Progress in Colloid & Polymer Science · Vol. 78

PROGRESS IN COLLOID & POLYMER SCIENCE
Editors: H.-G. Kilian (Ulm) and G. Lagaly (Kiel)

Volume 78 (1988)

Relationships of Polymeric Structure and Properties

Guest Editor: I. Chudáček (Prague)

Springer-Verlag
Berlin Heidelberg GmbH

ISBN 978-3-662-15682-7 ISBN 978-3-7985-1691-5 (eBook)
DOI 10.1007/978-3-7985-1691-5
ISSN 0340-255 X

Originally published by Dr. Dietrich Steinkopff Verlag GmbH & Co. KG, Darmstadt in 1988
Softcover reprint of the hardcover 1st edition 1988

Chemistry editor: Heidrun Sauer; Copy editor: James Willis; Production: Holger Frey.

Preface

The 5th International Seminar on Polymer Physics was organized by the Department of Polymer Physics at Charles University, Prague, 21–27 April, 1987. We present in this volume the main plenary lectures, short lectures, and posters from that seminar. The reader will obtain an overview of the principal scientific efforts in the field of polymer physics from many European universities and institutes. This book comprises two main scientific approaches:

1. The superstructure of polymer systems;
2. The electronic properties of thin polymer films.

It seems to me that the "heterogenous" state formed by these two appraoches represents a few inconveniences, but it has had a stimulating influence on each scientist present. The study of the polymeric state of condensed matter will in the future attract more scientists to its strange architecture, beautiful movements, and unexpected interactions. Also, new applications for polymers will emerge in future years and support existing ones.

In closing, I would like to thank each contributor, and the journal "Colloid and Polymer Science", which made this publication possible.

Ivo Chudáček

Contents

Equilibrium stress-strain properties and topological constraint release effects in end-linked model networks

G. Heinrich[1]), A. Havránek[2]), M. Ilavský[3]), and K. Dušek[3])

[1]) Technische Hochschule „Carl Schorlemmer“, Sektion Physik, Merseburg, G.D.R.
[2]) Faculty of Mathematics and Physics, Charles University, Prague, Czechoslovakia
[3]) Institute of Macromolecular Chemistry, Czechoslovakian Academy of Sciences, Prague, Czechoslovakia

Abstract: The interpretation of the equilibrium mechanical properties of polyurethane model networks in the large elongation region has been performed within a theory of tube-like topological constraints. This theory describes the small-strain properties as well as the large-strain properties of the networks in terms of the number of elastically active network chains (EANCs) and structural features contained in the structural parameters T_e (trapping factor) and w_s (sol content).

Key words: Model networks, topological constraints, network defects, network theory, stress-strain measurements.

1. Introduction

In a recent paper [1] a theory of polymer networks was proposed which uses a tube-like mean field concept for simulating the strength and deformation dependence of the topological constraints of the network chains. The effectiveness of these constraints depends on the external deformation. It has been explained that network defects (e.g., dangling ends) and the sol content, w_s, favor constraint release effects (CRE) of network chains which tend to return the configurational tubes to the configurations they had before deformation.

In the following we investigate experimental data of polyurethane model networks [2] within the theory of tube-like topological constraints. Attention is devoted to the prediction of large-strain Mooney-Rivlin parameters C_1 and C_2 in terms of molecular characteristics of the network.

The equilibrium mechanical properties (small-strain and large-strain behavior) of polyurethane model networks have been investigated in [2]. The networks were prepared from poly(oxypropylene) triols (PPT) and 4,4′-diphenylmethane diisocyanate (MDI) at various initial molar ratios of reactive groups, r_H. The experimental small-strain data have been compared with theoretical dependences of the equilibrium modulus, G, on the sol content, w_s. The calculations were based on the theory of branching processes [2, 3] and used the Langley-concept of trapped entanglements [4, 5] to correlate the equilibrium modulus with the concentration of elastically active network chains (EANCs), and with the so-called trapping factor, T_e, which characterizes the permanent topological interactions between segments in EANCs.

2. Theory

The statistical mechanics of polymer networks with deformation dependent tube-like topological constraints leads to the following dependence of the reduced stress σ_{Mg} on elongation λ [1]

$$\sigma_{Mg} \equiv \frac{\sigma_g(\lambda)}{\lambda - \lambda^{-2}} = G_{cg} + G_{Ng}\,\tilde{f}(\lambda) \tag{1}$$

where σ_g is the stress related to initial cross-section of the sample.

$$G_{cg} = \frac{f-2}{f}\,RT\,\nu_{eg} \tag{1a}$$

is the “chemical” contribution to the total shear modulus (f = functionality),

$$G_{Ng} = a\,G_N^0\,\beta_e^2 \tag{1b}$$

is the "topological" contribution to the total modulus and

$$\tilde{f}(\lambda, \nu, \beta_e) = \frac{1}{\nu\beta_e} \frac{\lambda^{\nu\beta_e} - \lambda^{-2\nu\beta_e}}{\lambda^2 - \lambda^{-1}}, \tilde{f}(\lambda = 1) = 1. \quad \text{(1c)}$$

The quantity ν_{eg} denotes the "chemical" number of EANCs in the unit volume of gel. In Eq. (1 c) the parameter ν describes the constraining mechanism ($-1 \leq \nu \leq +1$) of network chains. The model of trapped entanglements may be viewed as a special case of (tube-like) topological constraints which is mainly based on the assumption of dominating local constraints. This assumption fails increasingly if CREs allow local contraints to relax. Then, the effect of the remaining global entanglements can be described successfully by the trapping concept. This picture of trapped pure entanglements yields $\nu = -1$ [1].

The quantity β_e ($0 \leq \beta_e \leq 1$) characterizes the deformation dependence of the constraining tubes (deformed tube radius $d_\mu = d_0 \lambda_\mu^{\nu\beta_e}$, d_0 is the undeformed tube radius, $\mu = x, y, z$) and the relaxed microscopic deformation λ_{mic} of the *rms* end-to-end distance of network strands ($(\lambda_{mic})_\mu = \lambda_\mu^{\beta_e}$). The time dependence of this relaxation process to the final (equilibrium) state after a sudden external deformation will be discussed in a separate paper [10].

In Eq. (1 b), G_N^0 is the plateau modulus of the corresponding linear uncrosslinked polymer. The quantity $a \simeq 0.13\,(d_0^m/d_0)^2$ contains the ratio of the lateral tube radius in the melt (d_0^m) and in the corresponding network, respectively. The theory yields [1, 7, 9] $d_0^m/d_0 \simeq 2$–4 which leads to the relation

$$G_{Ng} \simeq (0.5 - 2)\, \beta_e^2\, G_N^0\,. \quad (2)$$

From comparison of Eq. (1 b) with the entanglement contribution to the shear modulus according to the Langley-concept it is necessary to write $T_{eg} \equiv T_e/w_g \simeq \beta_e^2$. T_e is the trapping factor and w_g is the weight fraction of the gel. The ratio $d_0^m/d_0 \simeq 2$–4 differs from the conclusions of Gaylord [11] and Marrucci [12] who assumed the relation $d_0^m/d_0 = T_e^{1/2} < 1$. Without discussing CRE they expected that a dilution effect of network defects leads to a cross-sectional area of the tube in the network larger than the cross-sectional area of the tube surrounding the same chain in the uncrosslinked system. Our model [1, 7, 9] gives an additional effect of the crosslinks to the strength of the topological constraints and, therefore, $d_0^m/d_0 > 1$. The strength of topological constraints depends on the space density of polymer contour length (or the local segment number density) and not on structural details of the system [7, 13].

In the following, we use the equality $\beta_e^2 = T_{eg}$. The quantities ν_{eg}, T_{eg}, and w_g were calculated for PUR model-networks using the theory of branching processes [2, 3]. The value $aG_N^0 \simeq 1.4\,MPa$ is taken from [2]. The Mooney-Rivlin parameters C_1 and C_2 are calculated by linear regression in the regime of moderate extensions ($\lambda = 1.1 - 2.5$) and use the relations

$$\sigma_{Mg}(\lambda) = G_{cg} + G_{Ng}\, \tilde{f}(\lambda, \nu = -1, \beta_e = T_{eg}^{1/2})$$
$$= C_{1g} + C_{2g}\lambda^{-1}\,. \quad (3)$$

The calculations were performed only in the case of networks characterized by the molar ratio $r_H \geq 1$ (samples No. 7–18 (PPT-LHT-240) and No. 28–40 (PPT-LG-56) of [2]). In the opposite case, $r_H < 1$, excess crosslinking took place, obviously due to the formation of allophanate groups. When determining C_{1g} and C_{2g} by Eq. (3), the correlation coefficients of all samples have been estimated to be in the range $r = 0.950$ and $r = 0.993$.

3. Results and Discussion

The calculated parameters C_{1g} and C_{2g} have been compared with the corresponding values $C_1^* \equiv C_1^{exp}/w_g$ and $C_2^* \equiv C_2^{exp}/w_g$. The quantities C_1^{exp} and C_2^{exp} denote the measured Mooney-Rivlin constants of the unextracted networks [2]. The quantities C_1^* and C_2^* consider the lowering of the elastic constants due to the volume fraction of the sol. The correlation between C_{1g} and C_1^* is reflected in the linear regression result $C_{1g}\,[MPa] = 0.985\, C_1^*\,[MPa] - 0.033\,[MPa]$ with a correlation coefficient $r = 0.997$.

When comparing the calculated values of C_{2g} with the corresponding values of C_2^*, we observe a non-linear decrease of $c = C_2^*/C_{2g}$ with increasing sol content, w_s (Fig. 1). This effect qualitatively agrees with the result of previous studies of dry extracted and non-extracted networks prepared from different polymers [14]. The similar effect is obtained in swollen networks in which the C_2-parameter at higher swelling degrees decreases non-linearly with increasing amount of solvent [1]. In the present paper the effect can be explained as the influence of the sol content on the CRE. Increasing amount of w_s (and, therefore, decreasing values of β_e) favors the knot rearrangement processes [15] which tend to return the configurations of tubes to those they had before deformation. The value of C_{2g} is determined by the amount of defects connected with the gel and their influence on the CRE. The value of C_2^* is additional determined by the sol content.

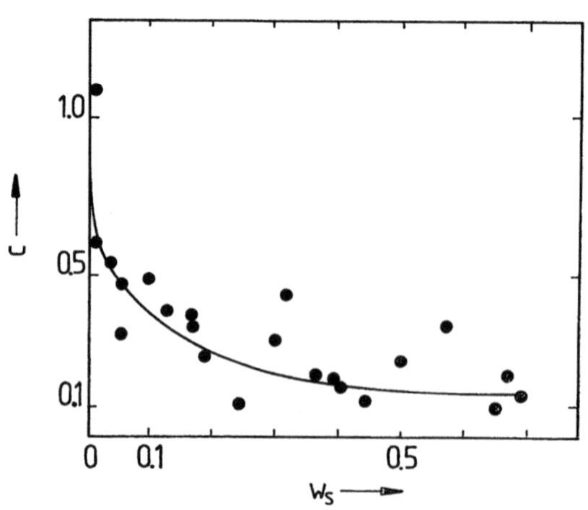

Fig. 1. The influence of sol content on the Mooney-Rivlin parameter C_2

Quantitatively, the curve in Fig. 1 is proposed to follow a power law dependence $c \sim (1 - w_s)^\alpha$. A similar result was found in [14]. A linear correlation analysis of a $\log c - \log w_g$ plot yields $\alpha \approx 1.42$ with a correlation coefficient $r = 0.660$.

On a molecular level the results can be interpreted as follows: The undeformed tube radius, d_0, mainly depends on the segment number density and not on structural details, whereas the deformation dependence of the tube radius ($d_\mu = d_0 \lambda_\mu^{-\beta_e}$) depends considerably on structural parameters. Physically, this effect has been explained by the concept of constraint release of network chains. A large content of sol fraction w_s (and, hence, a small value of β_e) leads to (nearly) deformation independent tube dimensions. It is well known that polymer networks in which the topological restrictions are nearly independent of deformation have small (in comparison with C_1) values of the C_2 constants [1]. Similar effects can be found in swollen networks.

Further experimental studies on extracted and nonextracted samples would be useful to clarify the quantitative dependence of the ratio c on the sol content w_s.

The theoretical basis of the interpretation of experimental data can be improved by additional consideration (to the configurational restrictions of network chain segments) of restrictions of junction fluctuations [16, 17]. In the case of restricted chains and partially restricted junction fluctuations one obtains additional contributions to the elastic free energy of the network; and the front-factor depends on d_0 and has values between the free-fluctuation limit and fixed junction limit, respectively [17, 18].

We note that Andrady and Llorente [19] recently interpreted stress-strain properties of poly(oxypropylene) model networks by the Mooney-Rivlin approach and the theory of restricted junction fluctuations.

References

1. Heinrich G, Straube E, Helmis G (1988) Adv Polym Sci 85:33
2. Ilavský M, Dušek K (1983) Polymer 24:981
3. Dušek K, Prins W (1969) Adv Polym Sci 6:1
4. Langley W (1968) Macromolecules 1:348
5. Pearson DS, Graessley WW (1980) Macromolecules 13:1001
6. Heinrich G, Straube E (1983) Acta Polymerica 34:589
7. Heinrich G, Straube E (1984) Acta Polymerica 35:115
8. Helmis G, Heinrich G, Straube E (1984) Wiss Zeitschr TH Leuna-Merseburg 26:461
9. Heinrich G, Straube E (1987) Polym Bull 17:247
10. Havránek A, Heinrich G (1987) Proceedings 5th intern Seminar on Polymer Physics, High Tatra 1987:63
11. Gaylord RJ (1982) Polym Bull 8:325; (1983) Polym Bull 9:186
12. Marrucci G (1981) Macromolecules 14:434
13. Graessley WW, Edwards SF (1981) Polymer 22:1329
14. Heinrich G, Hube H (1986) Plaste und Kautschuk 33:88
15. Bastide J, Picot C, Candau S (1981) J Macromol Sci, Phys B 19:13
16. Flory PJ (1977) J Chem Phys 66:5720
17. Kästner S (1981) Colloid Polym Sci 259:499
18. Heinrich G, Straube E, Helmis G (1987) Proceedings of IUPAC meeting Macro '87, Merseburg 1987
19. Andrady AL, Llorente MA (1987) J Polymer Sci, Polym Phys Ed 25:195

Received April 24, 1987;
accepted April 21, 1988

Authors' address:

G. Heinrich
Technische Hochschule „Carl Schorlemmer“ Leuna-Merseburg
Sektion Physik
DDR-4200 Merseburg, G.D.R.

Determination of mechanical parameters related to the kinetics of swelling in an electrically activated contractile gel

P. Chiarelli and D. De Rossi

Centro "E. Piaggio", Faculty of Engineering, University of Pisa, Italy

Abstract: We report on an analytical model and related experiments to determine the mechanical parameters governing the kinetics of mechanical readjustment of a thermally crosslinked polyacrylic acid (20% by weight) and polyvinylalcohol (80% by weight) (PAA-PVA) gel.

Either thin films on spherical samples have been used in the experiments. The characteristic time of swelling in the case of a thin film has been found to be proportional to the square of a linear dimension of the sample divided by a diffusion coefficient D, defined as $D = \mu/f$, where μ is the shear modulus of the gel network and f is the friction coefficient between the network and fluid.

This constitutes and essential difference between the swelling of spherical samples reported in literature in which the diffusion coefficient is $D^* = (k + 4/3)/f$, where k is the bulk modulus of the gel network.

We have derived the value of the bulk modulus k, the shear modulus μ and the friction coefficient f by free swelling experiments performed on spherical and planar thin samples, and by direct measurement of water permeability across the gel.

Mechanical readjustment time constant and diffusion-reaction time constant, which has been previously determined, are then compared to evaluate the limiting rate phenommena which govern the kinetics of gel deswelling under electrochemical stimuli in different pH ranges.

Key words: Polyelectrolyte gels, polyacrylic acid + polyvinylalcohol, thermally crosslinked, swelling kinetics, electromechanochemistry.

Introduction

Recently, experiments have been conducted aimed at eliciting contractile activity in polyelectrolyte networks by electrical excitation, to eventually develop a new kind of "soft actuator". Muscle-like motors could be instrumental, among other possible applications, in the design of innovative artificial organs and limb prostheses [1] and dexterousend effectors for advanced robots [2].

Although various physical effects can contribute to the observed electromechanical coupling [3], strongest contractile force density has been observed under provisions by which pH gradients are generated by electrode reactions and transported across the polymer network by electrodiffusion.

Several factors govern the electromechanochemical energy conversion performance of the system, such as gel material parameters, composition and solute concentration of the supernatant solution, geometrical arrangements, electrode material, and potential difference at the electrodes.

The purpose of this work is to discuss an electromechanochemical swelling model which has been formulated to describe the kinetics of electrically activated contractile phenomena observed in a polyelectrolyte gel and to determine, by free swelling experiments and mechanical measurements performed on spherical and planar samples, material parameters which are relevant to a phenomenological description.

Electromechanochemical swelling model

To analyze and describe the transient mechanical behavior of a polyelectrolyte gel element in response

PCPS HT 033

to proton and salt concentration gradients generated by electrode reactions and transported inside the gel by electrochemical potential differences, several rate processes should be taken into account, such as ion diffusion, diffusion limited chemical reaction, ion migration, and consequent electrical and mechanical readjustments of the polymer network [4].

Typically, proton diffusion-reaction and mechanical readjustment of the polymer network are the slowest phenomena, and a system of coupled differential equations is formulated to determine the kinetics of gel deswelling under electrochemical stimuli [5].
A comparison amoung proton diffusion and gel swelling characteristic time constants might prove very useful in decoupling the chemical and mechanical problems, particularly in the case of specific gel systems in which these limiting time constants are considerably different [6].

Analysis of the kinetics of free swelling

The kinetics of free swelling of a partially dehydrated gel is analyzed in the case of spherical and thin sheet samples; since no chemical change occurs in the gel during swelling; the kinetics is only governed by mechanical parameters.

An equation of motion of gel networks was formulated by Tanaka et al. [7] and proved to be adequate in describing free swelling phenomena either theoretically [8] on experimentally. Taking into account the chemical stress term β [9], which represents the zero-force strain generated by the proton concentration change, this equation is formulated

$$\partial u/\partial t = [(k + \mu/3)\, 1/f] \text{ grad div } u + \mu/f \text{ div grad } u + 1/f \text{ grad } \beta \,. \quad (1)$$

The stress-strain relationship becomes

$$\sigma_{ij} = k \text{ div } u\, \delta_{ij} + 2\mu(u_{ij} - 1/3 \text{ div } u\, \delta_{ij}) + \beta\, \delta_{ij} \quad (2)$$

where

$$u_{ij} = (\partial u_j/\partial x_i + \partial u_i/x_j)/2$$

where u is the displacement vector of gel elemental volume from its position when the swelling process has gone to completion, k is the bulk elastic modulus of the gel polymeric component, μ its shear modulus, and f is the friction coefficient between the polymer network and the liquid medium.

The chemical to mechanical coupling is provided by the proton concentration dependence of the phenomenological coefficients k, μ, f, and β. The kinetics of free swelling of a gel sample depends upon its geometry and material parameters as it is analyzed in the following in the case of spherical and thin planar samples.

Spherical samples

By referring to Eq. (1) in which the β term is omitteed since no chemical change occurs, the kinetic of swelling is described when the function $u(r,t)$, (r being the sample radius), is determined in the space domain occupied by the gel (in the state at $t = \infty$).

By solving Eq. (3) in spherical coordinates, it has been shown [10] how the characteristic relaxation time τ_s of the polymer network is proportional to the square of the radius a of the sphere when fully swollen and inversely proportional to the diffusion coefficient of the polymer network into the solution

$$\tau_s = a^2/(D_s \pi^2) \quad \text{where } D_s = (k + 4\mu/3)/f \,. \quad (3)$$

During the swelling process, $r(t) = a - \Delta r(t)$, where

$$\Delta r(t) = \Delta r_0 (6\pi)\, \Sigma_n\, n^{-2} \exp(-\, n^{-2}\, t/\tau_s) \quad (4)$$

and

$$\Delta r_0 = -\, u_{(r\,=\,a,\, t\,=\,0)} \,.$$

At $t > \tau_s/4$ in the time dependence of Δr the exponential term having $n = 1$ becomes preponderant.

Thin planar samples

We analyze the kinetics of swelling of a thin, planar sample whose thickness is negligible in respect to other dimensions, and we assume that the system is invariant in respect to translations along the planar coordinates x and y during the swelling process.

Under these assumptions, the strain components do not depend on the x and y coordinates; the following identities hence hold:

$$\partial u_{xx}/\partial x = \partial u_{yy}/\partial x = \partial u_{xx}/\partial y = \partial u_{yy}/\partial y = \partial u_{zz}/\partial x = \partial u_{zz}/\partial y = 0 \,. \quad (5)$$

By integrating the first four equations in relation to Eq. (5), we obtain $u_x = u(z)\, x$ and $u_y = u(z)\, y$.

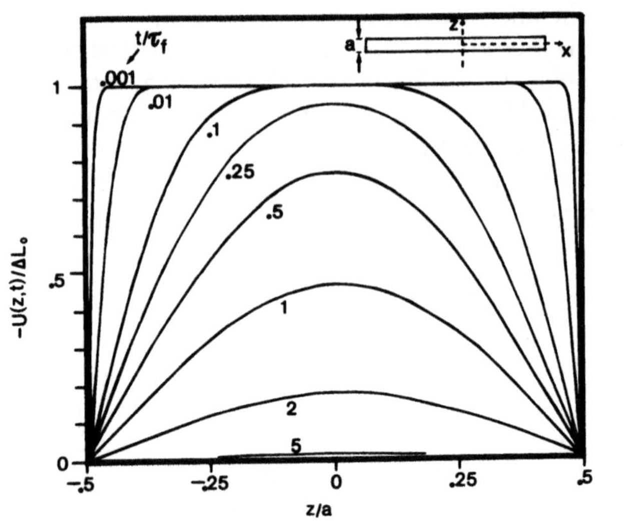

Fig. 1. Calculated fractional strain along x direction in a planar sample during the swelling process at various parametric values of the reduced time of swelling

Adhering to the definition of the strain tensor, it follows that $u_{xy} = 0$.

Since the system is invariant in interchanging x and y, the equality $u_{xx} = u_{yy}$ also holds. By introducing these relationships into the gel equation of motion we obtain

$$\partial u_x/\partial t = \mu/f\, \partial^2 u_x/\partial z^2 \quad (6)$$

$$\partial u_y/\partial t = \mu/f\, \partial^2 u_y/\partial z^2 \quad (7)$$

$$\partial u_z/\partial t = (k + \mu/3)/f\, \partial(\mathrm{div}\, u)/\partial z + \mu/f\, \partial^2 u_z/\partial z^2 \,. \quad (8)$$

We now are able to solve Eq. (6) by defining initial and boundary conditions which are congruent with the free swelling experiments.

Initial conditions, in the hypothesis of isotropic swelling, are defined as

$$u_i(x, y, z, t=0) = u_0\, i \qquad i = x, y, z \quad (9)$$

where u_0 indicates the percent change of sample dimentions referred to the final state at $t = \infty$.

Boundary conditions are defined by assuming that at time $t = 0^+$ the sample is immersed into the solution and "freed" from the virtual pressure which confines it to its acutal dimensions.

Hence, at any t, $\sigma_{xx} = \sigma_{yy} = \sigma_{zz} = 0$ at the sample surfaces perpendicular to the z axis at $z = \pm a/2$ (see Fig. 1 for the definition of the coordinate system). It follows

$$\sigma_{zz} = (k + 4/3\mu)u_{zz} + (k - 2/3\mu)(u_{xx} + u_{yy}) = 0$$

$$\text{at } t > 0 \text{ and } z = \pm a/2 \quad (10)$$

$$\sigma_{xx} = (k + 4/3\mu)u_{xx} + (k - 2/3\mu)(u_{yy} + u_{zz}) = 0.$$

By solving Eq. (10) in u, we obtain $u\,(\pm a/2, t) = 0$.

We can now find a solution of Eq. (6); introducing the initial and boundary conditions defined above

$$u = \Sigma_n \exp\left(-(2n+1)^2\, t/\tau_f\right)(-1)^n \left[4u_0/(2n + 1)\,\pi\right] \cos\left[(2n+1)\, z\,\pi/a\right], \quad (11)$$

where $\tau_f = a^2 f/\pi^2\, \mu = a^2/\pi^2 D_f$.

In Fig. 1 the strain in the planar sample during the swelling process is reported at various parametric values of the reduced time of swelling. The sample length along x coordinate $L(t)$ can be calculated

$$L(t) = L \cdot \{1 + 4u_0\, \Sigma_n \exp\left(-(2n + 1)^2\, t/\tau_f\right)(-1)^n/[(2n+1)\pi] \quad (12)$$

where L is the gel dimension when fully swollen.

It can be noted that for $t > \tau_f/9$, the term having $n = 0$ becomes predominant in the summation series.

Experimental

Sample preparation

Thermally crosslinked gel samples were prepared in thin sheet and spherical form using polyvinylalcohol (PVA) and polyacrylic acid (PAA). PVA with a degree of hydrolisis of 98% and an average molecular weight of 10000 (AnalytiCals, Carlo Erba, Milano, Italy) and PAA with an average molecular weight of 250000 (Aldrich Chemical Co., Milwaukee, Wisconsin USA) were dissolved separately into bidistilled water and subsequently mixed in the relative proportion of respectively, 80% and 20% by weight. After 20 min of stirring at 60 °C, the homogeneous solution was dehydrated in a thermostatic oven at 40 °C under mild vacuum. The sample was removed from the original container and thermally crosslinked at 130 °C for 45 min. The samples are then equilibrated with bidistilled water.

Free swelling experiments

Spherical samples (average diameter 0.05 cm) and square shape, thin sheets (average thickness 0.02 cm, 0.5 cm side) were partially dehydrated up to the point where linear dimensions reduced 8%. At $t = 0$, partially dehydrated samples were immersed into bidistilled water and allowed to free swell to equilibrium at 20 °C. Linear

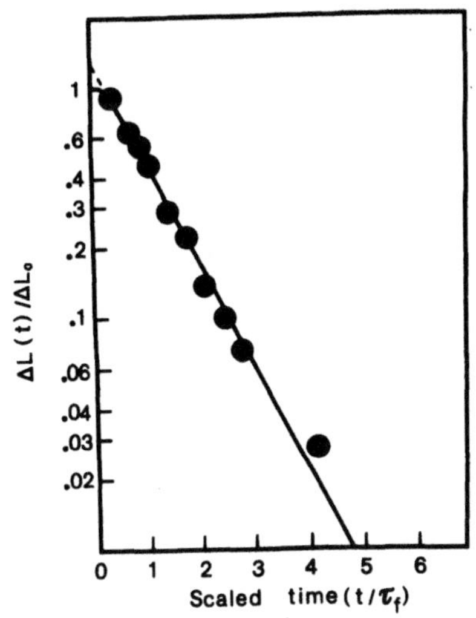

Fig. 2. The change in the length of a planar sample in respect to its lenth when fully swollen, at various scaled time of swelling. The solid curve is calculated using Eq. (11)

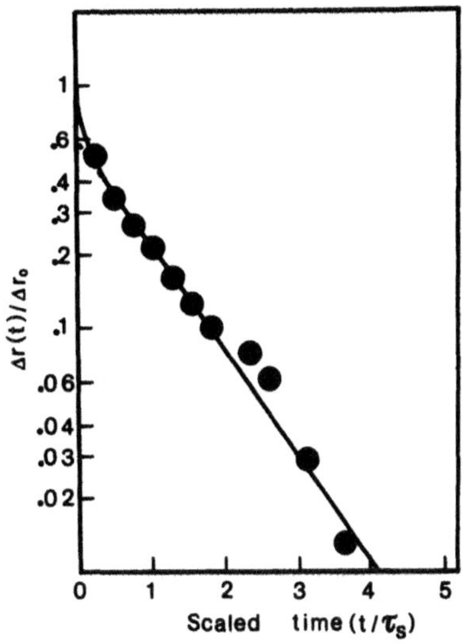

Fig. 3. The change in the radius of a spherical sample during swelling, plotted as a function of the scaled time of swelling. The solid curve is calculated using Eq. (4)

dimension changes were measured in time by using a graduated eyepiece. In Fig. 2 the change in the length of a planar gel sample $\Delta L(t)/\Delta L_0$ during swelling is reported, plotted as a function of relative time t/τ_f. The solid line represents the theoretical behavior calculated using Eq. (12). The extrapolated intercept with the ordinate axis of the line-fitting experimental data gives the value $4/\pi$, as theoretically predicted. A value of $\tau_f = 86 \pm 2$ sec has been obtained after averaging measurements performed on 10 gel samples. In Fig. 3. the change in the radius of a spherical sample is reported under the same condition illustrated for Fig. 2. Here again the theoretical prediction is quite accurate and the intercept at $t/\tau_s = 0$ gives the calculated value of $6/\pi^2$. A value of $D_f = 1/\pi^2 \cdot 4.7 \pm 0.4 \cdot 10^{-6}$ cm²/sec has been obtained after averaging measurements performed on 10 samples. A value of $D_s = 1/\pi^2 \cdot 2.0 \pm 0.3 \cdot 10^{-5}$ cm²/sec has been obtained for spherical samples.

Measurements of the friction coefficient and of the elastic moduli

Measurement of polymer-water friction coefficient was performed using the apparatus and technique indicated by Tanaka et al. [7]. A gel sample in the form of a disk (diameter = 0.8 cm; thickness = 0.018 cm) is contained in a glass apparatus as illustrated in Fig. 4 and located in a thermostatic oven at 20 °C. The time dependence of the capillary water column height is recorded using a cathetometer and it is shown in Fig. 5 in a logarithmic scale. Fitting of experimetal points permits us to obtain the value of $\tau = 2.6 \cdot 10^5$ sec for the exponential time dependence $h = h_0 \exp(-t/\tau)$.

The value of the friction coefficient f can be calculated [8] from the relationship

$$f = \varrho\, g\, A_g \tau / (\delta A_c) \tag{12}$$

where ϱ is the density of water, g is the acceleration of gravity, A_g is the gel surface area, δ is the gel thickness, and A_c is the crossectional area of the capillary column. The calculated value is $f = 3.1 \pm 0.8 \cdot 10^{13}$ dyne sec/cm⁴. Since, as it has been calculated in this paper, $f = \mu/D_f$ for planar samples, the combined measurement of D in free swelling experiments f and of the shear elastic modulus may provide a control of the theoretical arguments by comparing the independently measured and calculated values of f.

The measurement of the shear elastic modulus of the gel has been performed on planar samples under sinusoidal (3 Hz) uniaxial tensile stress using a commercially available instrument (Rheolograph Piezo, ToyoSeiki, Tokyo, Japan); a value of $\mu = 1.4 \cdot 10^7$ dyne/cm² has been obtained. Hence $f = \mu/D_f = 3.0 \pm 0.6 \cdot 10^{13}$ dyne sec/cm⁴, which is in reasonable agreement with the directly measured value of $f = 2.5 \pm 0.8 \cdot 10^{13}$ dyne sec/cm⁴. Finally, since $D_s = (k + 4\mu/3)/f$, the value of the elastic bulk modulus k is calculated to be $k = 3.5 \pm 1.5 \cdot 10^7$ dyne/cm².

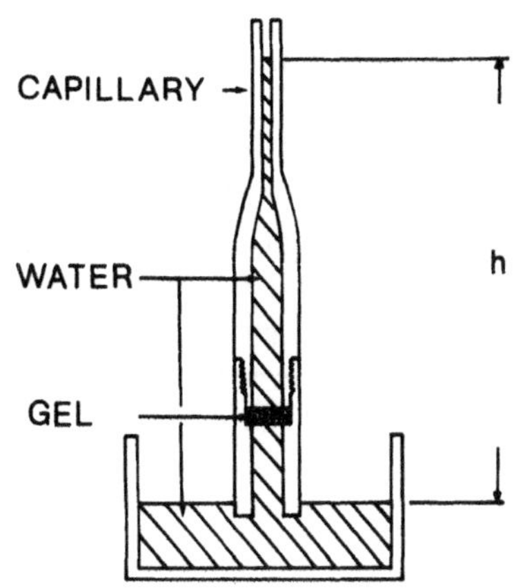

Fig. 4. Apparatus for measuring the friction coefficient f between the polymer network and water. After reference [7]

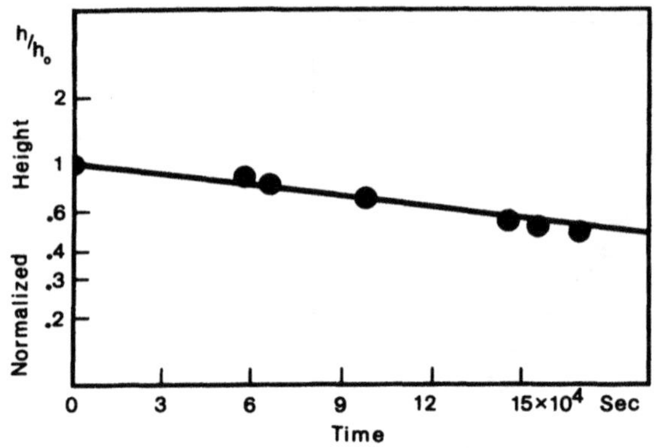

Fig. 5. Normalized height (logarithmic scale) of the capillary water column vs time of application of hydrostatic load

Conclusions

The time constant for free swelling of PAA-PVA gel samples having different geometry can be calculated using different combinations of the parameters k, μ and f.

A comparison between the proton diffusion-reaction time constant [5] and the mechanical readjustment time constant evaluated here shows that the kinetics of electrochemically activated PAA-PVA gels is limited by proton diffusion-reaction or by mechanical readjustment depending on the pH range. In the case of small pH increments, the proton diffusion-reaction time constant τ_c has been theoretically and experimentally found [5] to follow the relationship

$$\tau_c = (\pi^2 D_+)^{-1} [1 + K \cdot n/(K + c^D)^2] \quad (13)$$

where K is the intrinsic dissociation constant (mean value) of the carboxyl groups bound to the polymer network, c^D is the free proton concentration inside the gel, n is the total concentration of ionizable groups per liter of membrane water, and D_+ is the proton diffusion coefficient inside the gel.

The measured values of $K = 1.28 \cdot 10^{-5}$ mole/liter, $n = 0.568$ mole/liter, and D_+ comprised between $1.5 \cdot 10^{-9}$ and $3.8 \cdot 10^{-9}$ m^2/sec (depending on the gel water content) [5] allow us, by using Eq. (13), to assert that the mechanical readjustment time constant, as evaluated for a planar sample, is shorter than the diffusion-reaction time constant when the pH of the solution in equilibrium with the gel is higher than 2.9. At pH higher than 3.2 the kinetics are governed by the proton diffusion-reaction and the electromechanochemical problem can be decoupled. The polymer network, from a mechanical point of view, can be so considered at any time to be in equilibrium at the local proton concentration. In this context the mathematical description of gel kinetics under electrochemical stimuli is greatly simplified.

References

1. De Rossi D, Parrini P, Chiarelli P, Buzzigoli G (1985) Electrically induced contractile phenomena in charged polymer networks: preliminary study on the feasibility of muscle-like structures. Trans Am Soc Artif Int Organs XXXI: 60–65
2. De Rossi D, Domenici C, Chiarelli P (1988) In: Dario P (ed) Analogs of biological tissues for mechanoelectrical transduction: tactile sensors and muscle-like actuators, Sensors and sensory systems for advanced robots. Springer-Verlag, Berlin
3. De Rossi D, Chiarelli P, Buzzogoli G, Domenici C, Lazzeri L (1986) Contractile behavior of electrically activated mechanochemical polymer actuators. Trans Am Soc Artif Int Organs, vol XXXII, pp 157–162
4. Shoenfeld NA, Grodzinsky AJ (1980) Contribution of electrodiffusion to the dynamics of electrically stimulated changes in mechanical properties of collagen membranes. Biopolymers 19:241–262
5. De Rossi D, Chiarelli P, Genuini G, Dynamics of electrically driven muscle-like actuators, in preparation
6. Nussbaum JH, Grodzinsky AJ (1981) Proton diffusion reaction in a protein polyelectrolyte membrane and the kinetics of electromechanical forces. J Membrane Science 8:193–219
7. Tanaka T, Hocker LO, Benedek GB (1973) Spectrum of light scattered from a viscoelastic gel. J Chem Phys 59:5151–5159
8. Johnson DL (1982) Elastodynamics of gels. J Chem Phys 77:1531–1540
9. Eisemberg SR (1983) Nonequilibrium electromechanical interactions in cartilage: swelling and electrokinetics. Sc D thesis, Dep Elec Eng and Comp Science, MIT
10. Tanaka T, Fillmore DJ (1979) Kinetics of swelling of gels. J Chem Phys 70:1214–1218

Received June 25, 1987;
accepted April 21, 1988

Authors' address:

P. Chiarelli
Centro "E. Piaggio"
Facolta di Ingegneria
Universita di Pisa
Via Diotisalvi 2
I-56100 Pisa, Italy

Progress in Colloid & Polymer Science Progr Colloid Polym Sci 78:9–12 (1988)

Dielectric and thermal relaxations in amorphous and semicrystalline poly(ethylene terephthalate)

C. Schick[1]) and J. Nedbal[2])

[1]) Pedagogical College "Liselotte Herrmann", Department of Mathematics and Physics, Güstrow, GDR
[2]) Charles University, Faculty of Mathematics and Physics, Department of Polymer Physics, Prague, Czechoslovakia

Abstract: Undrawn PET was crystallized using several temperature programs to get samples of different crystalline structure, as indicated by a varying thickness of the amorphous layer between neighboring lamellae. The fluctuation model proposed by Donth was used to estimate the size (1–3 nm) of the cooperative, rearranging regions connected with the glass transition in these samples. There occurs a hindrance of the cooperative motions due to the crystalline structure, as shown by the correlation between the size of the cooperative, rearranging regions and the thickness of the layer of mobile amorphous material.

The WLF-curves are shown for the amorphous and the semicrystalline PET in the frequency range from 10^{-15} to 10^{8} Hz. A deviation from these curves, expected for the hindrance, is not observed.

Key words: Semicrystalline PET, amorphous fraction, glass transition, cooperative motion, characteristic length.

Introduction

The study of relaxation processes in semicrystalline polymers is a current topic of great scientific and technological interest. A large number of investigations have been undertaken to characterize relaxations in these materials, and there has been much scientific interest in the elucidation and detailed description of the molecular processes underlying them. Of particular interest is the main relaxation process or glass transition. There is some evidence that the molecular motion connected with this relaxation process is a cooperative one. The fluctuation model proposed by Donth [1, 2] is used to estimate the size of the cooperatively rearranging regions (CCR) at the glass transition:

$$V_a = k\,T_g^2\,(1/c_v)/\varrho(\delta T)^2 \qquad (1)$$

where k is Boltzmann's constant, T_g is glass transition temperature, $(1/c_v)$ is the step in the reciprocal specific heat at T_g, ϱ is the density of the amorphous material and δT is the mean temperature fluctuation. A great effort is necessary in order to extract δT from thermograms [2, 3]. Therefore, very rough approximations must be made [4]. From reference [3] a mean value of about 0.4 can be extracted for the ratio $\delta T/\Delta T$ (for three amorphous polymers), where ΔT is the temperature interval between 16% and 84% of the c_p step height for the heating thermogram. A variation, or shift, of this value is not believed to influence the qualitative results of the present paper.

Considering V_a as a sphere with a radius ξ, which is considered as a characteristic length, we obtain

$$\xi = (3\,V_a/4\pi)^{1/3}\,. \qquad (2)$$

The characteristic lengths ξ calculated according to Eqs. (1) and (2) for amorphous polymers are in the nm range (2–3 nm) [1–3]. If this is really the case, the question arises: What will happen if the dimension of the amorphous region is reduced to a value comparable with the characteristic length of cooperative motion? In semicrystalline polymers it should be possible to form this reduction which would hinder the cooperative motion.

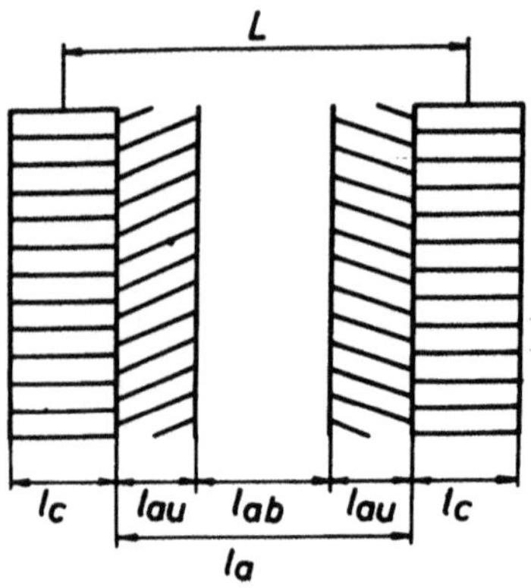

Fig. 1. Schematic representation of the arrangement of lamellae inside a regular crystalline structure (see text and Table 1)

Table 1. Parameters describing the crystalline structure 6 (see Fig. 1)

Material	Volume-fraction		Corresponding layer	
crystalline	α_v	$0.2 - 0.4$	$l_c = L \cdot \alpha_v$	2–4 nm
noncrystalline (amorphous)	$1 - \alpha_v$	$0.6 - 0.8$	$l_a = L \cdot (1 - \alpha_v)$	5–7 nm
mobile amorphous	β	$0.1 - 0.3$	$l_{ab} = L \cdot \beta$	1–3 nm
rigid amorphous	γ_v	$0.4 - 0.6$	$2 l_{au} = L \cdot \gamma$	3–4 nm

Experimental

We used commercial PET (ORWO K36, $\bar{M}_w \approx 23000$) as a model system since it allows us to estimate the size of the CRR in the totally amorphous material and in samples of different crystalline structure as well. In order to obtain the dimension of the amorphous regions inside the semicrystalline material we used the following simplified model (Fig. 1). The mean distance between the center of the lamellae equals the longperiod L, as it is determined by small angle x-ray scattering, in connection with the methods described by Vonk [5]. The longperiod varies from 7 to 9 nm for the samples investigated. From wide angle x-ray scattering we obtain the mass fraction of crystalline material and calculate the volume fraction α_v. Using this simple one-dimensional model it is possible to estimate the mean size of the lamella l_c and of the amorphous layer l_a (see Table 1). In addition to these crystallinity parameters we know from molecular mobility measurements (as NMR or calorimetric investigation of the step height of the specific heat, and by Raman spectroscopy as well) that the amorphous layer in the investigated PET samples is not a homogeneous one [6]. Here we are able to obtain the volume fraction β of the mobile amorphous material and the volume fraction γ of the rigid amorphous material with higher and lower molecular mobility, respectively. One possibility to estimate β and the thickness of the corresponding layer l_{ab} is given by

$$\beta = \Delta C_p / \Delta C_{pa} \qquad l_{ab} = L \cdot \beta \tag{3}$$

where ΔC_p is the step height of the specific heat at T_g for the observed semicrystalline sample and ΔC_{pa} is that for an amorphous one. The quantities we obtained [6] are given in Table 1. In our opinion the rigid amorphous material is located in the interfacial region. Its value of about 2 nm corresponds to the literature data for other polymers [7]. The figures shown in Table 1 correspond to three different crystallization regimes. These are:

1. Isothermal crystallization using temperatures between 120 °C and 220 °C;
2. Stepwise crystallization in the same temperature range in accordance with Groeninckx [8];
3. A long time secondary crystallization at 220 °C up to 500 h.

In this way we produced samples with different thicknesses of the layer of the mobile amorphous material. Table 1 reveals that they fall into the nm range so that they are of the same order of magnitude as the CRR.

Using Eqs. (1) and (2) it is possible to estimate the size of the CRR of these samples by analyzing a thermogram in the glass transition interval [9]. Since we investigated one polymer only, we think that the error in ξ of the order of 20% is a systematic one. Hence it does not influence the linearity shown in Fig. 2. Dielectric relaxation in these samples has been measured in the frequency range from 10^{-5} to 10^{8} Hz and mechanical relaxation at about 1 Hz as well as in the temperature range from −60 °C to 220 °C. The position of relaxation (as $\log f_{max}$ vs. $1/T$) for amorphous and semicrystalline PET are depicted in an Arrhenius-plot (Fig. 3). In order to include the thermal measurements, Eq. (4) [1,2] was used to transform the cooling rate $\dot{T}$ into a frequency

$$f \approx \dot{T} / 2\pi\delta T . \tag{4}$$

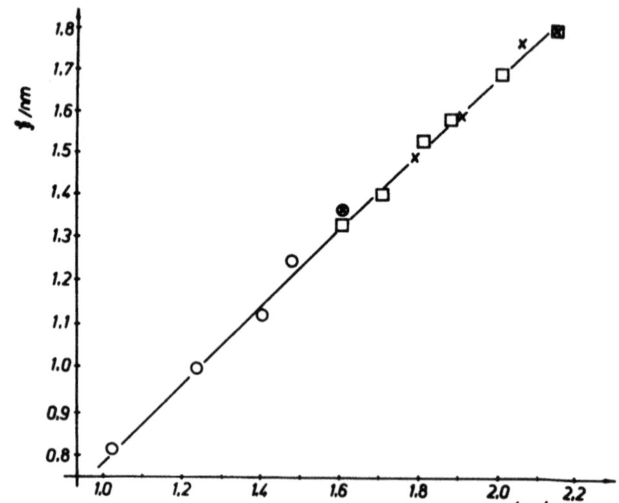

Fig. 2. Characteristic length of the cooperative motion versus thickness of the mobile amorphous layer for specimens of different thermal treatment; x: isothermal; O: stepwise; □: long time crystallization

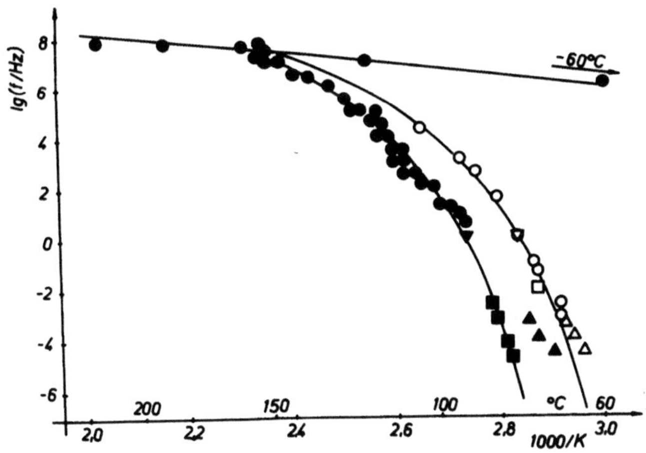

Fig. 3. Relaxation map of amorphous (open symbols) and semicrystalline (solid symbols) PET. ○, ●: dielectric; ▽, ▼: torsion pendulum; □, ■: glass temperature (Eq. (4)); △, ▲: enthalpy relaxation. ▼, ■, ▲: one sample crystallized at 120 °C for 16 h; ●: different crystallized samples

Results and discussion

Figure 2 depicts the dependence of the characteristic length (Eqs. (1) and (2)) for the cooperative motion in the semicrystalline samples on the thickness of the mobile amorphous layer (Eq. (3)). In any case the characteristic length is smaller than that determined for an amorphous sample (2.5 nm) and it is smaller than the width of the amorphous layer. It shows that the crystalline structure hinders the cooperative motions connected with the glass transition. Furthermore, it is a hint for the existence of CRR in the nm range as well. This hindrance of the cooperative motions should appear in the temperature-frequency relation for dynamic experiments, too. Since cooperativity of the motions connected with the glass transition increases as frequency decreases, a deviation from the WLF-curve will occur in the range where the hindering starts [10]. Figure 3 shows the WLF and Arrhenius behavior for the amorphous and semicrystalline samples. The secondary relaxation process was measured down to −60 °C and its location is the same for both samples. That means that there is no influence of crystalline structure on the local motions connected with this secondary relaxation process. The calculated activation parameters are $E_a = 43$ kJ/mol and the pre-exponential factor $A = 6 \cdot 10^{12}$ Hz. Near 150 °C for both samples, the main relaxation, i.e., the cooperative motion, separates from this local relaxation process.

The WLF equation (Eq. (5)) fits the curves using the parameters as

$$C_2 = 37\,\mathrm{K} \quad \text{and} \quad f_0 = 10^{-2}\,\mathrm{Hz},$$

$$C_1 = 14 \qquad T_0 = 347\,\mathrm{K}\ \text{(amorphous)}$$

$$C_1 = 14{,}5 \qquad T_0 = 361\,\mathrm{K}\ \text{(semicrystalline)}$$

$$\log(f/f_0) = C_1 \cdot (T - T_0)/(C_2 + T + T_0)\,. \qquad (5)$$

A deviation from the WLF-curve can be observed only for the enthalpy relaxation, but not in the frequency region around 10 Hz, where the hindering of the cooperative motions starts. In this region we only observe a drastic increase of *DC* conductivity so that it is not possible to perform dielectric measurements at lower frequencies. One might imagine that there exists a connection between charge transport and the hindering of the cooperative motions. Unfortunately, we have no experimental proof for this hypothesis at the moment and further investigations would be welcome.

Conclusions

PET proved to be a good model system for investigation of the influence of crystalline structure on the glass transition. On the basis of the fluctuation model of the glass transition proposed by Donth, it is possible to understand the correlation between the characteristic length of cooperative motions connected with the glass transition, and the dimensions of the crystalline structure. It is not clear why no deviation from the WLF-curve was found in the low frequency range for the semicrystalline samples. Probably there exists a small deviation but it is covered by the scatter of the measuring points which is relatively high and hence the fit of the WLF-curve does not reveal all details.

In our opinion these measurements support the picture of cooperatively rearanging regions in the nm range and the possibility to influence these motions by the dimensions of the crystalline structure.

Acknowledgements

This paper results from the cooperation organized by Prof. I. Chudacek and Prof. W. Mischok. The authors are also indepted to Doz. Dr. E. Donth, Dr. F. Fabry and Dr. W. Possart for helpful discussions.

References

1. Donth E (1981) Glasübergang, Akademie Verlag, Berlin
2. Donth E (1982) J Non-Crystalline Solids 53:325
3. Schneider K, Schönhals A, Donth E (1981) Acta Polym 32:471
4. Donth E (1984) Acta Polym 35:120
5. Vonk CG, Pijpers AP (1985) J Polym Sci Polym Phys Ed 23:2517
6. Schick C, Fabry F, Acta Polym in preparation
7. Varnell WD, Harrison IR, Kozminski SJ (1981) J Polym Sci Polym Phys Ed 19:1237
8. Groeninckx G, Reynaers H, Bergmans H, Smets G (1980) J Polym Sci Polym Phys Ed 18:1325
9. Wigger J (1986) Dissertation PH Güstrow
10. Schick C, Tanneberger H, Donth E (1980) Makromol Chem Rapid Commun 1:407

Received April 24, 1987;
accepted April 21, 1988

Authors' address:

C. Schick
Sektion Mathematik/Physik
Pädagogische Hochschule „Liselotte Herrmann“
Goldberger Str. 12
DDR-2600 Güstrow, G.D.R.

Progress in Colloid & Polymer Science Progr Colloid Polym Sci 78:13–16 (1988)

The relaxation of additives in polymer matrix

R. Bakule, J. Honskus, J. Nedbal, and M. Neubert

Department of Polymer Physics, Faculty of Mathematics and Physics, Charles University, Prague, Czechoslovakia

Abstract: The dielectric properties of synthetic polyisoprene (IR 305) samples with additives of acetonitrile were studied by means of the thermally stimulated depolarization current (TSDC) method. Acetonitrile in concentration from 0.02 to 5 wt% was added to synthetic polyisoprene samples IR 305. The dielectric measurements of natural rubber (NR) before and after extraction and measurements of IR 305 with extract from NR were studied by means of TSDC method and by means of temperature dependence of loss factor at fixed frequency. The measurements were carried out in the temperature range −150 °C to +130 °C

The high temperature absorption in NR and in IR 305 with extract was studied in detail and it can be seen that it is caused by orientational movement of additives in these materials.

The behavior of IR 305 with acetonitrile is more complicated. The main transition region is enlarged by the acetonitrile and we can observe its plasticizing effect in this region. The movement of free acetonitrile molecules in IR 305 appears at relatively high amounts of acetonitrile.

Key words: Dielectric relaxation, depolarization current, additives in polyisoprene.

1. Introduction

The following work extends the work of Bakule and Stoll who observed and described the high temperature maxima in natural rubber (NR) [1]. The peak height is approximately twice the dielectric loss value in the main transition region. The absorption curves are substantially broader and the peak position changes depending on the measurement temperature from 10^{-2} Hz at 0 °C up to 10^3 Hz at +140 °C. The peak shift could be described by the Doolittle formula [2]. A hypothesis was proposed [1] that the orientational polarization of low-molecular additives in NR, particularly proteins, may be considered an explanation.

Therefore, the measurements of synthetic rubber IR 305 (Cartiflex) with low-molecular additives were carried out [3]. Several liquid crystals (MW ~ 100) with the dipole moment along the molecule axis were chosen. Considerable secondary peaks obviously arising from these additives were observed and described [3]. The positions of maxima were one to three orders of frequency below the main transition region and this shift was constant in the whole temperature range −50 °C to +100 °C.

In the present work, we tried first to observe the influence of purifying the NR on the height and position of the above-mentioned high temperature peaks. Secondly, the extract from NR was added to IR 305 where the proteins are absent in principle. Finally, we searched for the dielectric absorption peak connected with the low-molecular additive that has a lower molecular weight than the polyisoprene (PI) monomeric unit (MW = 68). We selected the molecule of acetonitrile CH_3CN(MW = 42) with nearly spherical shape and substantially lower friction coefficient compared to the linear molecules of liquid crystals. Therefore, their orientational relaxation was expected to occur at a frequency above the glass transition region at constant temperature. This conclusion can be derived from the reasoning that the main transition region is caused by the mutual movement of PI monomeric units. The temperature dependence of additive relaxation is basically determined by the temperature dependence of the bulk internal viscosity.

2. Experimental

Two experimental methods were utilized. The method of thermostimulated depolarization currents (TSD) was preferred with regard to its high resolution (low equivalent frequency of 10^{-3} Hz). The homemade apparatus with temperature control logic unit ensured the rate of temperature rise of 4 K per min. The TSD current was measured with Keithley 610 C electrometer.

The second method was the complex permittivity measurement at a fixed frequency in the dependence on temperature. Two bridges were used during the measurements, the Tettex AG Zürich bridge for 50 Hz and the half-automatic bridge Tesla BM 484 working at 1 592 Hz. The measurements were carried out on sandwich-like samples composed of two equal layers (0.5 mm) on an aluminum plate, pressed together with a circular measuring electrode (25 mm) between them.

The conditions of extraction were as follows. The samples of partially purified crepe (PPC) had been extracted in acetone for 8 h. The extract, thickened by distilling off the bulk of acetone, was kept under room conditions until the constant weight was achieved. The mass fraction 1.5 % of the extract is lower by 0.4 % than it would be under optimal conditions.

The extract from PPC (1.5 %) was transferred to the foil of IR 305. The blend was homogenized by rolling on a laboratory mill at the temperature of + 60 °C. Results of TSD measurements for PPC, PPC after extraction in acetone, pure IR 305, and IR 305 with addition of the extract from PPC are shown in Fig. 1.

The main transition peaks for all measured samples have the same height and position. More essential differences can be seen in the high temperature region between − 40 °C up to + 40 °C. A distinct peak was observed near + 10 °C in all samples excluding pure IR 305 where peak can be detected near − 25 °C. It was shown elsewhere [4] that the latter peak is diminished during crystallization and a new clear maximum is created at a higher temperature near + 7 °C. The peak at − 25 °C was interpreted [4] as the relaxation of small arranged regions.

The peak at + 10 °C is diminished after extraction of PPC. On the other hand it appears in IR 305 after adding the extract of PPC. It can be concluded that the additive is responsible for this peak. As the peak is noticeable even in the extracted PPC, we may assume that the extraction was not complete.

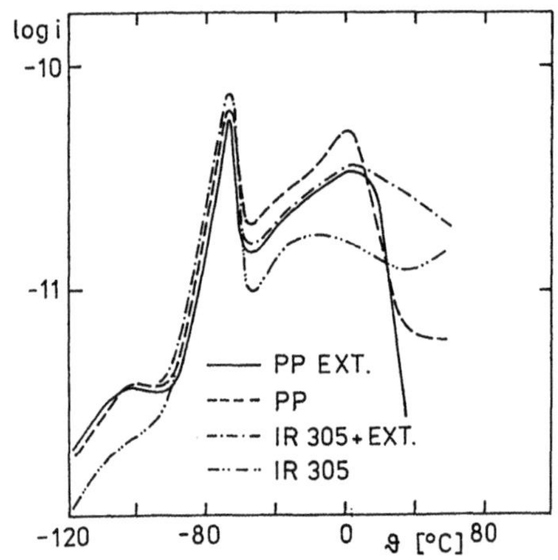

Fig. 1. TSD currents for PP crepe and synthetic rubber IR 305

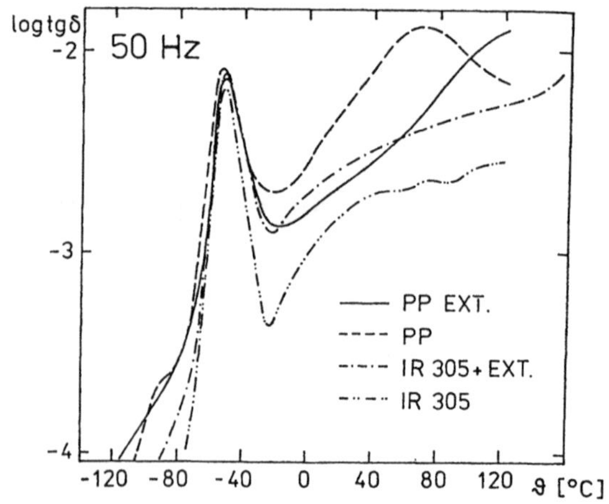

Fig. 2. Temperature dependence of the loss factor *tgδ* measured at 50 Hz for PP crepe and synthetic rubber IR 305

Thus we assume that two absorption mechanisms are superimposed in the range − 40 °C to + 40 °C − an already mentioned weak one observed separately in pure IR 305 (T_m = − 25 °C), and a strong one connected with the additives (T_m = + 15 °C).

The assumption is well supported by the temperature dependence of the loss factor *tgδ* at 50 Hz. The results for the same samples are presented in Fig. 2. The height and position of the main transition regions again do not differ. The high temperature peak was observed again for PPC at + 70 °C. The height of this peak is lowered by extraction and the addition of the extract to IR 305 manifests itself in the increase of the dielectric loss near + 50 °C. The high temperature peak for PPC after extraction and for IR 305 with extract is not marked. It is caused by the influence of the space charge that cannot be eliminated in AC measurements. The situation is worse in the case of 1 592 Hz because the peak is shifted to the space charge region and, therefore, this dependence is not shown here.

The results of our measurements on IR 305 with addition of 5 % (mass) of acetonitrile are presented in Fig. 3. A new low-temperature peak appeared in TSD at − 110 °C below the main transition region (− 75 °C). There is a weak peak near − 40 °C and a strong one at + 17 °C. Measurements of loss factor temperature dependences at 50 Hz and 1 592 Hz show both peaks at the lowest temperatures, too. Main transition temperature was − 60 °C at 50 Hz and − 50 °C at 1 592 Hz, the new peak position was − 90 °C at 50 Hz and − 80 °C at 1 592 Hz. In the high temperature region we only observed the dielectric loss increase caused by free charge motion.

The dependence of the position and the height of the absorption regions on the content of acetonitrile was measured especially by the TSD method, as the influence of the space charge could be eliminated by appropriate choice of the temperature at which the polarizing field was applied. Results of these measurements are collected in Fig. 4. Pure IR 305 has only two observable peaks. The height of the peaks is an order lower than in Fig. 1 as one order lower polarizing field was chosen for the concentration study. The addition of the acetonitrile causes the increase of the depolarization current practically in the whole temperature region.

Even a small addition of acetonitrile causes pronounced enlargement of the main transition peak. At 0.2 % of acetonitrile, the current maximum is two times higher and the region is shifted to lower

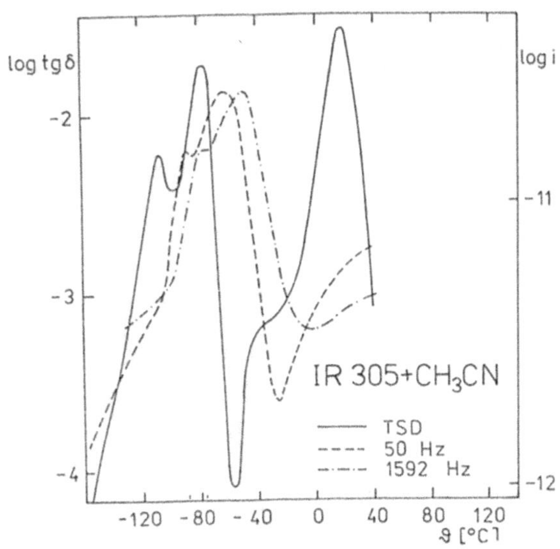

Fig. 3. Temperature dependence of the loss factor *tgδ* measured at 50 and 1592 Hz and TSD for IR 305 with acetonitrile

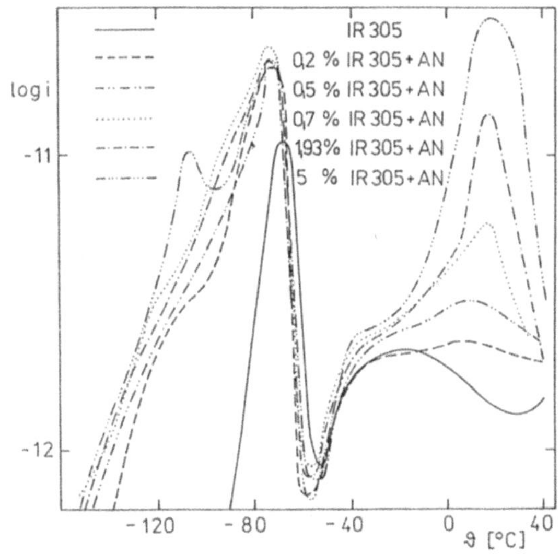

Fig. 4. TSD currents for the IR 305 samples with various content of acetonitrile

temperatures (approximately + 3 °C). The shift can be explained by the plasticizing effect of the acetonitrile and height increase by the cooperative motion of the complex polymer-acetonitrile.

The most pronounced influence of acetonitrile can be detected in the region above 0 °C. The height of a new peak near + 15 °C is proportional to the content of acetonitrile in the sample. It must be emphasized that the samples were polarized at + 40 °C. This temperature could not be very high in order not to release the acetonitrile from the polymer bulk.

In order to distinguish between the space charge effect and the orientational motion, measurements of the TSD thickness dependence were carried out.

The results for 1 % of acetonitrile in IR 305 are presented in Fig. 5. The polarization temperature was raised to + 60 °C. This may be the reason why no evident peak appears near + 17 °C. The polarizing voltage was selected to hold constant its relation to the sample thickness. Thus, in absence of the space charge the depolarization current should be the same. The space charge effects, on the contrary, should cause different responses in samples with different thickness. As there are apparent differences between the curves in the high temperature region, it may be concluded that the space charge plays an important role. The explanation of this behavior will be the topic of our next study.

3. Conclusions

Our results support the idea that the high temperature absorption region at + 10 °C in PPC and in IR 305 with extract is caused by the orientational movement of additives in the above mentioned materials. We assume that these rather large kinetic units are not bound to the main chains.

The measurement of IR 305 with acetonitrile showed more complicated behavior than we had expected.

The main transition region is enlarged by the acetonitrile addition, especially at concentrations lower than 0.1 %. It was very difficult for us to exactly determine the small quantities of acetonitrile. The height of the peak does not depend much on concentration

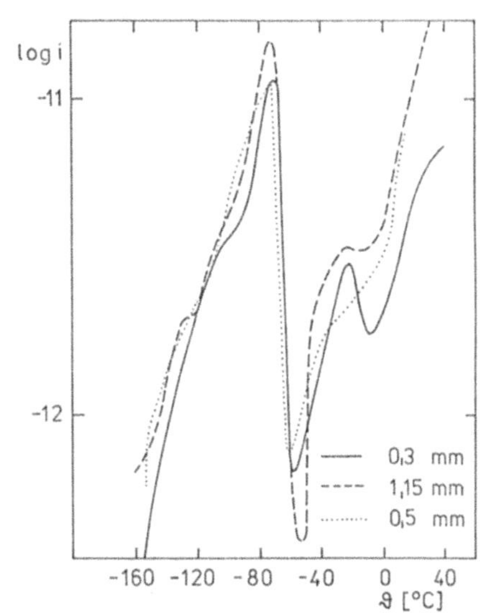

Fig. 5. TSD currents for samples of IR 305 with 1 % of acetonitrile and various thickness

above 0.2%. It may be supposed that there is only a limited number of acetonitrile molecules that can be firmly bound to the polymer. On the other hand, the plasticizing effect of the acetonitrile was observed in the whole concentration range.

The peak at −110 °C appears at relatively high amounts of acetonitrile. It may be connected with rotation of free acetonitrile molecules, while at the low concentrations (when the additive is predominantly bound to the polymer chains) the cooperative motion of the segments with acetonitrile is more probable.

References

1. Bakule R, Stoll B (1977) Colloid Polym Sci 255:1176
2. McCrumm NG, Read BE, Williams G (1976) Anelastic and Dielectric Effects in Polymeric Solids. Willey New York
3. Bakule R, Heinrich W, Pechhold W, Stoll B (1986) Colloid Polym Sci 264:185
4. Bakule R, Nedbal J, Štrof P (1986) In: Morphology of Polymers. Walter de Gruyter, Berlin, p 197

Received February 1, 1988;
accepted April 21, 1988

Authors' address:

J. Nedbal
Faculty of Mathematics and Physics
Charles University
V Holešovičkách 2
180 00 Praha 8, Czechoslovakia

Relaxation properties of butadiene-styrene copolymers

A. Havránek, J. Pospíšil, J. Buriánek, J. Honskus, and J. Nedbal

Faculty of Mathematics and Physics, Charles University, Prague, Czechoslovakia

Abstract: Dynamic mechanical, dielectric, luminescent, and thermodynamic relaxation spectra of butadiene-styrene copolymers are compared. Dynamic mechanical results were obtained by torsion pendulum measurements, dielectric results by thermally stimulated depolarization current (TSDC) method. Thermostimulated luminescence (TSL) and differential thermal analysis (DTA) were used in the luminescence and thermodynamic cases, respectively. The variations of the relaxation spectra with the copolymer composition (block or statistical, styrene content, 1,4 or 1,2-addition butadiene) and, especially, the possibilities of the comparison of different relaxation spectra as a tool for analysis of the structural origin of the observed spectral peaks are discussed.

Key words: Butadiene-styrene copolymers, viscoelasticity, luminescence, termostimulated currents, DTA.

1. Introduction

Copolymer systems are very variable in their structure. The relaxation mechanisms may, therefore, be very much influenced by the structural changes and the response of different relaxation measuring methods to these changes may be compared. The butadiene to styrene ratio, the statistical or block composition, the molecular masses of the blocks, the 1,2-addition butadiene content were varied in our samples. Relaxation mechanisms were detected by viscoelastic, dielectric, luminescent, and thermodynamic methods.

2. Measuring methods

The viscoelastic measurements were performed by torsion pendulum with a changeable moment of inertia which enables the measurements to be performed in a wide temperature range at a fixed frequency [1]. The measurements of the complex shear modulus $G' + iG''$ discussed in this article (Fig. 1 and 2b) were performed at a frequency of 1 Hz in the temperature range 130 K to 360 K.

The dielectric results were obtained by the thermally stimulated depolarization currents (TSDC) method. The electric field of intensity 30 kV/cm was applied to the samples at elevated temperature. The samples were then cooled down to approximately 150 K. At this low temperature the condenser electrodes were short-circuited over the picoampermeter and the depolarization currents (Fig. 2c) were measured while the samples were warmed up at a constant rate 4 K/min [2].

The thermostimulated luminescence (TSL) method consists of the ionization of a sample at low temperature and its subsequent heating during which trapped charges are released [3]. These charges after recombination emit light. This light is detected as an intensity of luminescence versus temperature curve, the so called glow curve (Fig. 2a).

The thermodynamic results were obtained by differential thermal analysis (DTA). These measurements were performed using the commercial equipment from Netzsch. The results of such measurements are the thermal difference ΔT against temperature curves (Fig. 2d).

3. Results

Typical results obtained during the butadiene-styrene copolymer study are given in Figs. 1 and 2. In Fig. 1 the imaginary part G'' of the complex shear moduli of three different butadiene-styrene copolymer samples are presented. The statistical copolymer (Krylene 1516) with 40% styrene content has one relaxation peak at 233 K (at high temperature this sample is so soft that it cannot be measured). The block copolymers (Solprene 411 and the laboratory prepared

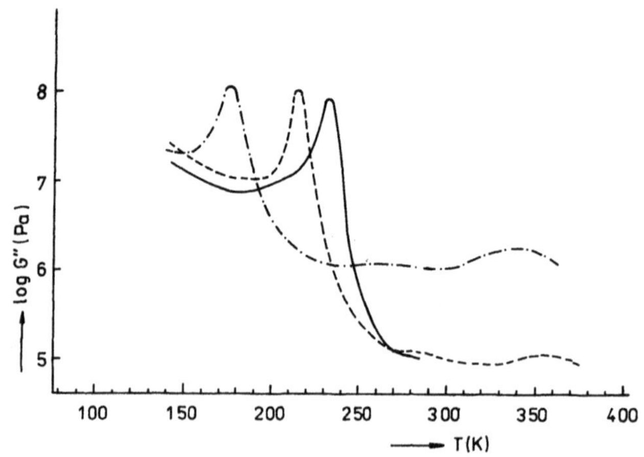

Fig. 1. Temperature dependence of G'' for Krylene (–), Solprene (–·–), and sample 5 of Table 1 (– – –)

sample denoted 5 in Table 1) have two main relaxation peaks, one at low temperature belonging to the transition zone of the butadiene blocks and the second at a temperature of approximately 350 K, reflecting the melting of styrene crystallites forming the network of this thermoplastic rubber. The butadiene transition zone of Solprene 411 is lower by 40 K than the butadiene transition of the sample 5, as in Solprene the butadiene is mainly of the 1,4-addition type (the glass transition temperature of this type of butadiene is $T_g =$ 166 K [4] and in the sample 5 the 1,2-addition content ($T_g = 258$ K) is nearly 50%.

In Fig. 2 the relaxation behaviour, as reflected by the described measuring methods of four block-copolymer samples, is given. We may compare both the different samples (composition is given in Table 1) and the different reflections of the relaxation mechanism in four utilised methods of measurement. Between 100 K and 150 K there is a broad intensity of luminescence maximum for all samples (in sample 3 this maximum is not separated as the TSL intensity decrease between this maximum and the maximum at 210 K is missing). These low temperature TSL maxima exhibit such values of activation energy and frequency factor that they are probably not connected with the mobility of a polymer chain, i. e. they are not connected with relaxation mechanisms [5]. Another maximum in the low frequency region is especially visible on the TSL curve of sample 4 at 160 K. This maximum, the symptoms of which are also visible on other TSL curves and on viscoelastic curves of samples 1 and 2, is of the secondary low temperature relaxation type (β-relaxation) [3]. The main transition zone of butadiene blocks is detected for all samples and by all methods in the vicinity of 210 K. This temperature, in accordance with the composition of butadiene blocks, follows the predictions given in [4]. In the range 240 K to 320 K large depolarization currents are detected. This relaxation activity is to some extent reflected also in viscoelastic (sample 1 and 2) and DTA measurements. The TSL method is insensitive in this temperature region as all the trapped charges had already recombined radiatively. This mainly dielectric relaxation has a complex character as may be seen from the TSDC and DTA curves. Some parts of it may be connected with the crystallization of polybutadiene and with the release of charges trapped between crystallites and amorphous regions [6]. The melting of crystal-type cross-links composed of styrene blocks is very probably the reason for the relaxation mechanism detected in the vicinity of 350 K by the viscoelastic, DTA, and TSDC measurements.

4. Discussion

The given results show that relaxation methods can be used very effectively in the study of the changes in the structure of butadiene-styrene copolymers. The

Table 1. Structure of the laboratory prepared S–B–S copolymers

Sample	Composition of B block (%) 1,4-addition trans	cis	1,2-addition	Molecular masses of blocks S	B	S	Modification
1	30	18	52	9000	44000	9000	–
2	30	18	52	14000	49000	14000	–
3	31	19	50	13000	64000	13000	oil added
4	31	19	50	13000	64000	13000	oligo-butadiene added
5	32	20	48	6300	75000	6300	–

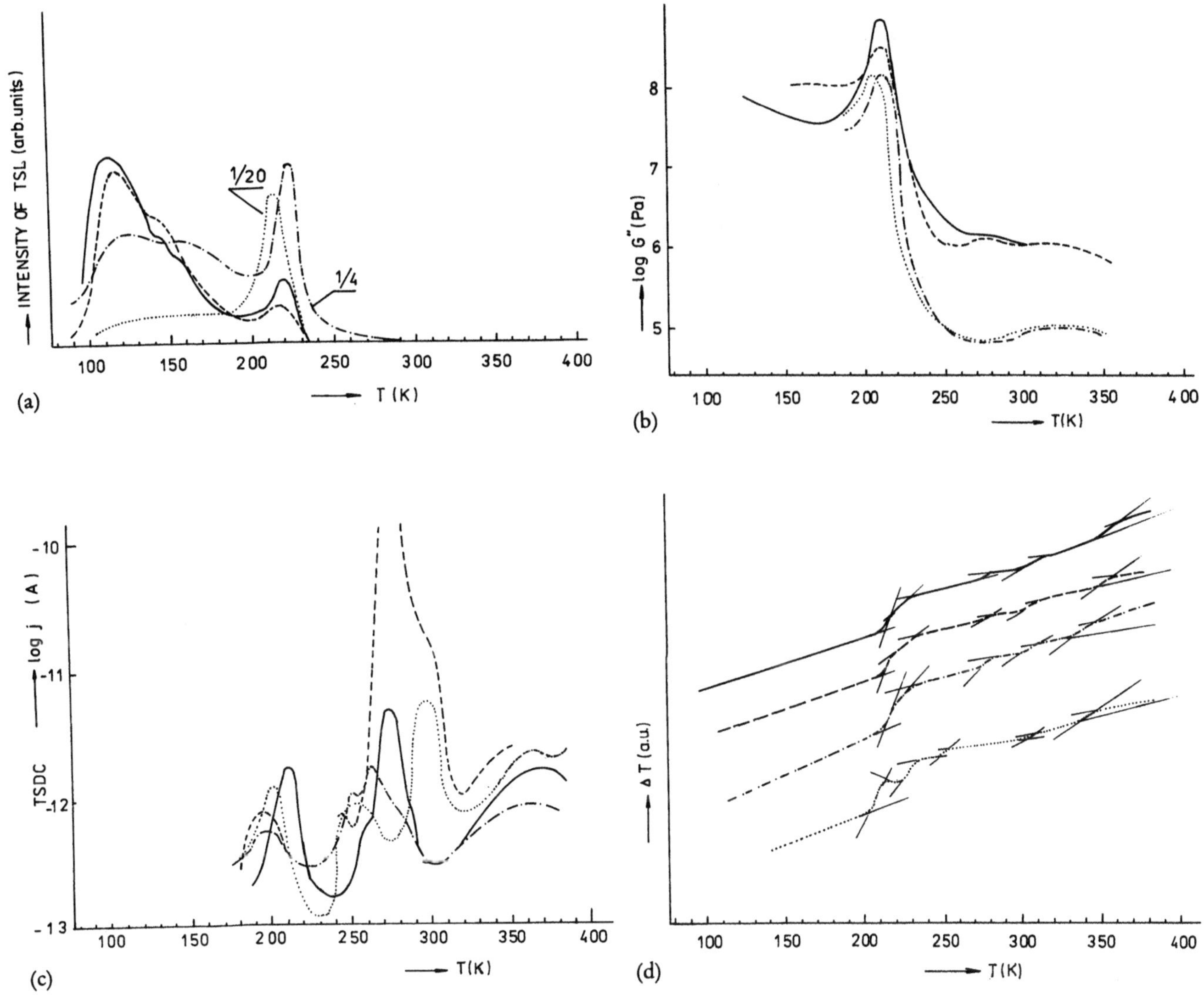

Fig. 2. Temperature dependence of luminescence intensity (a), G'' (b), depolarization current j (c) and thermal difference ΔT (d) of the four laboratory prepared samples; sample 1 (–), 2 (– – –), 3 (····), 4 (–·–). The ratios 1/20 and 1/4 in Fig. 2a denote the vertical reduction. The curves of Fig. 2d are vertically shifted for clarity

usefulness of the comparison is demonstrated in Fig. 2. We present some conclusions which come out from this comparison. The main transition zone of butadiene is reflected by all the methods, but the relaxation strength of the relaxation is different. The TSL maxima of samples with lower molecular weight additives (sample 3 and 4) are extreemly high, the maxima in TSDC are lower than the TSDC maxima in the 240 K to 320 K region. From this, it follows that the luminescence activity is very much favoured by the low molecular weight additives and this may be the key to a more detailed knowledge of the TSL mechanism. The small TSDC maxima reflect the low dielectric activity of butadiene. The very large TSDC maxima in the region 240 K to 320 K indicate the mainly dielectric-type origin of this relaxation mechanism. The softening activity of the low molecular weight additives is most evident in the DTA case, but also in the TSL and viscoelastic cases the butadiene maximum of the sample 3 (low molecular weight additive) lies at lowest temperature of all the measured samples.

Several more conclusions may be obtained from comparison of the curves in Figs. 1 and 2 but these are beyond the scope of this short article, in which we want to demonstrate the usefulness of a comparative study of relaxation behaviour of polymers.

References

1. Zimola B, Havránek A (1981) Plasty a kaučuk 18:337
2. Bakule R, Nedbal J, Štrof P (1986) In: Sedláček B (ed) Morphology of Polymers. Walter de Gruyter, Berlin, p 197
3. Pospíšil J, Havránek A (1986) In: Sedláček B (ed) Morphology of Polymers. Walter de Gruyter, Berlin, p 399
4. Saltman WM (ed) (1977) The Stereo Rubbers. Wiley, New York, p 592
5. Pospíšil J, Havránek A, Tale IA (1985) In: Sedláček B (ed) Optics of Dynamic Phenomena and Processes in Marcomolecular Systems. Walter de Gruyter, Berlin, p 245
6. Tkaczyk S, Swiatek A, Kwiatkovska A (1988) Progr Colloid Polym Sci 78:136

Received February 1, 1988;
accepted April 21, 1988

Authors' address:

Dr. A. Havránek, CSc
Department of Polymer Physics
MFF UK
V Holešovičkách 2
180 00 Praha 8, Czechoslovakia

Dielectric relaxation processes in poly (epoxypropylcarbazole) layers

I. Chudáček[1]), D. Slavínská[1]), V. Kriščiunas[2]), F. Kuliešius[2]), and P. Mačkus[2])

[1]) Charles University, Department of Polymer Physics, Prague, Č.S.S.R.
[2]) Kapsukas University, Department of Solid State Electronics, Vilnius, S.S.S.R.

Abstract: Polyepoxypropylcarbazole films were cast from solution on to an aluminium substrate and the upper Al-electrode was vacuum evaporated. The components of complex dielectric constants were measured in the frequency range from 10 to 10^5 Hz at temperatures 273 to 420 K. The experimental results lead to the conclusion that near T_g the observed relaxations are essentially associated with the cooperative nature of the carbazole units motions without the direct influence of the polymeric chain. The experimental data fit the nonexponential correlation function better than the functions obtained from the distribution of relaxation time concept.

Key words: Dielectric relaxation, poly (epoxypropylcarbazole), molecular correlation function, cooperative monomers motion.

1. Introduction

A very important goal of modern polymer physics is the unambiguous determination of the dynamics of molecular reorientation motions. The polymers with bulky side-groups are especially interesting in this sense.

The present paper covers the dielectric relaxation behaviour near the glass transition in poly (epoxypropylcarbazole) (PEPC), the main physical features of which are given in [1]. A choice of this material was influenced by need for fundamental comprehension of processes in this type of polymer, as well as by the widespread application of PEPC in imaging devices.

2. Experimental

PEPC from which the layers have been formed was purified by washing in a 10 % benzene solution in distilled water and fractionated later by consequent precipitations into benzene-hexane system. The molecular weight $\bar{M}_n$ alters from 760 to 1570. Films of thickness 10 to 100 μm were cast from solution on an aluminium substrate and the upper Al-electrod was then vacuum evaporated.

The impedance components were measured using General Radio 1621 and Wayne Kerr B 905 A capatitance assemblies for which Datron 1065 and Gould S 50000 served as output indicators. Measurements were performed in the frequency range from 10 to 10^5 Hz at temperatures 273 to 420 K.

PCPS HT 006

3. Results and discussion

Figures 1, 2 illustrate the dielectric loss tangent of PEPC layers versus temperature and frequency. In these layers a good correlation of absorption maxime location with glass transition temperature T_g [2] indicates the α segmental relaxations. Despite the pronounced dependence of relaxation region location on temperature and molecular weight, the shape of the relaxation curves is not affected by these parameters. Experimental data plotted in normalized coordinaves (see Fig. 3) demonstrated this. In this case the Havriliak-Negami [3] equation –

$$\varepsilon' + i\,\varepsilon'' = \varepsilon_\infty - \Delta\varepsilon\,[1 + (i\,\omega\tau_o)^\alpha]^{-\beta} \qquad (1)$$

(where ε', ε'' are the components of complex dielectric constant ε_o. ε_∞ are the in-phase component at zero and infinite frequencies ($\omega = 2\pi f$), respectively, $\Delta\varepsilon = \varepsilon_o - \varepsilon_\infty$, τ_o is the relaxation time) with fitting parameters $\alpha = 0.4$, $\beta = 0.9$ represents this loss curve well.

The dependence of the loss peak frequency f_{max} on reciprocal temperature compiled from temperature and frequency measurements is presented in Fig. 4. The activation energy E_a alters from 220 to 300 kJ/mol for the molecular weights under investigation. These

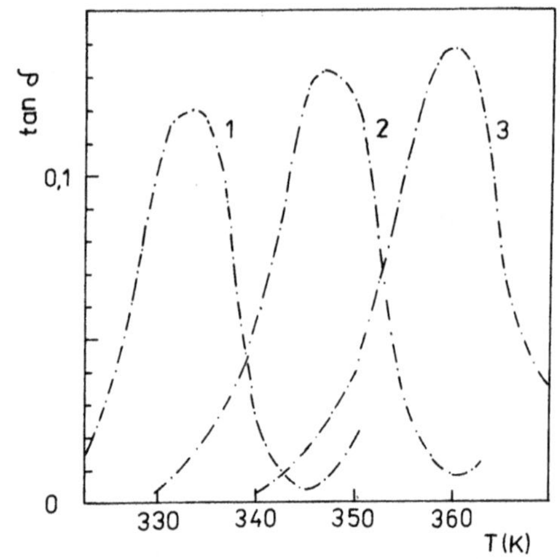

Fig. 1. Tan δ vs. temperature at 1 kHz for PEPC of different molecular weight. (Curve 1: $\bar{M}_n = 760$, 2: $\bar{M}_n = 1170$, 3: $\bar{M}_n = 1570$)

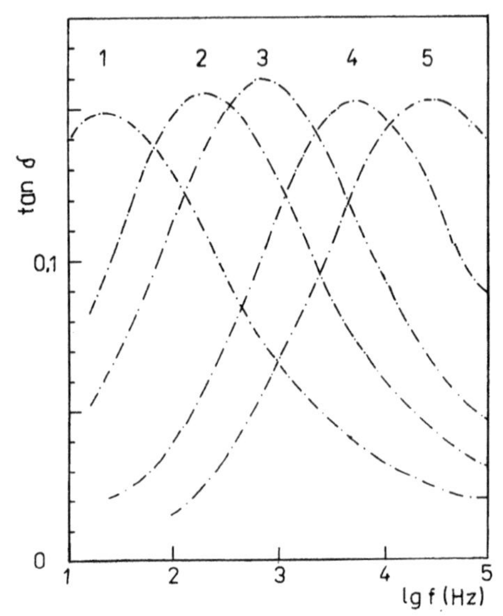

Fig. 2. Tan δ vs. $\log f$ for PEPC at T_g region. ($\bar{M}_n = 760$; curve 1: $T =$ 318 K, 2: 326,5 K, 3: 332 K, 4: 340 K, 5: 347,5 K)

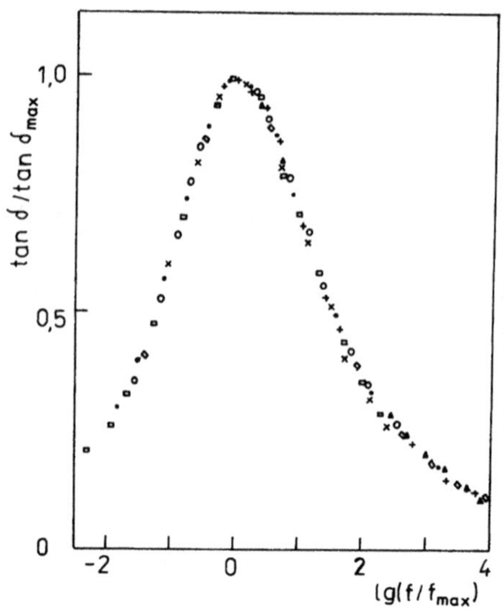

Fig. 3. Tan δ/tan δ_{max} vs. $\log (f/f_{max})$ for PEPC at different temperatures and molecular weight. ($\bar{M}_n = 760$; +: 318 K, ×: 326,5 K, ○: 329 K, •: 332 K, □: 340 K. $\bar{M}_n = 1570$; ▲: 342 K, ◇: 364 K)

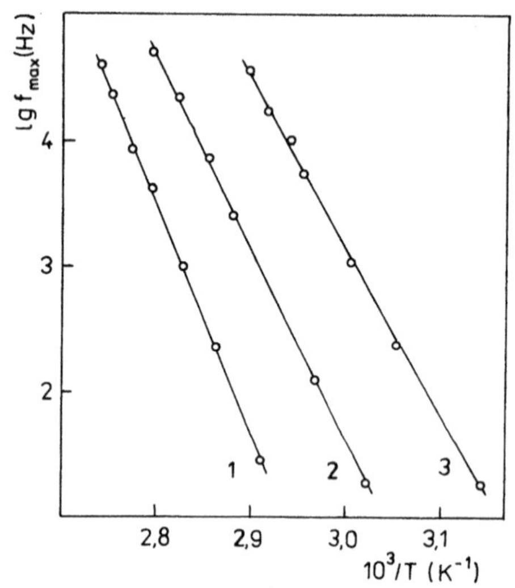

Fig. 4. Log f_{max} vs. reciprocal temperature for PEPC of different molecular weight. (Curve 1: $\bar{M}_n = 760$, 2: 1170, 3: 1570)

values are in accordance with the earliest PVK investigations [4] and subsequent studies [5, 6]. It seems interesting to note that for the polyphenyl ether case a value close to 270 kJ/mol for 9-acetylanthracene molecule, which is, regarding size, similar to the carbazole value, was obtained in [7]. At the temperature and frequency ranges under investigation experimental data could be fitted by an Arrhenius-plot, but these ranges are too narrow to exclude the possibility of a Williams-Landel-Ferry law.

The anomalously large activation energy, the independence of the relaxation curve shape from both temperature and molecular weight, as well as the value of the half-width near 2,5 decades of frequency, are typical for amorphous homopolymers and polar-nonpolar polymers [8]. Johari [9] emphasised that in supercooled liquids the rotations of simple molecules are similar to the relaxations in polymers. The assumptions mentioned above lead to the conclusion that in PEPC near T_g the observed relaxations are essentially associated with the cooperative nature of carbazole units motions and they are not the consequence of a chain connectivity phenomenon. The dominating factor, the carbazole side chain, which controls the main chain behaviour, was also underlined in works concerned with the PVK structural and charge transport properties [10]. Therefore, apart from hindered carbazole group rotation, the attribution of relaxation regions to fixed molecular unities on the polymeric chain [4, 5] is questionable.

In order to relate the experimental data with the material microparameters the macroscopic decay function [11]

$$\phi\left(\frac{t}{\tau_o}\right) = \frac{\int\limits_{-\infty}^{\infty} \left(\frac{\varepsilon''}{\varepsilon''_{\max}}\right) \cos\left[\left(\frac{f}{f_{\max}}\right)\left(\frac{t}{\tau_o}\right)\right] d\log\left(\frac{f}{f_{\max}}\right)}{\int\limits_{-\infty}^{\infty} \left(\frac{\varepsilon''}{\varepsilon''_{\max}}\right) d\log\left(\frac{f}{f_{\max}}\right)} \quad (2)$$

which is associated with molecular correlation function $\Gamma(t)$ was computed from dielectric relaxation data. $\Phi(t)$ received obeys the empirical Williams-Watts expression [12]

$$\Phi(t) = \exp\left[-\left(\frac{t}{\tau_o}\right)^{\beta}\right] \quad (3)$$

with the value of fitting parameter $\beta \simeq 0.5$. The loss spectrum involves asymmetry if expression (3) is valid. This is our case (see Fig. 3). For PEPC layers under investigation the proposition of nonexponential correlation function is preferred to an assumption of distributed relaxation times determined by (1). The available facts indicate that the model of conformational jumps [13] as well as of free volume [14] are preferred in our case. In order to confirm it, further investigations will be performed.

References

1. Gaidelis V, Montrimas E, Kriščiunas V (1976) Thin Solid Films 38:9
2. Kirščiunas V, Kuliešius F, Latakas R, Tamošiunas S (1986) Liet fiz rinkinys XXVII:233
3. Havriliak S, Negami S (1966) J Polym Sci C 14:99
4. Pochan JM, Hinman DF, Nash R (1975) J Appl Phys 46:4115
5. Pochan JM, Hinman DF (1976) J Polym Sci 14:2285
6. Block H, Cowden DR, Lord PW, Walker SM (1978) Polym Bull 1:137
7. Crossley J, Heravi A, Walker S (1981) J Chem Phys 75:418
8. Williams G (1982) Ann Rept Conf Electr Insul Diel Phenom, New York
9. Johari GP (1985) J Chem Phys 82:283
10. Williams DJ, Limburg NW, Pearson JM, Goedde AO, Yanus JF (1975) J Chem Phys 62:1501
11. Cook M, Watts DC, Williams G (1970) Trans Farad Soc 66:2503
12. Williams G, Watts DC (1970) Trans Farad Soc 66:80
13. Geny F, Monnerie L (1977) J Polym Sci, Polym Phys Ed 15:1
14. Chov TS (1985) J Non-Cryst Solids 75:209

Received November 13, 1987;
accepted April 21, 1988

Authors' address:

Prof. I. Chudáček
Department of Polymer Physics
Charles University
V Holešovičkách 2
180 00 Praha, C.S.S.R.

Dielectric relaxation of polyurethane systems Part II: Shape parameters

A. Schönhals and E. Schlosser

Academy of Sciences of GDR, Central Institute of Organic Chemistry, Berlin, G.D.R.

Abstract: We report on dielectric relaxation measurements carried out on a crosslinked polyurethane system. The shape of the main relaxation is characterized by two parameters of the model function of Havriliak and Negami. The dependence of the shape parameters on the urethane group concentration is presented and interpreted by a new model.

Key words: Dielectric relaxation, main transition, shape parameters, polyurethane network, theoretical model.

1. Introduction

Dielectric relaxation spectroscopy is a powerful method for the study of molecular mobility, an integral part of the microstructure of polymers.

In a preceding paper ([1]; part I) this method was described according to its application to crosslinked polyurethane systems. The intensity and the position of relaxation at the frequency scale could be elucidated semiquantitatively. It was one of the main results that dielectric behavior is much more influenced by the variation of the urethane concentration than by the change of the network density.

In this paper we extend that discussion to the interpretation of the shape of relaxational loss peak.

The dielectric relaxation behavior at constant temperature is characterized by the complex permittivity

$$\varepsilon^x(\omega) = \varepsilon'(\omega) - i\varepsilon''(\omega)\,. \qquad (1)$$

According to Havriliak and Negami [2] an isolated relaxation process can be described totally by the model function (HN-function)

$$\varepsilon^x(\omega) - \varepsilon_\infty \equiv \Delta\varepsilon[1 + (i\omega/\omega_0)^\beta]^{-\gamma}$$

$$\beta, \gamma < 1;\ \varepsilon_\infty = \varepsilon'(\omega),\ \omega >> \omega_0 \qquad (2)$$

with only four parameters, viz., the intensity $\Delta\varepsilon$, the position at the frequency scale ω_0, and the shape parameters β and γ.

The limiting dependencies of the imaginary part $\varepsilon''(\omega)$ is given by simple power laws [3]

$$\varepsilon''(\omega) \sim \omega^\beta \qquad \omega << \omega_0 \qquad (3)$$

$$\varepsilon''(\omega) \sim \omega^{-\beta\gamma} \qquad \omega >> \omega_0\,. \qquad (4)$$

Up to this time a discussion of the shape parameters has not been exhausted because of their unsatisfying accuracy.

However, now by means of computer-aided analysis of measurements the parameters can be determined precisely enough so that an interpretation seems worthwhile.

2. Experimental

Polymers under study (Fig. 1 and table 1):

Polyurethane, prepared with poly(propyleneglycol) (PPG), 2,4-toluylenediisocyanate (TDI) and trimethylolpropane (TMP) as crosslinking agent. The change of the structure is realized by variation of the molecular weight of the polyol which simultaneously alters the urethane concentration per network chain.

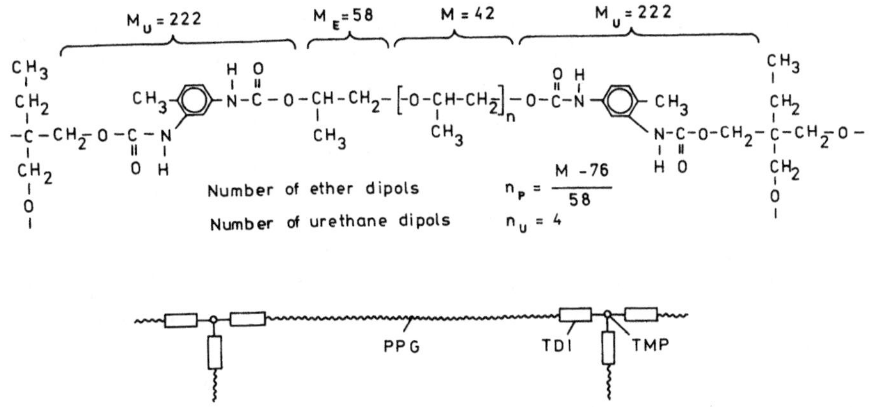

Fig. 1. Composition of the polyurethane model system

The complex permittivity $\varepsilon'(\omega)$ was determined in the frequency range $1 < f/\text{Hz} < 10^5$ at temperatures above the thermal glass transition (main relaxation).

The parameters $\Delta\varepsilon$, ω_0, β and γ were obtained by computer least-squares fitting [3] of the HN-model function to the experimental information at different temperatures. In order to compare them for the different samples a reduction to equal molecular mobility is necessary. We have chosen the values of the parameters at that temperature at which the position of the loss maximum $\varepsilon''_{\max} = \varepsilon''(2\pi \cdot f_m)$ *is* $f_m = 100\,\text{Hz}$.

With regard to the shape parameters β and γ results are given in Fig. 2. For convenience the parameters $m = \beta$ and $1 - n = \beta \cdot \gamma$ are plotted versus urethane concentration which is characterized by the ratio of the number of urethane n_U and ether groups n_P per network chain (cf. Fig. 1).

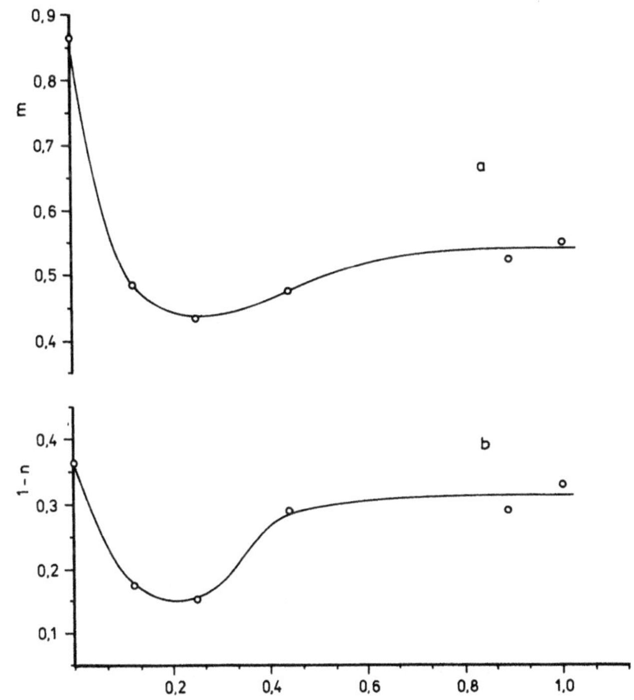

Fig. 2. Shape parameters m and $1 - n$ vs. urethane concentration

3. Model

The model proceeds from the general assumption that the molecular mobility in bulk polymers at the glass transition is always affected by intra- and intermolecular interactions. These types of interaction should be responsible for two different relaxation processes named m- and n-type. Both mechanisms contribute to the dielectric response and develop a uni-

Table 1. Composition of the PUR model system

Sample	Polyol	PUR 1	PUR 2	PUR 3	PUR 4	PUR 5
M of Polyol	∞	2055	1000	600	340	305
ur-concentration $= \frac{n_U}{n_P}$	0.0	0.12	0.25	0.44	0.89	1.

form relaxation process. But, because of their different relaxation times they are discernable at high and low frequencies (cf. Eqs. (2) and (3)) respectively. In our opinion, the n-type process is related to the local chain dynamics which can be studied by means of spectroscopic methods in solution [5, 6] as in bulk [6].

This motion can be treated theoretically as a damped diffusional propagation of the orientation $n(r,t)$ along the chain [7–9]. A generalization of this picture is possible using a time dependent diffusion coefficient $D(t)$. We get in the Laplace space [10]

$pN(x, p) = D(p)\, d^2 N/dx^2$ and assume

$$D(p) = K/p^{\alpha} \text{ with } -1 < \alpha < 1 \tag{5}$$

$-1 < \alpha < 0$ describes a hindered diffusion

$0 < \alpha < 1$ characterizes a motional process between diffusion and wave transfer.

From [10] follows

$$1 - n = (1 + \alpha)/2 . \tag{6}$$

The m-type process includes the relaxation of the surrounding of a test segment. Using the treatment of Nigmatullin and Hill [11] the parameter m is related to a decoupling parameter $\varkappa$ which increases with the correlation of the movement

$$m = (2 - \varkappa)/(\varkappa - 1) \qquad \text{with } 1.5 < \varkappa < 2 . \tag{7}$$

4. Interpretation of the shape parameters

– The parameter $\varkappa$ which describes the extent of correlation should increase with the strength of interaction, i.e., m should decrease according to Eq. (6).

– It is clearly seen from Fig. 1 that the interaction increases with the concentration of urethane groups most of all by formation of hydrogen bonds.

Therefore the level of m is diminished with urethane content as it is shown in Fig. 2a.

– At high urethane concentration the interaction possibilities will be saturated. For this reason m reaches a constant value.

– The minimum of m can be interpreted by the fact that the degree of saturation is gradually reduced and therefore m is raised up at higher crosslink density by steric reasons.

– Larger inhomogeneity of the chain and increased interaction with the surrounding caused by the urethane groups should hinder the local motional process.

Thus in Fig. 2b $1 - n$ decreases with the urethane content.

– The minimum of $1 - n$ vs. urethane concentration is also due to steric reasons.

References

1. Schlosser E (1985) Plaste u Kautschuk 32:20
2. Havriliak S, Negami S (1966) J Polym Sci C Polym Symp 14:99
3. Schlosser E, Kästner S, Friedland K (1981) Plaste u Kautschuk 28:77
4. Dissado LA, Hill RM (1983) Proc Roy Soc A 390:131
5. Viovy JL, Monnerie L, Brochon JC (1983) Macromolecules 16:1845
6. Laupretre F (1986) 8 Tagung Polymerphysik Leipzig, September 22–25
7. Hall CK, Helfand E (1982) J Chem Phys 77:3275
8. Valeur B, Jarry JR, Geny F, Monnerie L (1975) J Polym Sci, Polym Phys Ed 13:667, 675, 2251
9. Shore JE, Zwanzig R (1975) J Chem Phys 63:5445
10. Schönhals A, Schlosser E, Colloid Polym Sci, accepted
11. Nigmatullin RR, Hill RM (1983) phys stat sol (b) 118:769

Received June 25, 1987;
accepted April 21, 1988

Authors' address:

A. Schönhals
Academy of Sciences
Central Institute of Organic Chemistry
Rudower Chaussee 5
DDR-1199 Berlin, G.D.R.

Dielectric study of aqueous solutions and solid samples of methylcellulose

P. Pissis and D. Daoukaki-Diamanti

Physics Department, National Technical University Athens, Greece

Abstract: The dielectric properties of aqueous solutions and hydrated solid samples of methylcellulose were studied by means of the thermally stimulated depolarization current (TSDC) method in the temperature range 77–300 K and over a wide range of concentrations, 0.025–2.0 gr/lt, and water contents, 1.8–34.9 %. Two water-dependent complex dispersions were observed in the range of 77–155 K (low-temperature dispersion) and 155–300 K (high-temperature dispersion). The high-temperature dispersion was studied in detail. In the solutions the dispersion is due to thermal release of space charges trapped at the boundaries of the crystallites of the polycrystalline samples. In the solid samples three mechanisms contribute to the dispersion: water-assisted relaxation of local polar groups, space charge polarization connected with *dc* conductivity and accumulation of space charges at the electrodes. The results allow us to determine the critical water content for the completion of the primary hydration layer to 7–8 %.

Key words: Dielectric relaxation, hydration properties, depolarization current, space charges.

1. Introduction

We report in this paper on the dielectric behavior of aqueous solutions and hydrated solid samples of methylcellulose by means of the thermally stimulated depolarization current (TSDC) method in the temperature range 77–300 K. From a fundamental point of view it is interesting to compare the dielectric behavior of methylcellulose with that of cellulose. Another reason for choosing methylcellulose is that this polymer is water soluble, so both aqueous solutions and hydrated solid samples can be studied and the results of these studies can be compared. From a practical point of view the hydration properties of methylcellulose might be of interest for several of its applications, for example as a food additive, as a thickener, and as protective colloid. In contrast to cellulose, data on the hydration properties of methylcellulose and methylcellulose derivatives by means of dielectric as well as other methods are scarce [1, 2]. Most dielectric studies on methylcellulose are confined to dry samples [3].

2. Experimental

The TSDC method, its equations, the methods of evaluating the activation energy for the dipole reorientation, W, the pre-exponential factor in the Arrhenius equation, τ_0, and the contribution of a mechanism to the static permittivity, $\Delta\varepsilon$, as well as the experimental apparatus used have already been described [4, 5].

We used commercial grade methylcellulose (Hickock, USA). The aqueous solutions were prepared with doubly-distilled and deionized water (*dc* conductivity 1×10^{-6} ohm^{-1} cm^{-1}). The concentration c was varied between 0.025 and 2.0 gr/lt.

Solid samples were made by compressing methylcellulose powder to cylindrical pellets of 9.2 mm diameter and different heights between 2.0 and 2.5 mm. The water content h, defined as the ratio of weight of the sorbed water to the weight of dry methylcellulose, was varied between 1.8 and 34.9 %.

3. Results and discussion

In Fig. 1 we show TSDC plots measured on a solution with $c = 0.2$ gr/lt and on a solid sample with $h =$ 21.5 %. Each plot shows a dispersion in the range of 77–155 K (low temperature dispersion) and a second

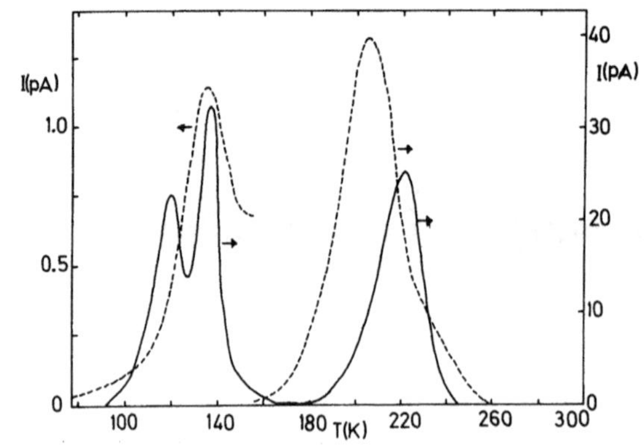

Fig. 1. TSDC plots measured on a methylcellulose solution with c = 0.2 gr/lt (——) and on a solid sample with h = 21.5% (- - -). The polarizing field was 3.7 kV/cm, the polarization temperature 295 K, the polarization time 5 min and the heating rate 3.0 K/min

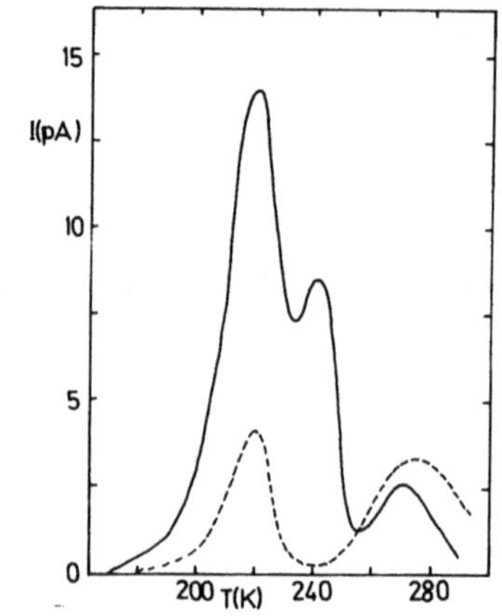

Fig. 2. TSDC plots measured on hydrated solid samples with h = 13.0% for two different electrode arrangements: direct contact to the brass electrodes (——) and teflon foils between the sample and the capacitor plates (- - -). Conditions as in Fig. 1

dispersion in the range of 155–300 K (high-temperature dispersion). Results to be reported elsewhere show that the low-temperature dispersion, both in solutions and in hydrated solid samples, is due to dielectric relaxation of free and loosely bound [5] water molecules. In the present work we studied the high-temperature dispersion in detail.

In the solutions the high-temperature dispersion consists of a single peak which can be satisfactorily described by a Debye dispersion. The peak temperature T_M, $T_M = 221 \pm 2$ K, and the activation energy W, $W = 0.43 \pm 0.02$ eV, are independent of concentration and within the limits of experimental errors equal to the corresponding values of the high-temperature TSDC peak in polycrystalline ice [6]. The contribution of the peak to the static permittivity, $\Delta\varepsilon$, on the contrary, as well as the dc conductivity of the solutions, increase with increasing c. These results reasonably indicate that the origin of the high-temperature dispersion in the solutions is the same as in pure polycrystalline ice, i.e., the thermal release of space charges generated inside the crystallites and trapped at the boundaries of the crystallites [6]. The solute itself does not contribute to this dispersion.

In contrast to the solutions, the high-temperature dispersion in solid samples is complex. The position, the shape, and the depolarization charge of the complex dispersion depend on the electrode arrangement for the same water content h (Fig. 2).

For samples in direct contact to the brass electrodes or with thin conducting foils (Au, Cu) between the sample and the brass electrodes the high-temperature dispersion consists of three peaks (Figs. 1 and 2) numbered I–III in order of increasing temperature. The position of the three peaks and $\Delta\varepsilon$ of the first two peaks are independent of the electrode material. $\Delta\varepsilon$ of peak III increases in the order brass-Cu-Au. Peak I shifts to lower temperatures with increasing h (Fig. 3). Its contribution to the static permittivity, $\Delta\varepsilon$, increases with increasing h and saturates at $h \simeq 10$%. The increase is very pronounced at h = 7–8%. Several tests offered by the TSDC method [4, 5] show that peak I is probably of dipolar origin: T_M is independent of T_p, $\Delta\varepsilon$ is independent of E_p. Peak II appears only for water contents higher than about 7–8%. Its position, $T_M = 242 \pm 3$ K, is independent of h, while $\Delta\varepsilon$ seems not to depend

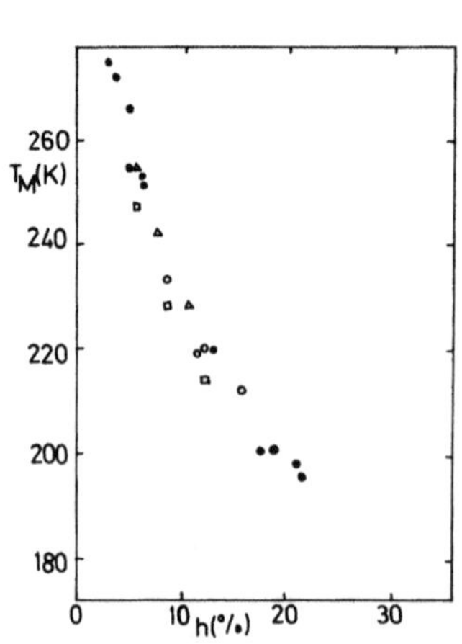

Fig. 3. Peak temperature T_M of peak I vs water content h. Symbols correspond to different electrode material: brass (●), Au (○), Cu (△) and teflon (□)

systematically on h. Peak III, a broad peak that shifts to lower temperatures with increasing h, was not studied in detail.

For samples in the MISIM contact configuration, i.e., with thin insulating foils (teflon, mica) between the sample and the capacitor plates, peak II is eliminated, while peaks I and III appear at the same position as with brass electrodes (Fig. 2).

The results support the hypothesis that peak I is due to dipolar polarization, peak II to space charge polarization connected with dc conductivity, and peak III to space charges accumulated at the electrodes. The dependence of the properties of the peaks on h makes reasonable that a critical water content of about 7–8% exists. Measurements to be reported elsewhere show that the primary hydration layer is completed at this water content.

Comparison of our results with TSDC measurements on hydrated cellulose [7], dielectric loss measuremens on dry methylcellulose [3], and mechanical and dielectric loss measurements on hydrated cellulose and cellulose derivatives [8–10] show that peak I is identical with a TSDC peak measured on hydrated cellulose and a mechanical and dielectric loss peak measured on cellulose and several cellulose derivatives. This peak could be attributed to motion of $CH_2OH{-}H_2O$ or glucopyranose rings relieved by water in the amorphous region. The results are less conclusive regarding peaks II and III. Mechanical and dielectric loss peaks observed in the temperature range of these peaks [3] have been attributed to Maxwell-Wagner interfacial polarization.

References

1. Allen DJ, Neale SM, Tait PJT (1972) J Polym Sci A2 10:433
2. Bordi F, Cametti C (1985) Ber Bunsenges Phys Chem 89:747
3. Algie JE (1979) Dielectric Properties of Keratin and Cellulose; in: Happey F (ed) Applied Fibre Science. vol 2, Academic, New York, pp 169–203
4. van Turnhout J (1980) Thermally Stimulated Discharge of Electrets; in: Sessler GM (ed) Electrets. Springer, Berlin, pp 81–215
5. Pissis P, Diamanti D, Boudouris G (1983) J Phys D 16:1311
6. Apekis L, Pissis P, Boudouris G (1983) J Phys Chem 87:4019
7. Pissis P, Anagnostopoulou-Konsta A (1985) In: Sessler GM, Gerhard-Multhaupt R (eds) Proceedings of the 5th International Symposium on Electrets, Heidelberg, IEEE Service Center, Piscataway, pp 842–847
8. Bradley SA, Carr SH (1976) J Polym Sci Phys Ed 14:111
9. Kimura M, Nakano J (1976) J Polym Sci Lett Ed 14:741
10. Seymour RW, Weinhold S, Haynes SK (1979) J Macromol Sci Phys B16:337

Received April 24, 1987;
accepted April 21, 1988

Authors' address:

P. Pissis
National Technical University, Athens
Physics Department
Zografou Campus
GR-15773 Athens, Greece

^{1}H-NMR transversal relaxation in crosslinked 1,4-cis-polybutadiene

G. Simon, H. Schneider, and K.-G. Häusler[1])

Department of Physics (Chemistry[1]), Technical University Leuna-Merseburg, Merseburg, GDR

Abstract: By means of ^{1}H-NMR relaxation molecular motion in crosslinked 1,4-cis-polybutadiene is studied. The relaxation function shows three clearly distinguishable components attributed to chain segements under different motion. Correlation times and the degree of anisotropy of segmental motion are obtained by a theoretical analysis of the form of magnetization decay.

Key words: Crosslinked polybutadiene, network dynamics, correlation time spectrum, ^{1}H-NMR relaxation, second moment.

1. Introduction

In the last decade transversal ^{1}H-NMR relaxation studies on polymeric networks were published by some authors [1–7] showing that this method is a useful tool for characterizing molecular dynamics above the glas temperature; thermal processes were characterized by their effect on averaging short-range interactions of nuclear magnetic dipols.

In general, two qualitatively different components were found in the relaxation function – a short one attributed to intercrosslink chains and a long one frequently attributed to free ends of chains. The aim of this work is to show that the form of relaxation function can be more complicated, in particular as far as the long relaxation times are concerned.

These investigations of the dynamics of polymeric networks by NMR are in part directed to creating a practical way of studying filled networks. In such systems the interpretation of the elastic modulus, which has an excellent theoretical background for unfilled systems [8], fails.

2. Experiments

The polybutadiene (BUNA cis 132) used is a technical product containing 96% of 1,4-cis-polybutadiene with a mean molar mass $\bar{M}_n = 62330$ g · mol^{-1} and a polydispersity of about 12. The original polybutadiene was exposed to different doses X of β^- irradiation (X = 0, 0.05, 0.10, 0.15, 0.30, 0.50, and 0.90 MGy). This gave crosslinked samples PB 0, PB.05, ..., and PB.90 which were characterized by sol-gel analysis, distribution of the molar masses of the sol fraction, determination of the degree of swelling, and linear compressibility. These and further characteristics from thermoanalytical methods can be found in [9].

The NMR experiments were carried out on a Bruker SXP4-100 spectrometer working at a proton resonance frequency $f_o = 88$ MHz. For the measurement of the transversal magnetization decay we used the common Hahn spin echo method (cf., e.g., [10]). Time-saving Carr-Purcell-Meiboom-Gill pulse trains (cf., e.g., [10]) could not be used because the decay depends on the pulse distance τ at $\tau < 2$ ms [3]. Relaxation curves of all samples are qualitatively similar (Fig. 1). Two long exponential components can be separated. A short component of about 85 proton-% is left in a form similar to a Gauss function; this is demonstrated for PB.15 in Fig. 2.

3. Data analysis

The interpretation of the relaxation function is carried out by a model which takes its pattern from results published in [1–4]. We will assume that near-static dipolar interactions are intragroup in origin at room temperature because of the liquid-like motion of polymers with low glass temperature (e.g., polybutadiene). The spectrum of correlation times should consist of three mean correlation times. Two very short times, τ_a and τ_c, are due to the local motion of statistical segments of a chain (τ_a), or are due to few, very

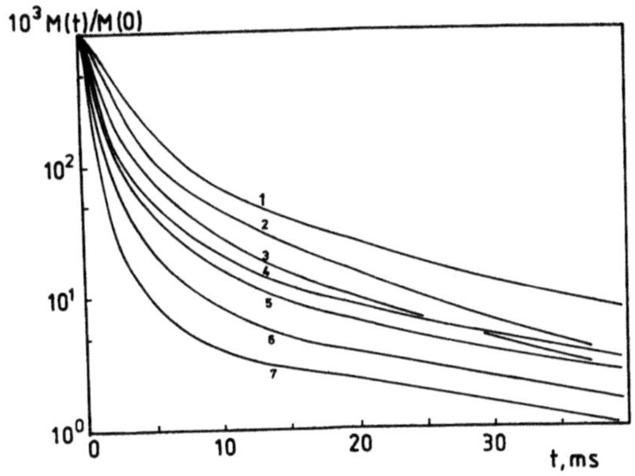

Fig. 1. Transversal magnetization decays of crosslinked polybutadiene (1–7 correspond to PB 0-PB.90), which are fitted to experimental decays. For all curves the following relation is valid

$$\left(\frac{\sum_{n=0}^{45}(y_n - f_n)^2}{\sum_{n=0}^{45}(y_n - \bar{y})^2}\right)^{1/2} < 0.06,$$

where y_n are the experimental values, $\bar{y}$ is the mean value of y_n, and f_n is the fit function

mobile segments at the end of a chain (τ_c) where $\tau_c < \tau_a$. The degree of anisotropy of local motion (τ_a) is mainly determined by certain restrictions – physical entanglements and chemical crosslinks. This anisotropy leads to a nonaverageable part of the second moment of the dipolar interaction, and consequently to a fast solid-like magnetization decay. The long time τ_b ($\tau_b \gg \tau_a, \tau_c$) is due to slow isotropic motion of an

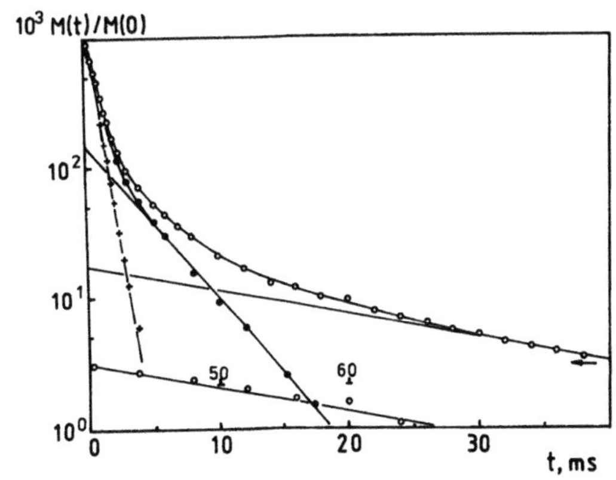

Fig. 2. Demonstration of separation of components for sample PB.15. Full curves correspond to the fit function. (○) initial experimental values, (●) after subtraction of the longest exponential component (C-component of Eq. (6)), (+) after subtraction of the B- and C-components. The initial curve is shown up to about 60 ms (⊥ 50, ⊥ 60 ms)

ensemble of chains which can average the remaining part of dipolar interaction.

That is why the short component of the relaxation function represents chains fixed between two points by chemical crosslinks or physical entanglements of a lifetime greater than the time of magnetization decay. The middle and the long components represent chain ends which are more or less mobile. This last conclusion is supported by analogous experiments on extracted and swollen poly(butadiene-styrene) networks [11]. The quality of the relaxation function is the same, but the amount of the middle and long component increases because of a decrease of physical restrictions close to the chain ends.

Further, it is assumed that all segments of each of the three dynamic groups can be characterized by mean quantities. A common correlation function of intercrosslink chains [4,14] is

$$G(\tau) = \{M_{2,a}\exp(-\tau/\tau_a) + M_{2,b}\}\exp(-\tau/\tau_b) \quad (1)$$

where $M_2 = M_{2,a} + M_{2,b}$ is the second moment of the rigid lattice; $M_{2,b}$ is the rest of the moment which cannot be averaged by the fast anisotropic motion (τ_a). Since $\tau_b \gg \tau_a$ and $M_{2,b} \ll M_{2,a} \simeq M_2$, Eq. (1) changes to

$$G(\tau) = M_2\exp(-\tau/\tau_a) + M_{2,b}\exp(-\tau/\tau_b). \quad (2)$$

For $\omega_0\tau = 2\pi f_o \tau \gg 1$ the relaxation function [12] can be written as

$$\phi(t) = \exp\left\{-\int_0^t (t-\tau)G(\tau)\,d\tau\right\} \quad (3)$$

or with Eq. (2)

$$\phi(t) = \exp\{-M_2(\tau_a^2 f(t/\tau_a) + q\tau_b^2 f(t/\tau_b))\} \quad (4)$$

where $q = M_{2,b}/M_2$ and $f(t/\tau) = \exp(-t/\tau) + t/\tau - 1$.

For the total relaxation function including the contribution of all polymeric net segments the following is valid

$$M(t) = A\phi(t)_{q\neq 0} + B\phi(t)_{q=0} + C\tilde{\phi}(t)_{q=0}. \quad (5)$$

intercrosslink chains — *less* / *more mobile chain ends*

In $\tilde{\phi}$ the time τ_a is replaced by τ_c. Since $M_2\tau_a^2, M_2\tau_c^2 \ll 1$ (correlation time is much smaller then the decay time)

Table 1. Fitted parameters of Eqs. (6) and (7); the M_c^+-values were determined by small linear compression [9]

Parameters	Samples with increasing doses of irradiation						
	PB 0	PB.05	PB.10	PB.15	PB.30	PB.50	PB.90
M_c^+, kg/mol	–	17.9	8.8	6.8	3.7	2.4	2.0
A	728	792	817	830	845	891	913
B	236	177	159	152	142	100	81
C	36	30	23	17	13	9	5
$q \cdot 10^4$	0.24	0.44	1.00	1.44	1.67	3.15	6.33
$\tau_a \cdot 10^8$	0.46	0.56	0.62	0.81	0.91	1.42	3.04
$\tau_b \cdot 10^3$	1.09	0.80	0.56	0.57	0.46	0.34	0.23
$\tau_c \cdot 10^9$	0.81	1.16	1.09	0.84	0.82	0.83	0.82

and $qM_2\tau_b^2$ is of the order of one [4], the resulting relaxation function is

$$M(t) = A \cdot \exp\{-t/T_{2,a} - qM_2\tau_b^2 f(t/\tau_b)\} + B \cdot \exp(-t/T_{2,a}) + C \cdot \exp(-t/T_{2,c}) \quad (6)$$

where $T_{2,x}^{-1} = M_2\tau_x$, and $T_{2,x}$ is the transversal relaxation time. The second moment $M_2 = 1.5 \cdot 10^{10}\ \mathrm{s}^{-2}$, was evaluated by an analysis of free induction decay which is an Abragam function [13]. Since $\omega_o \cdot \tau_a$ and $\omega_o\tau_c$ are of the order of one, the T_2 formula must be changed to [14]

$$T_{2,x}^{-1} = M_2\tau_x \left(1 + \frac{5/3}{1+\omega_o^2\tau_x^2} + \frac{2/3}{1+4\,\omega_o^2\tau_x^2}\right). \quad (7)$$

All the components of the magnetization decay can be easily separated because of the great differences in the time scale or the form of the decay. Fitted parameters are listed in Table 1; corresponding theoretical decays are shown in Figs. 1 and 2 by full curves. All curves accord well with experimental results. The change of most of the parameters is qualitatively correct with increasing crosslinking. The number of segments in chain ends decreases (B, C); the degree of anisotropy (q) increases; the correlation time τ_c of the final segments of chain ends does not change within the error limit. Up to now we haven't been able to sufficiently interpret the monotonous change in τ_a and τ_b. A further analysis of q according to [15] was not carried out. In [15] q is found to be a function of the number Z of intercrosslink segments in swollen networks. This function is not adequately satisfied here because the contribution of the physical network of unswollen samples has, up to now, proved impossible to describe theoretically.

References

1. Fedotov VD, Tschernov VM, Wolfson SI (1978) Vysokomol Soed B 20:679
2. Fedotov VD (1981) Thesis, KCTI Kasan, USSR
3. Müller R (1980) Thesis, University of Hamburg, FRG
4. Doskočilová D, Schneider B (1982) Pure & Appl Chem 54:575
5. English AD (1985) Macromolecules 18:178
6. Cohen-Addad JP, Dupeyre R (1983) Polymer 24:400
7. Charlesby A, Bridges BJ (1981) Europ Polym J 17:645
8. Heinrich G, Straube E, Helmis G (1987) Adv Polym Sci 85:33
9. Häusler KG, Zander P, Arndt KF, Stepto RFT (1987) Acta Polymerica 38:519
10. Michel D (1981) Grundlagen und Methoden der kernmagnetischen Resonanz. Akademie Verlag Berlin, GDR
11. Götschmann B (1987) Diploma Work, Technical University Leuna-Merseburg, Merseburg, GDR
12. Anderson PW, Weiss PR (1953) Rev Mod Phys 25:269
13. Simon G, Schneider H, Häusler KG, Pohlent A, Leitgebel J (1987) Poster, IUPAC 31st Macromolecular Symposium Merseburg, GDR
14. Abragam A (1961) The Principles of Nuclear Magnetism. Clarendon Press Oxford
15. Gotlib YuYa, Lifshitz MI, Shevelev VA, Lishanskij IS, Balanina IV (1976) Vysokomol Soed A 18:2299

Received June 25, 1987;
accepted April 21, 1988

Authors' address:

G. Simon
Sektion Physik
TH "Carl Schorlemmer"
O.-Nuschke-Str.
DDR-4200 Merseburg, G.D.R.

Progress in Colloid & Polymer Science Progr Colloid Polym Sci 78:33–36 (1988)

Flexoelectric effect in elastomers

M. Marvan and A. Havránek

Department of Polymer Physics, Charles University, Prague, Czechoslovakia

Abstract: Electric polarization induced by a nonhomogeneous deformation field – flexoelectricity – has been investigated in elastomers. A simple model of dipole orientation mechanism controlled by anisotropic free volume is given. The order of magnitude of the polarization to the deformation gradient ratio β predicted by the model is $\beta \sim 10^{-11}$ C/m. Deformation gradient was experimentally realized in a truncated pyramid cut from elastomers stressed by axial pressure. Voltage difference between the bases of the pyramids approximately proportional to the deformation gradient and depending on the direction of the gradient has been observed. The β values calculated from these experiments have the order of magnitude 10^{-11} or 10^{-10} C/m.

Key words: Electric polarization, nonhomogeneous deformation, flexoelectricity, bending piezoelectricity, elastomers.

1. Introduction

The specific type of interaction between electrical and mechanical fields is studied. The best known effect of this type is the piezoelectric effect (e.g., [1]) in which homogeneous mechanical deformation induces electric polarization. This effect is possible only in crystals in which a centre of symmetry is missing. In crystals with a centre of symmetry, electric polarization may be induced by a nonhomogeneous deformation field [2]. Indenbom et al. [3] called this effect "flexoelectric" in analogy to similar effects in liquid crystals. Tagantsev gave a critical review of the previous works, introduced a more proper definition of the flexoelectric effect, and elaborated on its microscopic theory for ionic crystals [4, 5]. He found that this effect is combined from three contributions: dynamic, static volume, and static surface flexoelectric effects. In crystals these three contributions are of the same order of magnitude. The flexoelectric coefficient – the polarization to deformation gradient ratio – is predicted to be of the same order of magnitude as the particle charge e to the lattice constant a ratio; e/a.

Many experiments were described (e.g., [6, 7]) in which deformation of polymeric materials led to the occurence of non-zero voltage on the sample. The interpretation of these experiments is often vague. For example, "bending piezoelectricity" measured at unpoled PP films [8] may be possibly explained by flexoelectricity. We think that also some other experiments may be explained by flexoelectric effects. This is the reason why we have started the study of flexoelectric effects in polymers.

2. Phenomenology

In a material with an arbitrary symmetry the electric polarization vector is

$$P_i = \gamma_{ijkl} \frac{\partial e_{kl}}{\partial x_j} . \tag{1}$$

In (1) γ_{ijkl} are the flexoelectric coefficients, e_{ij} is the tensor of small deformations, and x_i ($i = 1, 2, 3$) are the space coordinates. In an isotropic material Eq. (1) reduces to the form

$$P_i = \alpha \frac{\partial}{\partial x_i} (e_{11} + e_{22} + e_{33}) + \beta \frac{\partial e_{ij}}{\partial x_j} \tag{2}$$

and only two flexoelectric coefficients α, β remain. The studied materials are elastomers and in them it is reasonable to assume that no volume deformation occurs,

$$\Delta V/V = e_{11} + e_{22} + e_{33} = 0 \tag{3}$$

and so Eq. (2) for elastomers obtains the form

$$P_i = \beta \frac{\partial e_{ij}}{\partial x_j} \,. \tag{4}$$

If, e.g., the deformation field has the form

$$e_{33} = f(x_3),\ e_{ij} = 0 \text{ for } i \neq j \tag{5}$$

(according to (3) in this case $e_{22} = e_{33} = -f(x_3)/2$) then the only non-zero partial derivative from (4) is $\partial e_{33}/\partial x_3$ and

$$P_3 = \beta \frac{\partial e_{33}}{\partial x_3} = \beta \frac{df}{dx_3}\,,\ P_1 = P_2 = 0\,. \tag{6}$$

In this way we obtain a non-zero electric polarization P_3 if the flexoelectric coefficient β is not zero.

3. Microscopic model

The microscopic theory of flexoelectric effects in crystals starts from the theory of crystal lattice vibrations [4]. The static volume effect is explained by intrinsic deformation of a unit cell caused by the deformation gradient. This intrinsic deformation differs from the macroscopic deformation.

The rotation of chain links is the dominant type of motion in elastomers. The flexoelectric effect may occur only if the behavior of the + and − charge carriers differ in the nonhomogeneous deformation field. The different masses of the opposite sign carriers may induce the dynamic flexoelectric effect. The different way of binding to the main polymer chain of the two opposite sign carriers may induce the static volume flexoelectric effect. We shall put down a model of such an effect and in the framework of this model we shall estimate the order of magnitude of the induced flexoelectric coefficient β.

The dipole moment p directed at fixed angle α to the main chain link occurs because the binding of the two opposite sign charge carriers to the main chain is different (compare [9]). In our model we simplify the real situation by the assumption that the dipole moment p is connected with one side-group and has the same

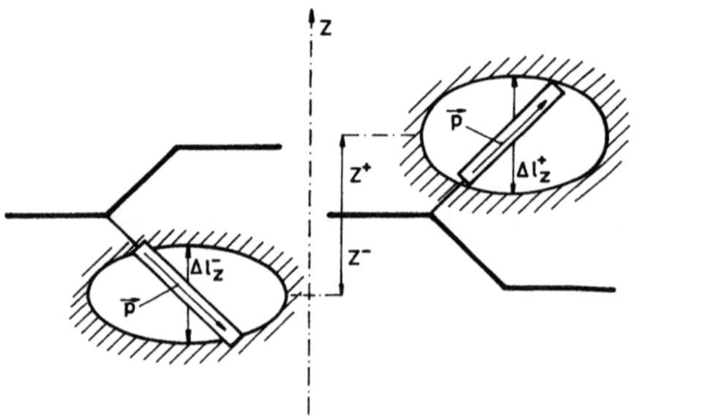

Fig. 1. The positions of polymeric main chain and side-group assumed in the microscopic model. The deformation of free volume is illustrated

direction as this group (see Fig. 1). The deformation is changed in the $x_3 \equiv z$ direction as in Eq. (5). The free volume available for the side-group is deformed in accordance with the deformation field. Although the whole free volume (see Eq. (3)) cannot be changed, its form is deformed and its mean characteristic dimension in z-direction Δl_z depends on the z- coordinate. This Δl_z dimension has a decisive role in the probability with which the side group bearing the dipole moment p will occupy the free volume (compare Fig. 1). We assume that the main chain link is oriented perpendicular to z-direction as this is the representative orientation for the solved problem. For the probability w we suppose the dependence

$$w \sim e^{-\Delta l_c/\Delta l_z} \tag{7}$$

where Δl_c is the critical dimension and (7) has the form of the well known expression

$$w \sim e^{-V_c/V} \tag{8}$$

of the free volume theory (e.g., [10]). The ratio of the probabilities (7) in the z^+ and opposite z^- direction is

$$\frac{w^+}{w^-} = e^{-\Delta l_c\left(\frac{1}{\Delta l_z^+} - \frac{1}{\Delta l_z^-}\right)} = e^{\frac{\Delta l_c}{\Delta l_o^2}(\Delta l_z^+ - \Delta l_z^-)} \tag{9}$$

where Δl_z^+ and Δl_z^- are the Δl_z dimensions in z^+ resp., z^- distance from the main chain link and

$$\Delta l_o = \sqrt{\Delta l_z^+ \,\Delta l_z^-}\,. \tag{10}$$

The Δl_z dimensions deform in the same way as the macroscopic dimensions, therefore

$$\frac{\Delta l_z^+}{\Delta l_o} - \frac{\Delta l_z^-}{\Delta l_o} = \frac{\partial e_{zz}}{\partial z}\,\delta \tag{11}$$

where

$$\delta = z^+ - z^- \tag{12}$$

is the distance of the centres of the side group in its opposite positions (see Fig. 1). From Eqs. (9) and (11) it follows

$$\frac{w^+}{w^-} = e^{\frac{\Delta l_c}{\Delta l_o}\frac{\partial e_{zz}}{\partial z}\delta}\,. \tag{13}$$

A simple expression

$$P_z = N p_z\,(w^+ - w^-) \tag{14}$$

for the polarization P_z may be written in the approximation supposing that only the positions with probabilities w^+ and w^- of the side group are possible,

$$w^+ + w^- = 1\,. \tag{15}$$

The symbol p_z denotes the absolute value of the z-component of the dipole moment p, and N is the number of dipoles in unit volume. The expression

$$\beta = \frac{\Delta l_c}{2\Delta l_o} N p_z\,\delta \tag{16}$$

for the flexoelectric coefficient β follows from Eqs. (6), (13) – (15).

Equation (16) is used for the order of magnitude estimate of the β value. The dipole moment $p_z \sim 10^{-29}$ cm. The value of $N \sim 10^{28}\,\mathrm{m}^{-3}$ if we assume that one dipole belongs to one monomer unit. The order of magnitude of $\delta \sim 10^{-10}$ m. The ratio 1/40 is used in the free volume theory [10] as the ratio of the mean free volume v belonging to one monomer unit to the critical volume v_c necessary for the elementary segment jump, $v/v_c = 1/40$. We use this ratio in the one dimensional modification, $l \sim v^{1/3}$, for the estimate of the $\Delta l_c/\Delta l_o$ ratio, $\Delta l_c/\Delta l_o = 3.4$. With the values just given we obtain

$$\beta \sim 10^{-11}\ \mathrm{C/m}\,. \tag{17}$$

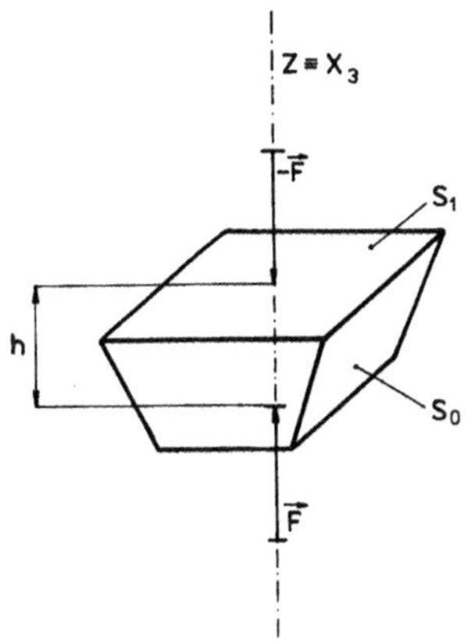

Fig. 2. The form of the samples and the mode of their deformation

The evident geometric and other simplifications used in the derivation of (17) may hardly cause deviation of the final value larger than one order of magnitude.

4. Experimental

Preliminary measurements were performed with the aim to test the possible existence of flexoelectric effect in elastomers. The samples were chosen in the form of a truncated pyramid and this pyramid was deformed by axial pressure (Fig. 2). So the nonhomogeneous deformation field of the Eq. (5)-type is obtained with

$$f(x_3) = f(z) = e_{zz} = \frac{F}{E\,S_o(1+kz)^2}\,. \tag{18}$$

In Eq. (18) F is the applied force, E the Young modulus of the deformed elastomer, $k = (1/h)\,(\sqrt{S_1/S_o} - 1)$ and S_o, S_1, h are the sample dimensions. The areas of the bases S_o, S_1 are in the vicinity of 10^{-4} m^2 and the heights h are several millimeters in our samples.

According to Eq. (6) polarization $P_3 = P_z$ arise in the ample during deformation. The corresponding voltage difference ΔU between the two bases S_o, S_1 of a pyramid was measured by an electrometer (the Keithley electrometers 604 and 610 C were used in our experiments). From the elementary electrostatic field theory and Eq. (6) it follows

$$U = \int_o^h \frac{P_z}{\varepsilon_o \varepsilon_z}\,dz = \frac{1}{\varepsilon_o \varepsilon_z}\int_o^h \beta\,\frac{de_{zz}}{dz}\,dz = \frac{1}{\varepsilon_o \varepsilon_z}\,\beta \Delta e \tag{19}$$

and therefore

$$\beta = \varepsilon_0 \varepsilon_r \frac{\Delta U}{\Delta e} \,. \qquad (20)$$

In Eqs. (19), (20) $\varepsilon_0\,\varepsilon_r$ is the sample permitivity and $\Delta e = e_{zz}(0) - e_{zz}(h)$ is the e_{zz}-deformation difference between the two bases S_0 and S_1. The Δe value has been calculated from the measured overall deformation of the sample using Eq. (18). The maximum Δe values reached in the samples were about 0.5. Six samples gave reasonable results. These were: three samples of the liquid rubber type crosslinked by epoxy resins, two of them with butadiene main chain, one with polyoxypropylene main chain (Jeffamine), two samples of ethylene-propylene rubber type (Keltan) crosslinked by sulphur, and one natural rubber with carbon black sample crosslinked by sulphur. In these samples ΔU was nearly proportional to Δe, maximum ΔU values were of the order of volts, and the $\Delta U/\Delta e$ ratios vary from 3V to 5V. The β values obtained from Eq. (20) were in the limits 7.10^{-11} C/m to 15.10^{-11} C/m (relative permitivity ε_r of the samples is between 2 and 3), with the exception of the carbon black sample, which due to the large and not well defined ε_r value gave β of approximately 50.10^{-11} C/m. The low capacitance ($\sim pF$) of the measuring condenser might cause some uncertainties in the ΔU voltage statement. These uncertainties may enhance the actual β values by a factor less than 5.

Some tests were made to reduce the probability that the observed ΔU values come from other sources than the flexoelectric effect. Two samples were cut from sheets in such a way that their smaller bases lay on opposite surfaces of the sheets. The sign of ΔU of these samples depends only on the position of the smaller base in the measuring condenser and not on the way they were cut from the sheet. Liquid grafite was sprayed on the pyramid bases to change the friction between the sample and condenser desks and to change the electric contact between desks and the sample. The effect of grafite didn't change the measured $\Delta U/\Delta e$ ratios significantly.

In samples which did not give reasonable results (four samples) the measured ΔU values were too small to give reproducible values. More details of the experiments are in the dissertation of Prokšová [11] who has done most of the experimental work.

5. Discussion

The results of the work seem to be positive. The voltage difference on some nonhomogeneously deformed elastomers has been detected in accordance with the theoretical predictions. The presented microscopic theoretical model of flexoelectricity in polymers gives reasonable order of magnitude of the flexoelectric coefficient β.

Some objections may arise. It may be thought that the measured effect comes out from other sources than is the flexoelectric effect, e. g., it may be the result of triboelectricity, pressure dependent contact potential, or some effect of volume and surface charges. The microscopic model can be said to be too artificial. Our further work will be oriented on overcoming these objections.

Nevertheless, we think that the preliminary results obtained are encouraging.

Acknowledgement

The authors thank Dr. J. Fähnrich for helpful discussions which led to improvement of the paper.

References

1. Mason WP (1950) Piezoelectric Crystals and Their Application to Ultrasonic. van Nostrand, New York Toronto London
2. Kogan ShM (1963) Fiz Tverd Tela (Leningrad) 5:2829, Sov Phys Solid State 5:2069
3. Indenbom VL, Loginov EB, Osipov MA (1981) Kristalografiya (in Russian) 26:1157
4. Tagantsev AK (1985) Zh Exp Teor Fiz 88:2108, Sov Phys YETP 61:1246
5. Tagantsev AK (1986) Phys Rev B 34:588
6. Lushcheikin GA (1984) Polymernye Elektrety (in Russian), Khimiya, Moscow
7. Sessler GM (ed) (1981) Electrets. Springer, Berlin
8. Takamatsu T, Tian RW, Sasabe H (1985) In: Sesseler GM, Gerhardt-Multhaupt R (eds) Proceedings of the 5th International Symposium on Electrets, Heidelberg 1985 (Available from IEEE Service Center, Piscataway, NJ, USA), p 942
9. Flory PJ (1969) Statistical Mechanics of Chain Molecules, Wiley, New York London
10. Bueche F (1962) Physical Properties of Polymers. Wiley, New York London
11. Prokšová J (1986) Thesis, Faculty of Mathematics and Physics, Charles University, Prague

Received February 1, 1988;
accepted April 21, 1988

Authors' address:

Dr. M. Marvan, CSc
Department of Polymer Physics
MFF UK
V Holešovičkách 2
180 00 Praha 8, Czechoslovakia

Progress in Colloid & Polymer Science Progr Colloid Polym Sci 78:37–53 (1988)

Description of the freezing-in process in poly(vinyl acetate) based on the meander model

W. Heinrich and B. Stoll

Abteilung Angewandte Physik, Universität Ulm, Ulm, F.R.G.

Abstract: The meander model was used to derive molecular parameters from the pressure dependence of the main dielectric relaxation in PVAc. The specific volume and the specific heat capacity of the melt were calculated with the same parameters in agreement with experimental data from the literature and additional experimental data presented here. The validity of the Ehrenfest relations is discussed. For constant cooling rate, it is shown by model calculations, that the product of cooling rate and relaxation time is about 15 K at T_g. The relation between the dielectric and the thermodynamic relaxation times are investigated and it is found that they are equal within an uncertainty factor of approximately 2. Mechanical measurements are also presented.

Key words: Poly(vinyl acetate), meander model, glass temperature, dielectric relaxation, mechanical relaxation, specific heat capacity, specific volume.

Introduction

If we compare the molten state of an amorphous polymer with the glassy state, the most prominent difference is the relaxation behaviour, which is detected by dynamic mechanical, dielectric and NMR measurements. In addition, there are differences in the thermal expansion coefficient, the compressibility and the specific heat capacity and other properties. For many years, efforts have been made to understand the nature of the transition between the molten and the glassy state and thus to correlate various properties. The most widely known of these is the theory of the free volume [1]. A new assessment was made in 1979 with the dislocation concept based on the meander model [2], which is a special case of the free volume theory, defining quasi-dislocations that represent the "holes". The dislocation concept presents a relatively simple description of molecular mobility above the glass temperature T_g, since it contains only one variable – in addition to temperature and pressure – to determine the state of the polymer system and the mean relaxation time: the density of quasi-dislocations. A relaxation time distribution is not included in the model until now; therefore some effects (e. g. the broadening of the dielectric loss peak or the memory effect) cannot be discussed quantitatively. But, of course, the model may be extended in this direction in the future. Presumably, it will be possible to give a physical meaning to the relaxation time distribution in this model.

The aim of this paper is to use the dislocation concept based on the meander model to describe various properties of a typical amorphous polymer, poly(vinyl acetate). Especially the influence of hydrostatic pressure will be analysed and used to derive molecular parameters.

Simple WLF-theory

The pressure dependence of the dielectric relaxation in PVAc was measured in [3]. To test simple WLF-theory, the following form was chosen (see for instance [1, 4])

$$f_m = f_o \exp\left(-\frac{BV_o}{V_f}\right) \tag{1}$$

where f_m is the maximum frequency of the dielectric loss peak, f_o is a fixed frequency (specific for each polymer), V_o is the total volume of the sample (taken approximately as a constant) and V_f is the free volume, B is a parameter (≈ 1). The temperature and pressure dependence of V_f was expressed by its first derivatives at the glass temperature T_{go} at atmospheric pressure. In this section, of the paper, the glass temperature T_g is defined arbitrarily at constant relaxation time by (τ_d means dielectric relaxation time)

$$1/(2\pi\,\tau_d) = f_m = f_g = 10^{-3}\ \text{Hz} \tag{2}$$

$$V_f = V_g\,[1 + \Delta\alpha(T - T_{go}) - \Delta\varkappa(p - p_o)] \tag{3}$$

V_g is the free volume at T_{go}, $\Delta\alpha$ resp. $\Delta\varkappa$ are the differences in the volume expansion coefficient resp. compressibility between melt and glass. These were measured by McKinney and Goldstein [5] as $\Delta\alpha = 4.2 \cdot 10^{-4}$ K^{-1} and $\Delta\varkappa = 2.1 \cdot 10^{-5}$ bar^{-1}. The reference pressure $p_o = 1$ bar was neglected here. This leads to the equivalent form

$$\log\frac{f_m}{f_g} = \frac{C_1\left(T - T_{go} - \dfrac{\Delta\varkappa}{\Delta\alpha}\,p\right)}{C_2 + T - T_{go} - \dfrac{\Delta\varkappa}{\Delta\alpha}\,p}\,. \tag{4}$$

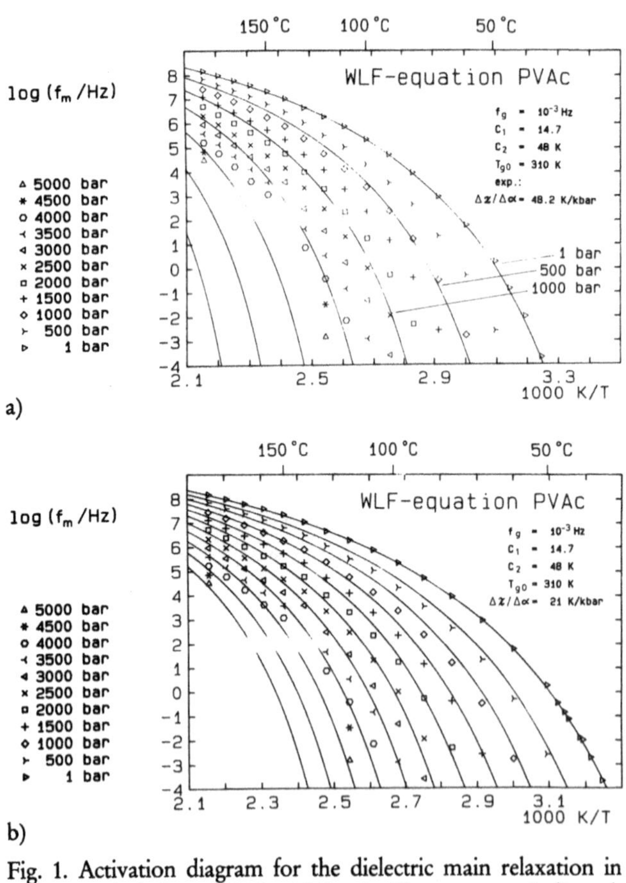

Fig. 1. Activation diagram for the dielectric main relaxation in PVAc. Symbols from [3]; Eq. (4) fitted with parameters indicated. Experimental data for $\Delta\varkappa/\Delta\alpha$ from Ref. [5]

First, Eq. (4) was fitted to the experimental data for atmospheric pressure in Fig. 1a with the parameters indicated in Fig. 1a.

We obtain

$$B = 2.303\ C_1 C_2 \Delta\alpha = 0.68$$

$$V_g/V_o = C_2\Delta\alpha = 0.020$$

$$f_o = f_g \exp(2.303\ C_1) = 5 \cdot 10^{11}\ \text{Hz}\,. \tag{5}$$

The other curves were drawn for 500 bar, 1000 bar, 1500 bar, and so on, without additional fitting parameters using $\Delta\alpha$ and $\Delta\varkappa$ from [5]. The difference between theory and experiment is very large at high pressures. In our opinion, this discrepency can only be substantially reduced by choosing $\Delta\varkappa/\Delta\alpha$ about one half of its experimental value, as is shown in Fig. 1b.

Adam-Gibbs-Theory

Havlicek et al. [6] claimed that the Adam-Gibbs-Theory, when expanded to high pressures, describes the pressure dependence of the dielectric relaxation in PVAc well. However, they did not take the correct experimental value of Δc_p (which is 0.46 $J\,g^{-1}\,K^{-1}$, see Ref. [7] and Fig. 11 of this paper) for their calculation. The formula given by them is (the index zero refers to atmospheric pressure)

$$\log\frac{f_m}{f_o} = -\,0.4343\ C\ \{[T_{go}\Delta c_{po}\ln(T_{go}/T_{20})]^{-1} - [T\Delta c_{po}\ln(T/T_{20}) - TV_o(\Delta\alpha_o p - a p^2/2)]^{-1}\}\,. \tag{6}$$

This formula was used again here but with $\Delta c_{po} = 0.46$ $J\,g^{-1}\,K^{-1}$ and fitted to the experimental data from [3] for atmospheric pressure (Fig. 2a).

The fitting parameters are indicated in Fig. 2a. The curves for higher pressures (500 bar, 1000 bar, 1500 bar ...) were drawn without additional fitting parameters. In spite of using a second order term, the high pressures in Fig. 2a are poorly described by the theory. Perhaps, other second order terms (which were not considered by Havliček et al. [6]) are of importance here. The pressure dependence of $\Delta\alpha$ is represented by

$$\Delta\alpha = \Delta\alpha_o - ap \tag{7}$$

with $a = 1.16 \cdot 10^{-7}$ $K^{-1}\,bar^{-1}$ in Fig. 2a in agreement with Refs. [5] and [6]. To test the influence of this

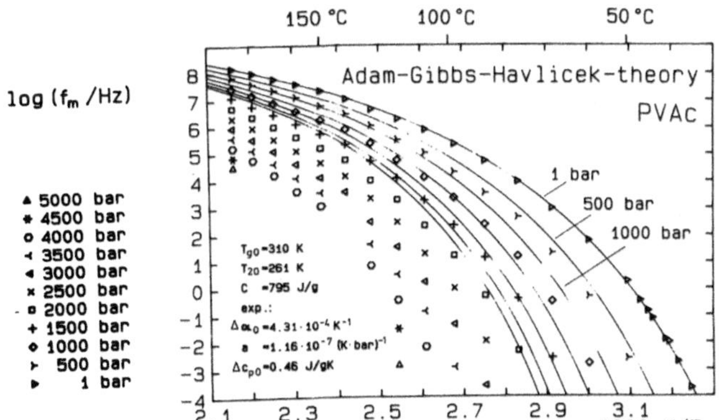

a)

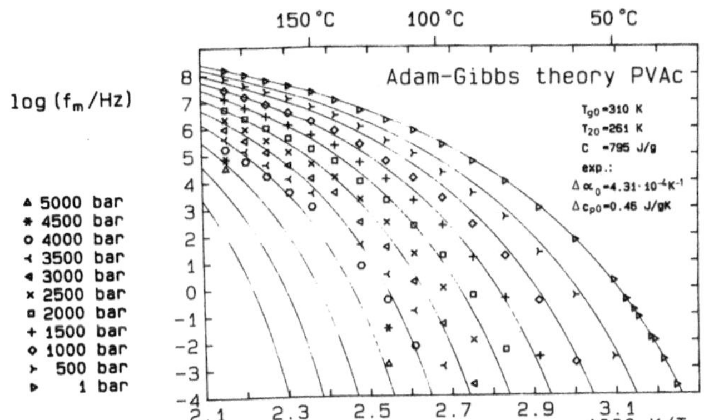

b)

Fig. 2. Activation diagram as Fig. 1, but the theory of Adam-Gibbs in the form proposed by [6], which is rewritten in Eq. (6) fitted to the experiment with the specific heat measured by Höhne (Fig. 11) and parameters indicated (tentatively $a = 0$ in Fig. 2b)

paramter, we calculated theoretical curves in Fig. 2b with the same paramters as Fig. 2a, but with $a = 0$. The result looks even better than Fig. 2a.

Dislocation concept

The dislocation concept is described in [2, 8] and was compared there with experimental data at atmospheric pressure for many amorphous polymers. The formula for the maximum frequency of dielectric loss is

$$\log \frac{f_m}{f_o} = -\log \pi - \frac{Q_y}{2.303RT} + 3\left(\frac{3r}{d}\right)^2 \frac{d}{s} \log\left[1 - (1-x)^{3r/d}\right]. \quad (8)$$

Here f_o = local vibration frequency; Q_y = local activation energy; d = mean distance between neighbouring chains (= 0.76 nm for PVAc); s = segment length (= 0.42 nm for PVAc); r = mean diameter of a bundle of macromolecules; $x = \exp[-\varepsilon_s/(RT)]$ is the number of quasi-dislocations of a certain type per segment (compared with the notation in [8], $x = \Sigma x_i$); ε_s = energy of a quasi-dislocation wall per segment

$$\varepsilon_s = \frac{0.3\, Gb^2\, d}{4\pi\,(1-\nu)} \quad (9)$$

G = shear modulus of the structure in which dislocations are introduced (= shear modulus of the glass); b = Burgers vector; ν = Poisson's ratio.

To calculate the temperature and pressure dependence of f_m, we neglect the temperature and pressure dependence of all parameters except ε_s and Q_y. Assuming now, the Poisson's ratio is constant, we may use Slater's expression [9]

$$\frac{d \ln G}{d \ln V} = \frac{d \ln K}{d \ln V} = -2\gamma \quad (10)$$

where K is the bulk modulus, γ the Grüneisen parameter and V the volume of the glass. This implies, that K depends only on the volume and not explicitly on temperature and pressure. This assumption is supported by the fact that nearly the same Grüneisen parameter was obtained from the pressure dependence of K and from the temperature dependence of K in several polymers [10–14]. We should however note that rather different Grüneisen parameters are obtained by different experimental methods: static bulk modulus, ultrasonic wave velocity and Brillouin scattering yield $3 < \gamma < 10$ [10–16] because they involve low frequency modes; specific heat capacity and infrared frequency shift yield $0.1 < \gamma < 2$, because in this case high frequency modes play a dominant role [17–19]. One might think that the dislocation energy is caused only by intermolecular interaction and, therefore, one should take the γ for low frequency modes. The macromolecular chains are, however, not rigid rods, the formation of quasi-dislocations discussed here may thus also involve intramolecular conformation changes. For these reasons, we will take γ_{disl} as a Grüneisen parameter for quasi-dislocations, and consider it as an unknown parameter between 0.5 and 5. In Eq. (9), not only G, but also b and d, will depend on the volume. Because b is proportional to d (see Ref. [2]), we write

$$\frac{d \ln \varepsilon_s}{d \ln V} \approx \frac{d \ln G}{d \ln V} + \frac{d \ln V}{d \ln V} = -2\gamma_{\text{disl}} + 1 \quad (11)$$

$$(\varepsilon_s/\varepsilon_{so}) = (V/V_o)^{1-2\gamma_{\text{disl}}}. \tag{12}$$

Now, to find $V(T, p)$ of the glass, we can simply use the compressibility $\varkappa = 1/K$ and the expansion coefficient α measured by McKinney and Goldstein [5]. Alternatively, according to W. Pechhold [20] we may use Eq. (10) to establish an equation of state for the glass and fit this to the data in Ref. [5]. The advantage of the latter procedure is it permits higher order terms in T and p, and thus an extended range of validity. The derivative of K follows from Eq. (10)

$$\frac{\partial K}{\partial p} = K\left(\frac{\partial \ln K}{\partial \ln V}\right)\left(\frac{\partial \ln V}{\partial p}\right) = 2\gamma \tag{13}$$

$$\frac{\partial K}{\partial T} = K\left(\frac{\partial \ln K}{\partial \ln V}\right)\left(\frac{\partial \ln V}{\partial T}\right) = -2\gamma K\alpha. \tag{14}$$

We now assume, that $\gamma =$ const (independent of p and T) and that $K\alpha =$ const (in the range measured by McKinney and Goldstein [5], $K\alpha$ is indeed independent of p and T). We may then integrate and obtain:

$$K - K_o = 2\gamma[p - K_o\alpha_o(T - T_o)] \tag{15}$$

where $p_o = 1$ bar was neglected. The derivatives of V are

$$\frac{\partial V}{\partial p} = -\frac{V}{K} \tag{16}$$

$$\frac{\partial V}{\partial T} = V\alpha = VK_o\alpha_o/K. \tag{17}$$

Integration yields

$$\frac{V}{V_o} = \left\{1 + \frac{2\gamma}{K_o}[p - K_o\alpha_o(T - T_o)]\right\}^{-1/(2\gamma)}. \tag{18}$$

As we consider volume changes in the glass without moving dislocations and negligible local relaxation, we should use the low frequency mode Grüneisen parameter Eq. (18), which was measured by Brillouin scattering [16] as $\gamma = 5$ in PVAc. The data from [5] are: $K_o\alpha_o = 10.5$ bar K^{-1}, $K_o = 26\,100$ bar and $V_o = 0.87$ cm^3/g at $T_o = 360$ K. We now consider the local activation energy Q_y which may contain an unknown intermolcular part; we therefore split Q_y into an intramolecular part and an intermolecular part. The latter should depend on pressure and temperature in the same way as ε_s. At the reference temperature T_o, the intermolecular part is assumed to be the fraction ζ of the Q_{yo}. This leads to the form

$$Q_y = (1-\zeta)\,Q_{yo} + \zeta Q_{yo}\,\varepsilon_s/\varepsilon_{so}. \tag{19}$$

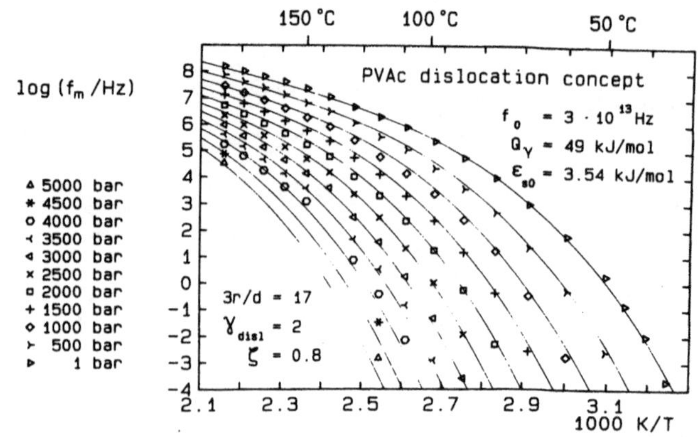

Fig. 3. Activation diagram as Fig. 1, but curves calculated from the dislocation concept (Eqs. (8), (12), (18), (19)) fitted to the experiment with data from [5], calculation parameters are indicated

Theoretical curves using Eqs. (8), (12), (18), (19) were fitted to the experimental results for 1, 500, 1000, 1500 ... bar in Fig. 3. The fitting parameters are indicated in Fig. 3.

The values of f_o and Q_{yo} are as expected from secondary relaxation data [21]; ε_{so} and $3r/d$ agree roughly with values already obtained for atmospheric pressure in Ref. [2, 8]. ε_s was checked in [2, 8] by a formula from dislocation theory (Eq. (9)) to be of the correct order of magnitude. $3r/d$ agrees satisfactorily with the evaluation [2] from the rubbery plateau compliance J_{eN}. Dynamic mechanical measurements of the same PVAc as used in [3] are shown in Figs. 4 and 5. The rubbery plateau compliance J_{eN} and the maximum frequencies of J'' for the glass relaxation were calculated, using a computer, from a best fit of superimposed relaxation peaks to the experimental curve. This fit yielded $J_{eN} = 1.5 \cdot 10^{-6}$ Pa^{-1} for $M_v = 170\,000$ g/mol resp. $J_{eN} = 2.1 \cdot 10^{-6}$ Pa^{-1} for $M_v = 435\,000$ g/mol. Using the formula

$$J_{eN} = \frac{1}{9}\frac{d^3}{RT}\left(\frac{3r}{d}\right)^2 \tag{20}$$

for the meander model [2], we calculated $3r/d = 12$ and 14 for $M_v = 170\,000$ and $435\,000$ g/mol, resp., which is slightly lower than $3r/d = 17$, obtained in Ref. [2, 8] on the basis of other (literature) data. The Grüneisen parameter $\gamma_{\text{disl}} = 2$ turned out to be considerably lower than the low frequency mode Grüneisen param-

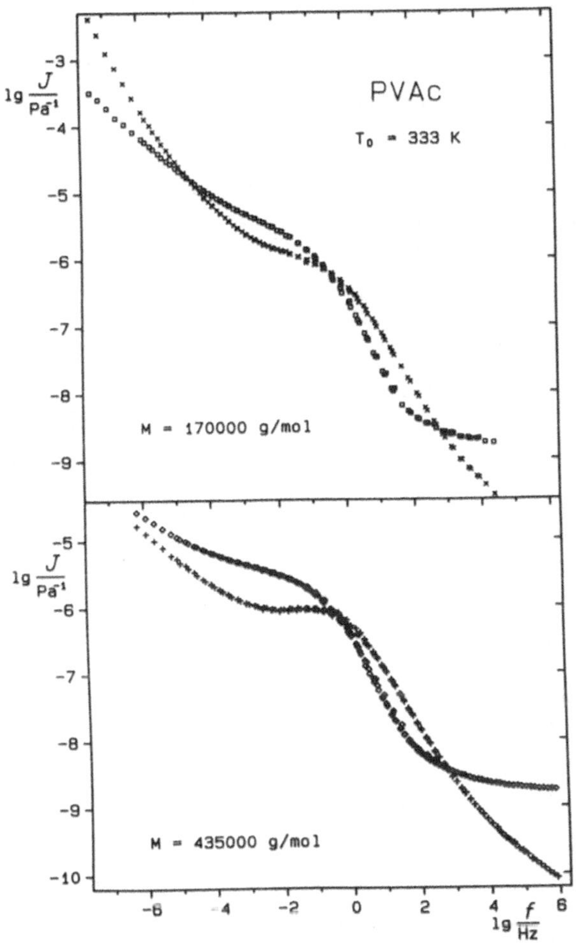

Fig. 4. Master curves for the complex shear compliance of two PVAc samples from Wacker Chemie, Burghausen. The frequency scale refers to the reference temperature $T_o = 333$ K. The apparatus and the construction of mastercurves are described in [22]. The shape of the measured curves changes slightly with temperature, therefore no ideal mastercurves could be constructed. The authors wish to thank Dipl.-Phys. W. v. Soden et al.; University of Ulm, for carrying out the measurements and their evaluation. Squares: real part J'; Crosses: imaginary part J''

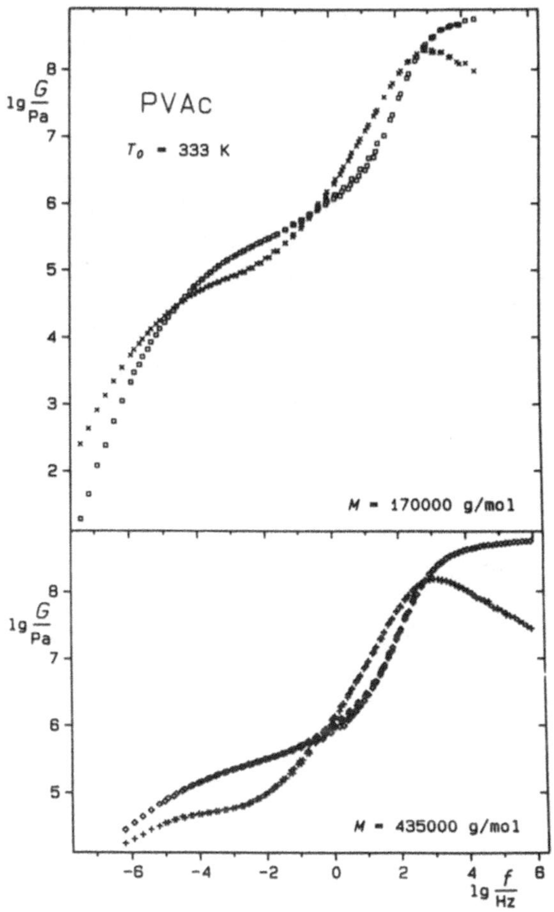

Fig. 5. The complex shear modulus calculated from Fig. 4 by $G' + iG'' = 1/(J' - iJ'')$. Squares: real part G'; crosses: imaginary part G''. Details of the measurement are described in [22]

eter $y = 5$. It follows that ε_s contains a considerable portion of intramolecular energy. The behaviour of Q_y and ε_s are not very different (this follows from $\zeta = 0.8$).

Free volume and specific heat capacity

In the preceding section we determined some parameters for our model. It is now possible to calculate the contribution of the quasi-dislocations to the thermal properties without further information. We shall do this using the formulae

$$\Delta \varkappa = -(1/V_o)(\partial V/\partial p) = -(1/V_o)(\partial^2 g^*/\partial p^2) \quad (21)$$

$$\Delta \alpha = (1/V_o)(\partial V/\partial T) = (1/V_o)(\partial^2 g^*/\partial p\, \partial T) \quad (22)$$

$$\Delta c_p = (1/m_o)(\partial h^*/\partial T) = -(T/m_o)(\partial^2 g^*/\partial T^2) \quad (23)$$

where g^* is the free energy due to all quasi-dislocations for one mol of segments, V_o resp. m_o are the volume resp. mass of one mol of segments. We prefer the following form:

$$g^* = 8\, x\, \varepsilon_s + 8\, RT\, (x \ln x - x)\,. \quad (24)$$

This yields the same total number $x = \Sigma x_i$ of quasi-dislocation of one type per segment at equilibrium as the form used in [2, 8]; however, it has the advantage of reducing the enthalpy

$$h^* = g^* - T\left(\frac{\partial g^*}{\partial T}\right) \quad (25)$$

by a factor $(1-x)$ compared with [2, 8], which is probably more realistic (considering the interactions between quasi-dislocations at high concentration). Besides, there is no explicit term with pressure in Eq. (24) because ε_s is pressure dependent here – in contrast to [2, 8]. The factor 8 indicates, that additional types of quasi-dislocations are taken into account (see Fig. 6) which were not considered in [2, 8]. Using the parameters from the preceding section, we calculated the curves in Figs. 7–9 for the pressures 1, 1 000, 2 000, 3 000 bar. The good agreement with the experimental results shows that the differences in the thermal properties between melt and glass may be ascribed to the quasi-dislocations in this model. The relative free volume V_{disl}/V_o attributed to the quasi-dislocations, calculated by

$$V_{disl} = (\partial g^* / \partial p) \tag{26}$$

turned out to be $V_{disl}/V_o = 0.040$ at $T_{go} = 310$ K.

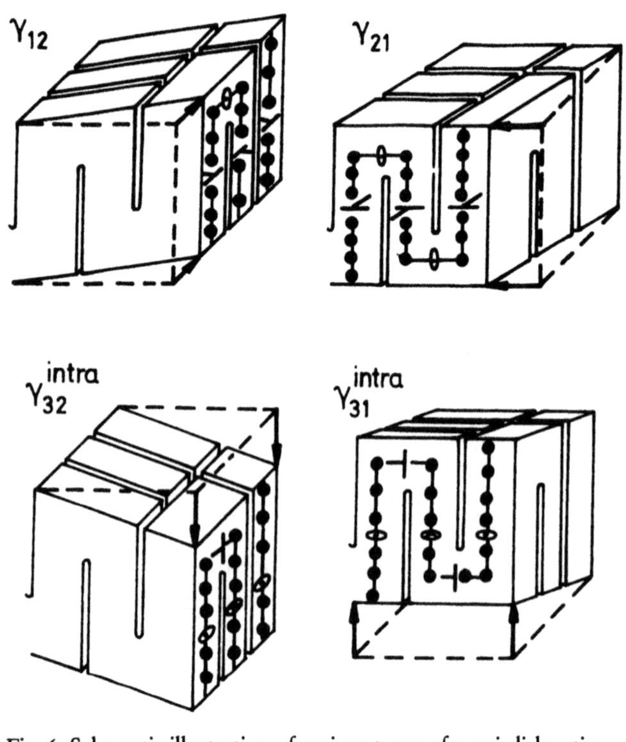

Fig. 6. Schematic illustration of various types of quasi-dislocations in a meander cube performing various shear deformations

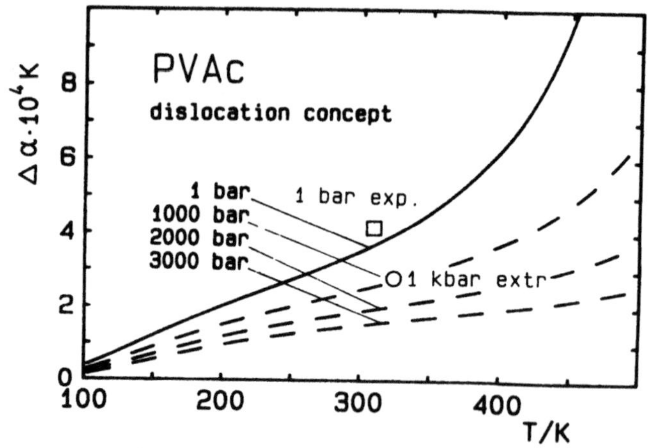

Fig. 7. Part of the volume expansion coefficient of PVAc which is due to quasi-dislocations, calculated with Eq. (24) and parameters of the preceding section. The symbols indicate the difference between melt and glass measured in [5] for atmospheric pressure and extrapolated from the data in [5] to 1 kbar

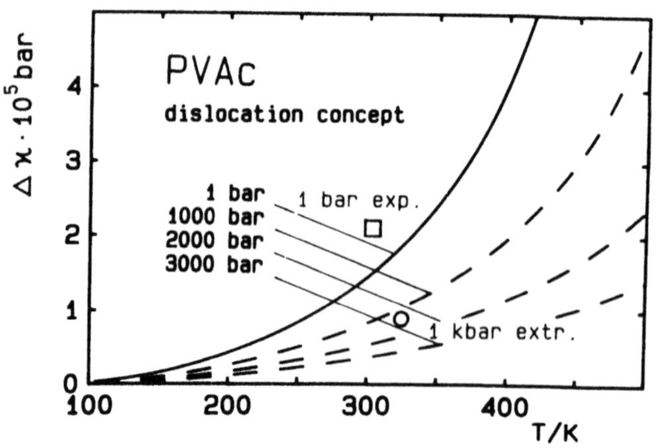

Fig. 8. Part of the isothermal compressibility analogous to Fig. 7

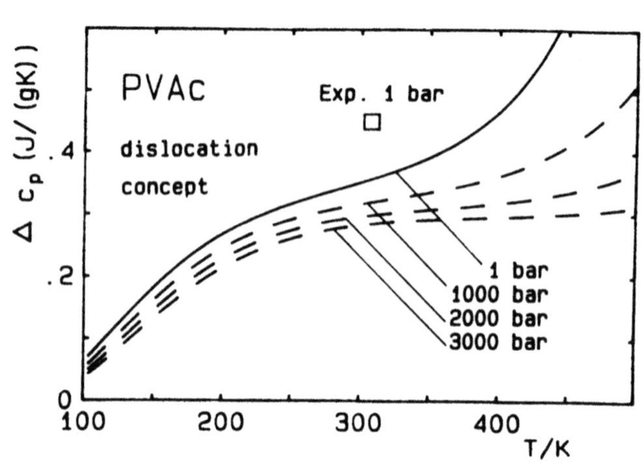

Fig. 9. Part of the specific heat capacity analogous to Fig. 7. The square symbol represents experimental data (see Fig. 11) of Höhne et al.

Ehrenfest's relations

$$\frac{dT_g}{dp} = \frac{\Delta\varkappa}{\Delta\alpha} \tag{27}$$

$$\frac{dT_g}{dp} = \frac{TV\Delta\alpha}{\Delta c_p} . \tag{28}$$

These relations were originally derived in a very general way for a thermodynamic transition of second order (see for instance [23]). However, the experimentally observable (in a realistic time scale) "glass transition" is of a different nature [23–26]. Several arguments indicate, that it is a freezing-in process; that means, the glass is not in thermal equilibrium and the freezing-in line is controlled by kinetic effects. Therefore, before discussing Eqs. (27), (28), it is necessary to adopt a criterion for the "glass-transition" temperature T_g. Some of the theoretical criteria used in the literature (see for instance [6, 23, 25]) are the following: T_g is defined by a critical value (independent of temperature and pressure) of

a) free volume or
b) configurational entropy or
c) an internal variable.

Criterion a) leads directly to Eq. (27). Criterion b) leads directly to Eq. (28) (see for instance [25]). Criterion c) refers to irreversible thermodynamics, where a non-equilibrium state of the polymer is described by one or more internal variables (sometimes called "order parameters"). It was first shown by Davies and Jones [27] and clearly pointed out by Breuer and Rehage [23] that both Ehrenfest relations (27), (28) are formally valid for a system with a single internal variable, if criterion c) is applied.

In addition, there are numerous experimental criteria. Some of them define T_g by

d) the break in volume-temperature-curves measured at constant pressure and constant cooling rate, or
e) a critical viscosity or
f) a critical dielectric relaxation time, corresponding to a critical maximum frequency of the dielectric loss or
g) a critical relaxation time from measurements of the shear compliance or
h) a critical relaxation time from measurements of the shear modulus.

There is experimental evidence for polyisobutylene (see activation diagram in [2]) and PVAc (see Fig. 16 of this paper) – and we hope it can be generalized – that the dielectric and mechanical relaxation times in the "glass transition region" differ by a constant factor which is independent on temperature (and presumably also independent on pressure). If this is true, criteria e), f), g), h) are virtually equivalent, because (for instance) a certain critical maximum frequency of the imaginary part of the shear compliance corresponds to another critical maximum frequency of the dielectric loss, independent of pressure. At first glance, criterion d) seems to be different from the others, but it is known [28–30] from experiment, that

$$dT_g/d\,(\log q) = dT_g/d\,(\log f_m) \tag{29}$$

where q is the cooling rate. In addition, it will be shown by model calculations in a later section of this paper, that in PVAc there is a nearly constant factor between cooling rate and reciprocal equilibrium dielectric relaxation time at T_g (defined by criterion d)). Therefore, taking T_g for a certain constant cooling rate from volume measurements at various pressures should lead to the same dT_g/dp as evaluating T_g from dielectric (or mechanical) measurements at the corresponding constant maximum frequency. Consequently we consider criterion d) as equivalent to criteria e), f), g), h).

To allow an experimental proof of Eqs. (27) and (28) the T_g therein should be defined by an experimental criterion; since the above mentioned experimental criteria are considered to be equivalent, we choose here criterion f). The T_g were taken from dielectric data in Ref. [3] at the point where $f_m = 10^{-3}$ Hz and plotted in Fig. 10. For $p = 1$ bar, $dT_g/dp = 25$ K/kbar. For comparison, volume measurements [5] with criterion d) yield $dT_g/dp = 27$ K/kbar; both dT_g/dp agree within experimental uncertainty. The other experimental terms in Table 1 are calculated from volume data in Ref. [5] and specific heat capacity measure-

Table 1. Ehrenfest's relations for $T = 310$ K and $p = 1$ bar

all units: K/kbar	dT_g/dp	$\Delta\varkappa/\Delta\alpha$	$TV\Delta\alpha/\Delta c_p$
exp.	25	50	24
WLF (simple form)	50	50 (fitted)	–
Adam-Gibbs-Havlicek	21	–	24 (fitted)
dislocation concept	23 (fitted)	43	27

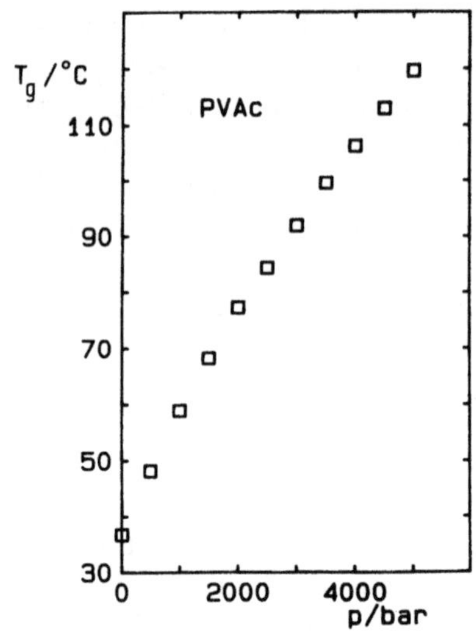

Fig. 10. Glass temperatures T_g (defined by $f_m = 10^{-3}$ Hz) from dielectric measurments [3] plotted as a function of pressure

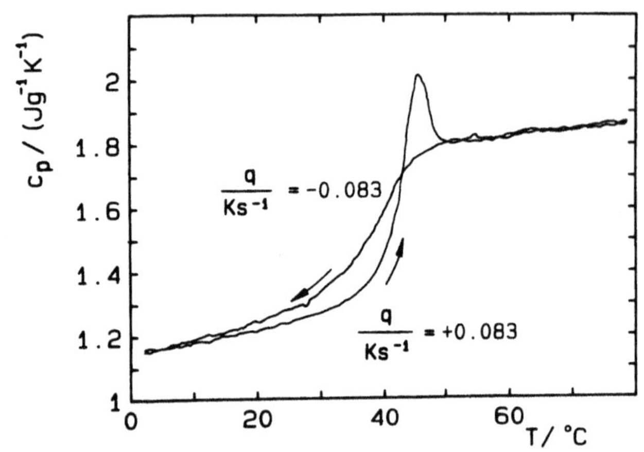

Fig. 11. Experimental specific heat capacity curves obtained on cooling and reheating with the same rate q with a Perkin-Elmer DSC-2 for a PVAc sample ($M_v = 170\,000$ g/mol) from Wacker Chemie Burghausen. Measurements by Höhne et al.

ments presented here in Fig. 11. Obviously Eq. (28) holds, but Eq. (27) not. This agrees with data from some other polymers, collected in Ref. [31], but is in contrast to data reported for polystyrene [24] and polyisoprene [4].

Looking at the simple form of the free volume theory discussed in this paper, we see from Eq. (1) that criterion f) is equivalent to criterion a). Therefore this theory predicts that Eq. (27) should hold. If we accept experimental values of $\Delta\alpha$ and $\Delta\varkappa$, we arrive at the predictions in the second line of table 1.

Looking at the Adam-Gibbs theory in the form which was proposed in Ref. [6], we realise that criterion f) is not equivalent to criterion b). This was already shown by Havlićek et al. [6]. Therefore it is predicted that Eq. (28) will not hold. Accepting experimental values for $\Delta\alpha$ and Δc_p, we have to read T_g from the theoretical curves in Fig. 2a at the point, where $f_m = 10^{-3}$ Hz. Thus we calculated $dT_g/dp = 21$ K/kbar (third line of Table 1).

Looking at the dislocation concept, it follows from Eqs. (8), (12), (18), (19) that criterion f) is not equivalent to criterion c). Therefore it is predicted that Eqs. (27) and (28) will not hold. In the presentation in this paper, the dislocation concept was fitted to the activation diagram (Fig. 3). The aim was to describe the whole range of data and not only the low frequencies in particular; therefore the theoretical dT_g/dp (calculated using criterion f)) coincides only approximately with the experimental dT_g/dp (see Table 1). The other terms in the fourth line of Table 1 are calculated without further fit parameters, representing the contribution of the quasi-dislocations to the thermal properties.

Determination of T_g by DSC

Considering the preceding section, it is very desirable to have a calorimetric definition of T_g which is equivalent to the other experimental criteria mentioned there. An excellent treatment of this problem was given by Richardson and Savill [32]. They pointed out that DSC-curves obtained by heating depend not only on the heating rate, but also on the thermal history of the glass. In contrast, cooling curves depend only on the cooling rate. Heating curves of samples annealed below T_g show a marked peak, the so-called enthalpy relaxation, whereas cooling curves show a simple step function. The enthalpy relaxation is seen even in samples cooled and reheated at the same rate (measurements of Richardson and Savill [32] and of Höhne (Fig. 11 of this paper)). Richardson and Savill [32] eliminated the effect of enthalpy relaxation by the following procedure: They determined T_g from heating curves by construction of the corresponding enthalpy curves and by taking the intersection point of the extrapolated liquid and glassy line. This intersection point should be the same for cooling curves leading to the same glass. Fig. 13 shows a construction of the enthalpy curve corresponding to a cooling curve measured in PVAc by Höhne (Fig. 12). We note that T_g, defined by the above mentioned intersection point, may alternatively be found by the intersection of the half angle line indicated in Fig. 12 with the cooling

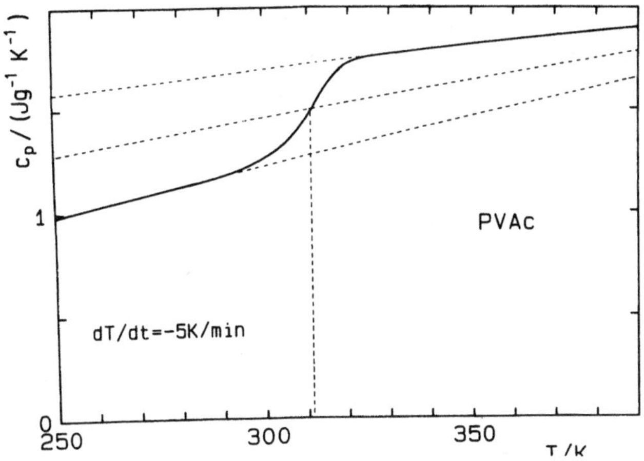

Fig. 12. Smoothed DSC-curve, measured with the cooling rate 5 K/min showing the determination of T_g (measured by Höhne et al.)

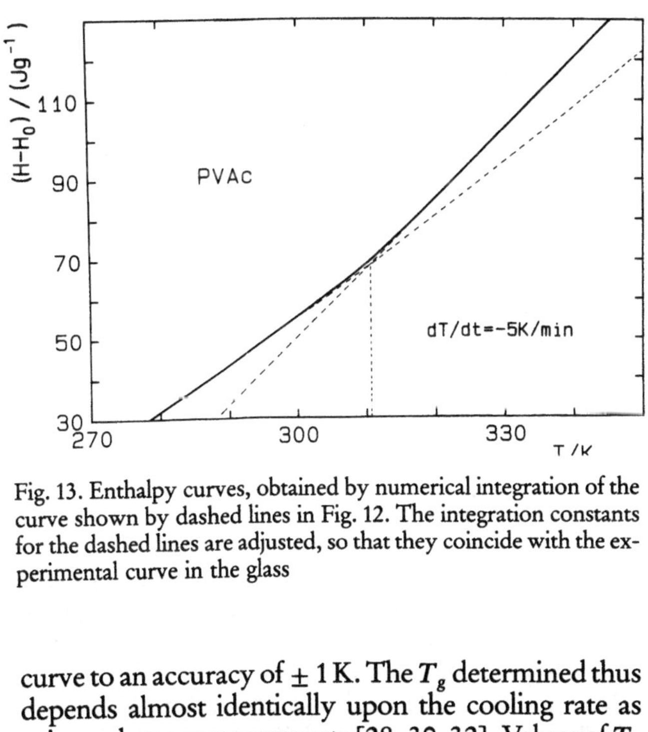

Fig. 13. Enthalpy curves, obtained by numerical integration of the curve shown by dashed lines in Fig. 12. The integration constants for the dashed lines are adjusted, so that they coincide with the experimental curve in the glass

curve to an accuracy of ± 1 K. The T_g determined thus depends almost identically upon the cooling rate as using volume measurements [28–30, 32]. Values of T_g were determined by the method shown in Fig. 12 from measurements by Höhne et al. and listed in Table 2 (the same PVAc as in [3] was dried 30 min at 150 °C in vacuum; then discs were cut 0.3 mm thick, 5 mm ∅ from the films and measured by cooling from 420 K to 240 K in a DSC-2 from Perkin-Elmer). A correction was made to eliminate the influence of heat conduction. The estimated accuracy of T_g is ± 2 K. The mean value of Δc_p at $T = T_g$ (taken from several cooling curves) is $\Delta c_p = 0.456$ Jg^{-1} K^{-1}.

Table 2. Calorimetric glass temperatures for PVAc

cooling rate K/min	5	10	20	40
M_v = 170000	312	312.5	313.5	
M_v = 1100000	312	313.0	313.5	313.5

Structural relaxation time at T_g

It is very important for modelling the relaxation behaviour of a polymer melt to know not only the dielectric and dynamic mechanical relaxation times, but also those of enthalpy and volume, and especially the relations between them. Dielectric and mechanical relaxation are very often investigated by dynamic methods, and it is easy to calculate relaxation times from the frequency of maximum loss. Volume and enthalpy relaxation, on the other hand, are scarcely observed in dynamic experiments. A lot of measurements were carried out, however, at constant cooling rate or constant heating rate. To get information about the relaxation time from these measurements, it is necessary to perform model calculations and compare them to experimental curves. A part of this information would be easily accessible, if the product of relaxation time and cooling rate at the glass transition temperature T_g was constant (specific for each polymer and known from other experiments). Volkenstein and Ptitsyn [33] performed model calculations for a system of independent segments with two possible conformations assuming an Arrhenius law

$$\tau_v = \tau_\infty \exp (U/RT) \tag{30}$$

for the thermodynamic relaxation time τ_v. They showed that at the temperature T_q (close to T_g) the following equation should hold

$$q\tau_v = - RT_q^2/U \tag{31}$$

where the righthand side is nearly independent of q. Bakule [34] applied these calculations to his volume measurements on polyisoprene. Wunderlich et al. [35, 36] starting from the same basis, calculated theoretical heat capacity curves for constant heating rate and compared them with DTA-experiments on polystyrene. Thus they obtained numerical values for the temperature dependence of τ_v. In addition, they measured cooling curves and determined $q\tau_v$ at T_g to be 6.6 K.

The T_g in [35] was defined in the same way as in Fig. 12 here.

An essential feature of the above mentioned model calculations [33, 35] is the assumption that τ_v is a function of temperature only (not dependent of the actual free volume); thus the difference between equilibrium and non-equilibrium state of the polymer is not taken into account. The heating curves in Ref. [35] are well described, in spite of the fact that no relaxation time distribution is involved in the theory. We suppose that both simplifying assumptions in the theory compensate for each other, leading to agreement with experimental heating curves. Kovacs and coworkers [37] calculated heating and cooling curves for a typical amorphous polymer numerically, without a relaxation time distribution, but with τ_v as a function of volume and temperature. Their theoretical heating curves show very narrow peaks. By introducing a relaxation time distribution in Ref. [38] they reached good agreement with experimental results. We did not find, however, numerical values of $q\tau_v$ at T_g in these papers.

We present here some calculations of $q\tau_v$ from our laboratory. First a simple derivation on the basis of the WLF equation. The simple form of the free volume theory is outlined in Eqs. (1), (3), (5). We start with the relaxation equation for the actual free volume V_f, which was used by Kovacs [37, 38]. Here, V_{feq} is the free volume in thermal equilibrium, t is the time

$$\tau_v(dV_f/dt) = V_{feq} - V_f \tag{32}$$

$$\tau_v(d^2V_f/dt^2) + (d\tau_v/dt)(dV_f/dt) = V_o q \Delta\alpha - (dV_f/dt)\,. \tag{33}$$

We now use Eq. (33) at T_g, which is defined by (see preceding section)

$$(dV_f/dt) = \frac{1}{2} V_o q \Delta\alpha \tag{34}$$

and neglected the term $\tau_v(d^2V_f/dt^2)$ (certainly a very rough approximation).

At T_g we obtain

$$d\tau_v/dt = 1. \tag{35}$$

Now we have to distinguish between the equilibrium relaxation time τ_{veq} and the actual relaxation time τ_v in a cooling experiment. We shall assume that there is a constant factor A between the dielectric (τ_d) and the volume (τ_v) relaxation time and use Eq. (1):

$$\tau_v = A\tau_o \exp(BV_o/V_f) \tag{36}$$

$$d\tau_v/dt = -\tau_v B(V_o/V_f^2)(dV_f/dt)\,. \tag{37}$$

Equations (34), (35), (37) yield at $T = T_g$

$$q\tau_v = -\frac{2}{B\Delta\alpha}(V_f/V_o)^2 \approx -\frac{2\,C_2}{2.303\,C_1}\,. \tag{38}$$

Now we can calculate the actual volume relaxation time at T_g; but, if we want a comparison with the dielectric relaxation time, we should know the equilibrium volume relaxation time at T_g. Therefore we use

$$\tau_{veq} = A\tau_o \exp(B\,V_o/V_{feq}) \tag{39}$$

$$\frac{\tau_{veq}}{\tau_v} = \exp\left[B\,V_o\,\frac{V_f - V_{feq}}{V_f\,V_{feq}}\right]. \tag{40}$$

Equations (32), (34), (38) yield at $T = T_g$

$$\frac{V_f - V_{feq}}{V_o} = \frac{1}{B}\left(\frac{V_f}{V_o}\right)^2 \approx \frac{C_2\Delta\alpha}{2.303\,C_1} \tag{41}$$

$$\frac{\tau_{veq}}{\tau_v} = \exp\left(\frac{V_f}{V_{feq}}\right) \approx 2.72 \tag{42}$$

$$q\tau_{veq} \approx -2.36\,C_2/C_1\,. \tag{43}$$

This result suggests that the product $q\tau_{veq}$ is independent of q and is of the order of 10 K for many polymers. Of course, this result is uncertain because of the rough approximation used in its derivation. In order to obtain more reliable results, we solved Eq. (32) numerically using a computer for PVAc at atmospheric pressure. We chose four different expressions for τ_v. First the case where $\tau_v = \tau_{veq}$ with the temperature dependence given by the WLF equation. The result for $q = 0.1$ K/s and $A = 1$ was a curve $V_f(T)$ and its numerical derivative is the lowest curve in Fig. 14. The second curve (named "WLF") was computed using Eqs. (32), (36), (5). Third, we wanted to show the influence of an additional Arrhenius term which was used by Macedo and Litowitz [1]. Therefore, Eq. (36) was replaced by

$$\tau_v = A\tau_o[\exp(BV_o/V_f)]\exp(Q/RT) \tag{44}$$

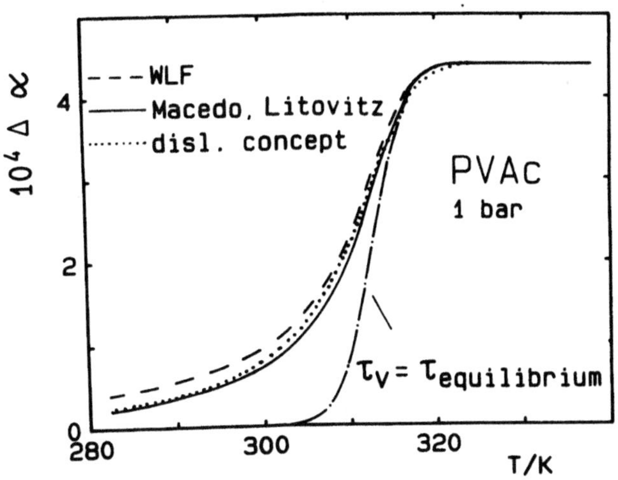

Fig. 14. Theoretical cooling curves, representing the expansion coefficient due to the free volume, multiplied by 10^4. Cooling rate $q = -0.1$ K/s and $A = 1$ in Eq. (36)

with $Q = 42$ kJ/mol. New parameters were used to fit to the experimentally known temperature dependence of the equilibrium relaxation time. The fourth curve (named "dislocation concept") represents dx/dT which is approximately proportional to $\Delta\alpha$ (normalized to compare it with the others). We used Eqs. (8), (12), (18), (19) and

$$\tau_v \frac{dx}{dt} = x_{eq} - x\,. \tag{45}$$

There is little difference between the freezing-in curves of these theories except the case $\tau_v = \tau_{veq}$. We should note that all curves in Fig. 14 involve no relaxation time distribution. Experimental cooling curves should therefore show a smoother step. This is especially important considering the heating curve in Fig. 15.

At the point, where $\Delta\alpha$ reaches half of its equilibrium value ($T = T_g$), $q\tau_v$ and $q\tau_{veq}$ were calculated and listed in Table 3 to compare them with the algebraic approximation. Also included in Table 3 are calculations for cooling rates varied by a factor 100 in both directions. Figure 15 illustrates this effect. It turned out that varying A has the same effect as varying q. Looking at Table 3, we note that $q\tau_{veq}$ is, indeed, nearly independent of q. All numerical methods (except the unrealistic "$\tau_v = \tau_{veq}$" method) give nearly the same result. Even a simulation of other polymers by drastically changing ε_{so} resp. $3r/d$ in the dislocation concept results in nearly the same $q\tau_{veq}$. Therefore we can use

$$q\tau_{veq} = 15\ \mathrm{K} \tag{46}$$

to determine volume (resp. enthalpy) relaxation times at $T = T_g$ from experimental cooling curves. Considering the uncertainty in the experimental determination of T_g and the strong temperature dependence of τ_{veq}, it is not necessary to know $q\tau_{veq}$ more accurate than to ± 50%. We did not consider a relaxation time distribution in this paper. We hope, however, that the above calculations give correct values for the mean relaxation time.

Table 3. Calculations of the product of cooling rate and volumetric relaxation time at T_g for different models (equation numbers in parentheses) with $A = 1$

	$q\tau_v$			$q\tau_{veq}$			T_g		
cooling rate q/Ks^{-1}	0.001	0.1	10	0.001	0.1	10	0.001	0.1	10
algebraic approximation (PVAc)									
$\tau_v = \tau_{veq}$ (30), (31)	1.7	1.7	1.7	1.7	1.7	1.7	306.6	313.0	319.7
WLF (36), (38), (43)	2.7	2.7	2.7	7.3	7.3	7.3	305.5	311.0	317.7
numerical solution (PVAc)									
$\tau_v = \tau_{veq}$ (31), (32)	2.5	2.5	2.5	2.5	2.5	2.5	306.1	312.5	319.1
WLF (5), (32), (36)	3.4	4.3	5.9	15.3	19.0	27.9	304.1	309.6	316.3
Macedo + Litovitz (32), (44)	3.0	4.0	5.5	11.5	15.0	20.2	305.6	310.6	317.3
dislocation concept (8), (45)	3.7	4.4	5.5	12.1	14.5	17.1	303.4	310.0	317.9
simulation of other polymers with the dislocation concept									
$\varepsilon_{so} = 2.8$ kJ/mol, $3r/d = 17$	3.9	4.8	6.2	11.5	14.2	16.6	258.7	265.7	274.6
$\varepsilon_{so} = 4.48$ kJ/mol, $3r/d = 17$	3.5	4.2	5.1	11.8	14.7	17.3	355.3	361.4	368.7
$\varepsilon_{so} = 4.48$ kJ/mol, $3r/d = 34$	2.6	3.2	4.0	9.7	11.6	13.3	306.2	310.6	316.1

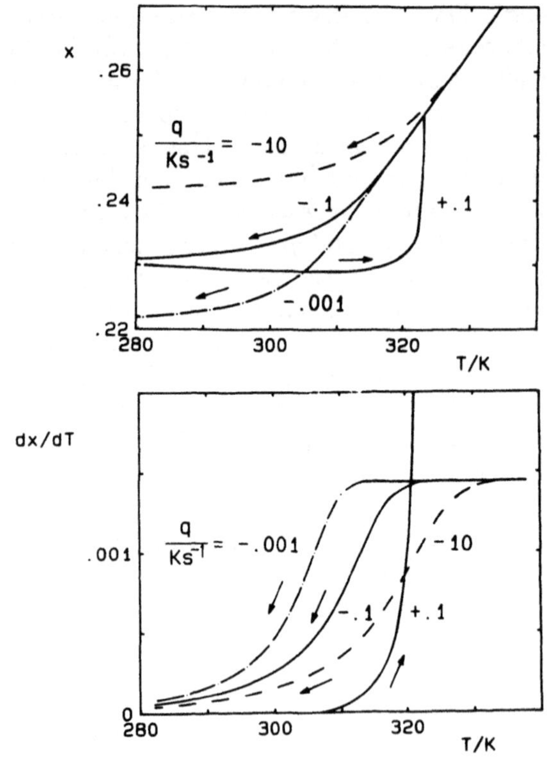

Fig. 15. Upper diagram: The number of quasi-dislocations of a certain type per segment in PVAc, as a function of temperature, calculated for the cooling rates 10, 0.1, 0.001 K/s. One heating curve for 0.1 K/s is also shown. The lower diagram shows the derivative dx/dT

Comparison of the dielectric, calorimetric and dilatometric T_g

Using Eq. (46) and the well-known relation between relaxation time and maximum frequency in dynamic experiments

$$2\pi f_v \tau_{veq} = 1 \tag{47}$$

for enthalpy instead of volume, the calorimetric experiments (Table 2) are compared directly with the dielectric and mechanical results in Fig. 16. All measurements were made on the same material (Vinnapas from Wacker). The difference in maximum frequencies at a given temperature for samples with M_v = 170 000 to M_v = 1 100 000 g/mol was not greater than a factor 2.

Lower molecular weights than M_v = 170 000 were not included in Fig. 16. All samples were dried before measurements. It could be asked "are measurements of high molecular weight samples (which cannot flow near T_g) influenced by internal stress and orientation

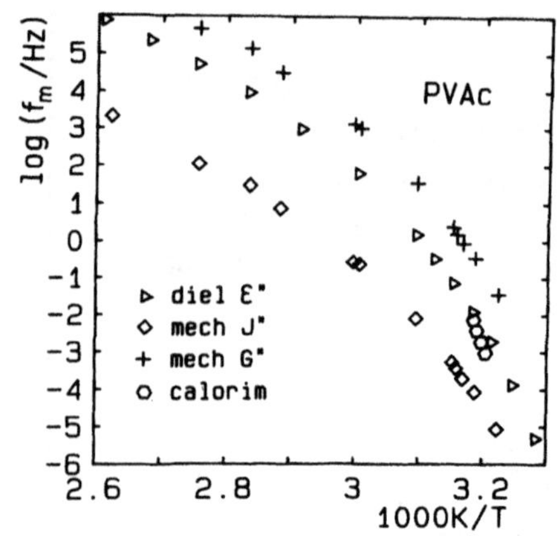

Fig. 16. Activation diagram for PVAc (Vinnapas from Wacker Chemie, Burghausen)

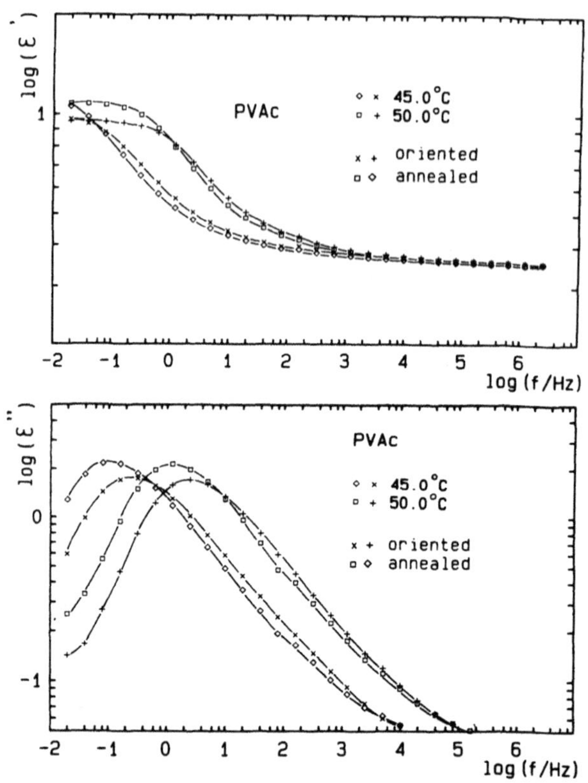

Fig. 17. The real and imaginary part of the permittivity as a function of fequency for a PVAc sample (M_v = 1 100 000 g/mol) which was formed at 170 °C in vacuo to obtain the thickness 2 mm, then cooled to 80 °C, then pressed between stainless steel plates in vacuo (with negligible flow) to obtain the new thickness 0.3 mm and a biaxial orientation, then measured at 45 °C and 50 °C, then heated in vacuo to 170 °C and relaxed 2 hours at fixed thickness, then measured again at 45 °C and 50 °C. The difference between oriented and relaxed state is small and might perhaps be explained by a poorer adhesion of the oriented sample to the stainless steel plates

of the macromolecules?" An investigation of this is presented in Fig. 17. We conclude that this influence is negligible for the discussion in this paper. From Fig. 16 we conclude that dielectric and enthalpy relaxation times are nearly equal in PVAc, and that the mechanical relaxation times for shear modulus, resp. shear compliance differ from them by a constant factor. Volume measurements on PVAc are being carried out in our laboratory; they will be presented in another paper. Dilatometric data from the literature are plotted in Fig. 18 for PVAc and in Fig. 19 for PS, using Eqs. (46), (47). Here we note, that Sasabe and Moynihan [39] used the same material as McKinney and Goldstein [5]. Similarly, we were lucky to find, that Brather et al. [41] used the same polystyrene as Schwarzl and Zahradnik [28]. We conclude from Figs. 18 and 19 that the volume relaxation time is nearly equal to the dielectric relaxation time for PVAc and PS. Matsuoka et al. [44, 45] arrive at a similar conclusion for frequencies higher than about 1 Hz, but note differences for lower frequencies (resp. temperatures).

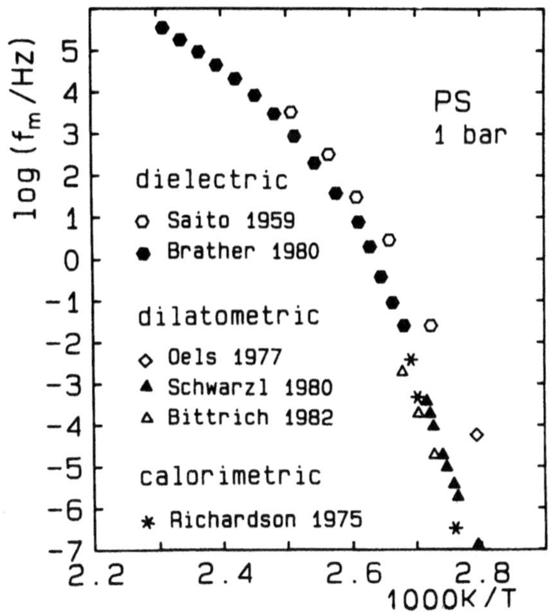

Fig. 19. Activation diagram for polystyrene at atmospheric pressure. Dielectric maximum frequencies from Brather et al. [41] (filled hexagons) and Saito [43] (open hexagons); dielectric measurements of Broens and Müller [42] above 100 Hz coincide with Brather's [41]. Dilatometric frequencies, using Eqs. (46), (47) were calculated from Oels and Rehage [24] open square), from Schwarzl and Zahradnik [28] (filled triangles), and from Bittrich et al. [29] (open triangles). Calorimetric frequencies are from Richardson and Savill [32] (stars)

Dielectric observation of freezing-in

Since the dielectric relaxation time is controlled by the free volume (WLF theory) or the density of quasi-dislocations (dislocation concept), the freezing-in point in cooling curves should be observable by dielectric measurements too. Therefore we cooled a PVAc sample with evaporated gold electrodes from 44 °C in steps of 5 °C per 90 minutes and made measurements on each step (Fig. 20). For comparison, measurments on a very slowly cooled sample are presented in Fig. 21. Assuming, that the shape of the dielectric loss curve is constant, we extrapolated the position of the maximum frequency and plotted it in Fig. 22. Another run was carried out with continuous cooling (rate 0.01 K/s), while monitoring only the dielectric loss at 0.1 Hz and extrapolating the peak frequency as before. We note, that the points in Fig. 22 deviate from the extrapolated equilibrium curve at low temperatures. A similar behaviour was found by Koppelmann [46] and Schlosser [47]. Theoretical curves are calculated for the dislocation concept, using Eqs. (8)–(19), (45), (47) and the parameters from Fig. 3 assuming $\tau_d = \tau_v$, and are plotted in Fig. 22. The Arrhenius term in Eq. (8) is here an important factor at low temperatures. The theory describes the experiments rather well, consi-

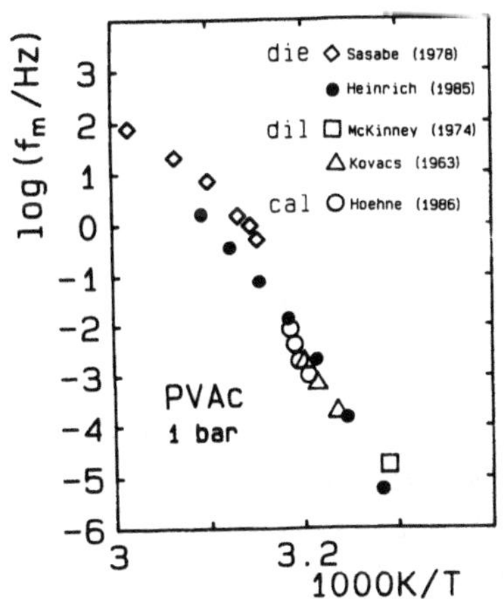

Fig. 18. Dielectric maximum frequencies from [3] (filled circles) and [39] (open squares on the tip) together with corresponding maximum frequencies from calorimetry (this paper, open circles) and volume measurements [5] (open square, flat) using Eqs. (46), (47). In addition, frequencies were calculated with Eq. (47) from volume relaxation times, which were roughly taken from isothermal measurements of Kovacs [40] and plotted as triangles. Only those curves of Kovacs [40] were used, which had nearly an exponential form (that means: the temperature step before measurement should be sufficiently small in order to be not far from equilibrium)

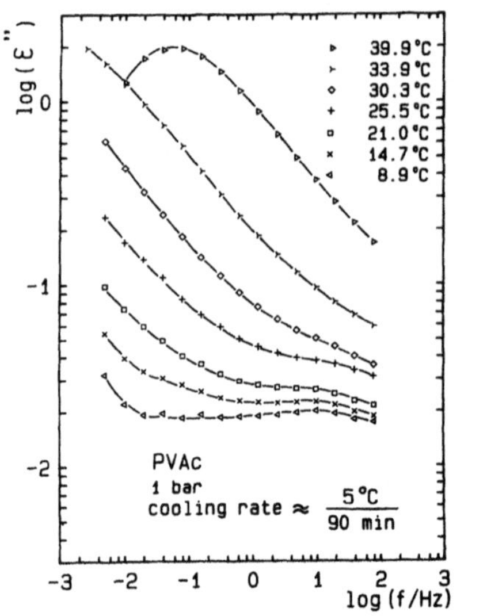

Fig. 20. The dielectric loss of PVAc at atmospheric pressure (M_v = 1 100 000 g/mol) cooled in steps of 5 °C per 90 minutes and measured at the temperatures indicated as a function of frequency

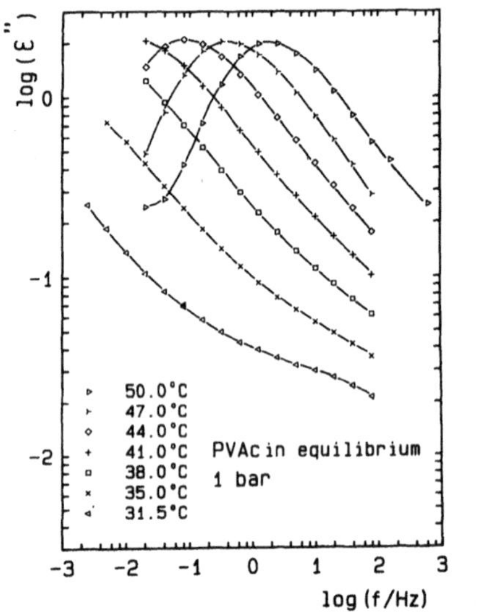

Fig. 21. The dielectric loss of PVAc at atmospheric pressure (M_v = 1 100 000 g/mol) cooled slowly enough to remain in thermal equilibrium, measured at the temperatures indicated as a function of frequency

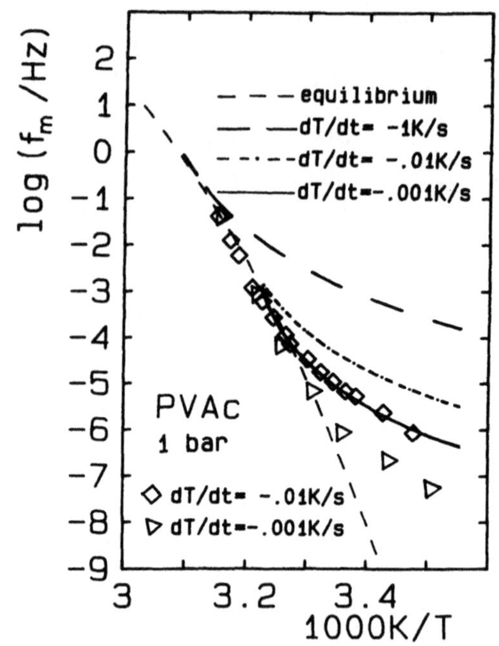

Fig. 22. Activation diagram for PVAc at atmospheric pressure from dielectric measurements on samples cooled at constant cooling rate. Symbols: experimental. Lines: Theory based on the dislocation concept for cooling rates indicated

dering the large extrapolation from experimental frequencies (10^{-2} resp. 10^{-1} Hz) to f_m in Fig. 22.

Relaxation after pressure steps

Having found that we can observe the freezing-in of the free volume (resp. the quasi-dislocation density (Fig. 22)) in dielectric measurements, we observed the behaviour of a PVAc sample after a sudden change of hydrostatic pressure. The relaxation time needed for the dielectric loss peak to reach a new equilibrium position is presumably equal to the volume relaxation time. The temperature 40 °C and the reference pressure 210 bar were chosen in order to have a sufficiently long volume relaxation time, compared with the time $\Delta t_p \leq 3$ s necessary to achieve a pressure jump of about 100 bar. This pressure jump is large enough to observe the change of the dielectric loss easily and is small enough to avoid non-linear effects caused by the change of volume relaxation time in the course of the measurement. The sample (0.5 mm thick, 17 cm^2 area) had evaporated gold electrodes (≈ 0.1 μm thick) and was clamped with a spring between stainless steel plates in order to allow free expansion resp. contraction. The pressure was transmitted by silicone oil [3]. The dielectric loss was monitored at 0.1 Hz and the maximum frequency f_m of the dielectric loss peak was extrapolated using its known shape. Log f_m is plotted in Fig. 23. Another experiment with another sequence of pressures is presented in Fig. 24. The arrows indicate the log f_m immediately before a pressure change. One might interpret Figs. 23 and 24 so: an instantaneous response of the sample is present in addition to the

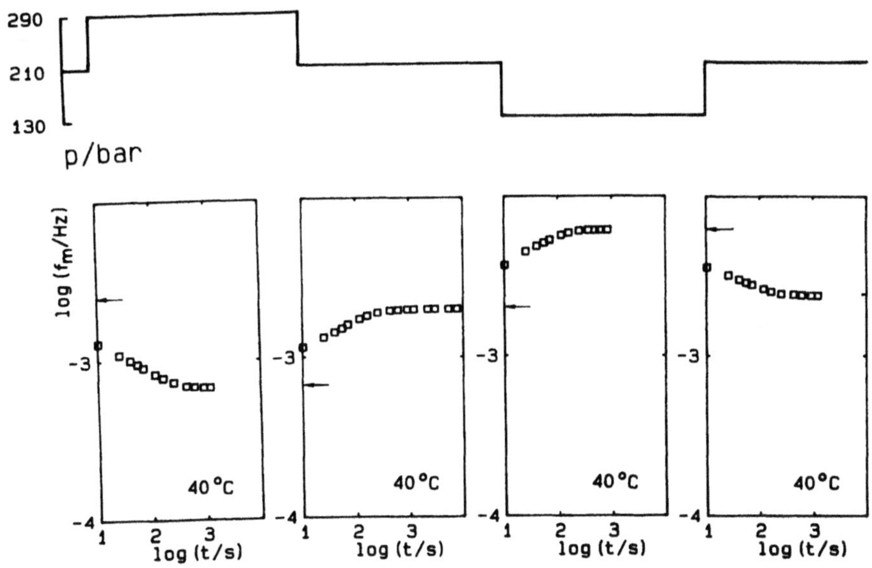

Fig. 23. Extrapolated maximum frequency of the dielectric loss peak as a function of the logarithm of time after a sudden change of pressure. Arrows indicate the equilibrium value before the pressure change. The pressure belonging to each of the four relaxation experiments is indicated above. The temperature was constant 40 °C

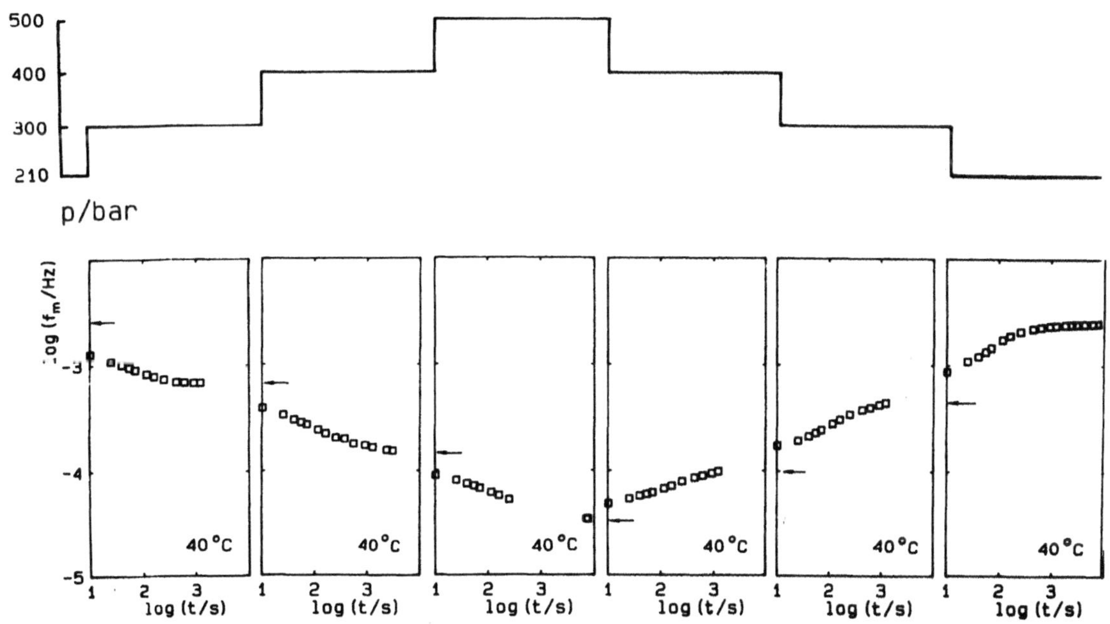

Fig. 24. Data as in Fig. 23, but with another sequence of pressure

relaxation response; this instantaneous response may be explained by the pressure dependence of Q_v in Eq. (8). In the following, however, we neglected the instantaneous response and, although the observed curves did not follow a pure exponential law, a mean relaxation time τ_v was determined roughly by reading $\log t = \log \tau_v$ a the point where

$$\log f_m = \log f_m\,(\text{new}) + \frac{1}{e}\,[\log f_m\,(\text{old}) - \log f_m\,(\text{new})] \tag{48}$$

("old" and "new" refer here to the equilibrium states observed in Figs. 23 and 24 before and after a pressure change). The $\log f_m$ value was taken at the same point from Figs. 23 and 24 and plotted in Fig. 25 versus $\log f_v$ using Eq. (47) and assuming $\tau_v = \tau_{veq}$ (because the pressure jumps are small). If the relaxation times of volume and dipole orientation are the same, the points in Fig. 25 should lie on the 45° degree line. The deviation of experimental points from it is greater, the lower the frequency (the larger the extrapolation from the experimental frequency 0.1 Hz to the maximum fre-

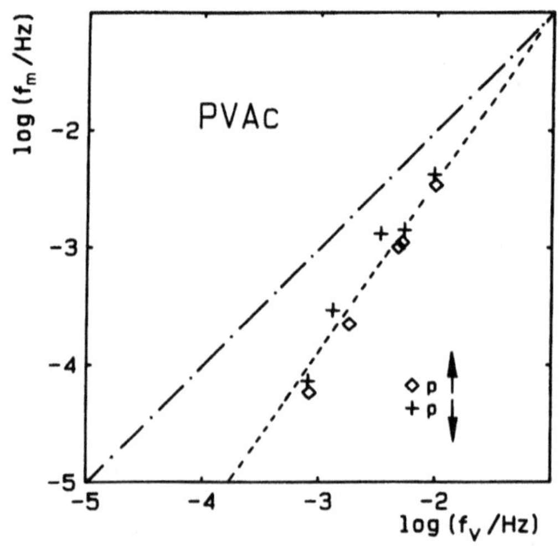

Fig. 25. Extrapolated maximum frequency of the dielectric loss peak plotted versus the structural relaxation frequency (calculated by $f_v = 1/(2\pi\tau_v)$ from the relaxation time τ_v needed for the dielectric loss peak to reach a new equilibrium position after a small pressure change)

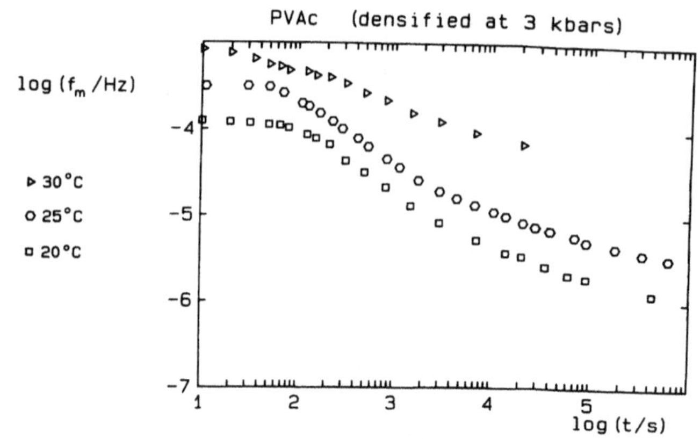

Fig. 26. Extrapolated maximum frequency of the dielectric loss peak as function of the logarithm of the time after pressure release for the same PVAc sample as used in Figs. 23 and 24. The sample was heated to 130 °C and cooled under a pressure $p = 3\,000$ bar with the rate 10 °C/h to the temperatures indicated; then the pressure was released suddenly and the dielectric loss was observed at 0.1 Hz at constant temperature. The position of the loss peak was extrapolated assuming that its shape remains constant

quency f_m). If we realise, that the position of the dielectric loss peak is estimated more accurately at higher frequencies than at lower frequencies, we should extrapolate the data in Fig. 25 to 0.1 Hz. We conclude that τ_v is equal to τ_d. The differences between them in Fig. 25 are probably due to small changes of the shape of the dielectric losss peak with pressure.

Further problems

It has been shown in the preceding section that the dislocation concept is able to describe the temperature and pressure dependence of dielectric relaxation times in PVAc, as well as that of volume and enthalpy in the melt, using molecularly defined parameters, which can (in principle) be checked by other methods.

Some interesting points, however, have not yet been discussed (half-width and asymmetry of dielectric and mechanical loss peaks, relation between mechanical and dielectric maximum frequencies, memory effect [40], and the "densified" glasses [48–51]). The latter are especially interesting because volume and dipole relaxation show a time dependence opposite from that to be expected from the free volume theory. This is illustrated in Fig. 26. Volume measurements after pressure release [24, 48, 49, 51] show an increase in volume (not only the sudden response, but also the relaxation shows increasing volume). It is not clear, whether a relaxation time distribution, or fluctuations in density, can explain this behaviour.

It is planned to apply the dislocation concept to other amorphous polymers and to polymer blends. To this aim many additional measurements are necessary. It is desirable to describe with a molecular model not only the above mentioned effects, but also the influence of plasticiser, crosslinking, crystallinity and fillers on the dielectric and mechanical relaxation of the melt. Little has yet been done in this direction.

Some interesting problems in relation to the relaxation behaviour of a polymer melt are represented by liquid crystalline polymers and by mixtures of polar and non-polar components.

Acknowledgement

The authors wish to express their thanks to Prof. Dr. W. Pechhold for his interest in this work, to Dr. G. Höhne and Dipl.-Phys. W. von Soden for their experimental contributions (calorimetric and mechanical measurements) and to the Deutsche Forschungsgemeinschaft for the financial support.

References

1. Ferry JD (1980) Viscoelastic Properties of Polymers. 3rd Ed, Wiley, New York
2. Pechhold W, Sautter E, v Soden W, Stoll B, Grossmann HP (1979) Makromol Chemistry Suppl 3:247
3. Heinrich W, Stoll B (1985) Colloid Polym Sci 263:873
4. Dalal EN, Phillips PJ (1983) Macromolecules 16:890

5. McKinney JE, Goldstein M (1974) J Res Nat Bur Standards 78A:331
6. Havlíček I, Ilavsky M, Hrouz J (1982) J Macromol Sci-Phys B21:425
7. Kanig G (1969) Kolloid Z Z Polym 233:829
8. Pechhold W, Stoll B (1982) Polymer Bull 7:413
9. Slater JC (1939) Introduction to Chemical Physics. McGraw Hill, New York
10. Hartmann B (1976/77) Acustica 36:24
11. Warfield RW (1974) Makromol Chem 175:3285
12. Asay JR, Lamberson DL, Guenther AH (1969) J Appl Phys 40:1768
13. Leute U, Grossmann HP (1981) Polymer 22:1335
14. Brody EM, Lubell CJ, Beatty CL (1975) J Polym Sci, Polym Phys Ed 13:295
15. Roberts R (1985) Dissertation, Saarbrücken
16. Weeger R (1986) University of Ulm, Abt Exp Physik, unpublished results
17. Gilmor I, Trainor A, Haward RN (1978) J Polym Sci, Polym Phys Ed 16:1291
18. Carlone C, Hota NK, Stolz HJ, Elbert M, Hochheimer HD (1981) J Chem Phys 75:3220
19. Boehler R, Kennedy GC (1977) J Appl Phys 48:4183
20. Pechhold W (1986) University of Ulm, personal communication
21. McCrum NG, Read BE, Williams G (1967) Anelastic and Dielectric Effects in Polymeric Solids. Wiley, New York
22. Havranek A, Ilavsky M, Nedbal J, Böhm M, v Soden W, Stoll B (1987) Colloid Polym Sci 265:8
23. Breuer H, Rehage G (1967) Kolloid Z Z Polym 216/217:159
24. Oels HJ, Rehage G (1977) Macromolecules 10:1036
25. Stavermann AJ (1966) Rheol Acta 5:283
26. Donth EJ (1981) Glasübergang. Akademie-Verlag, Berlin
27. Davies RO, Jones GO (1953) Adv Phys 2:370
28. Schwarzl FR, Zahradnik F (1980) Rheol Acta 19:137
29. Bittrich HJ, Schad HJ, Tanneberger H (1982) Acta Polymerica 33:736
30. Greiner R, Schwarzl FR (1984) Rheol Acta 23:378
31. Havlíček I, Vojta V, Ilavsky M, Hrouz J (1980) Macromolecules 13:357
32. Richardson MJ, Savill NG (1975) Polymer 16:753
33. Volkenstein MV, Ptitsyn OB (1956) zh techn fiz 26:2204
34. Bakule R (1987) Charles University of Prague, personal communication
35. Wunderlich B, Bodily DM, Kaplan MH (1964) J Appl Phys 35:95
36. Wolpert SM, Weitz A, Wunderlich B (1971) J Polym Sci A-2 9:1887
37. Hutchinson JM, Kovacs AJ (1976) J Polym Sci, Polym Phys Ed 14:1575
38. Kovacs AJ, Aklonis JJ, Hutchinson JM (1979) J Polym Sci, Polym Phys Ed 17:1097
39. Sasabe H, Moynihan CT (1978) J Polym Sci, Polym Phys Ed 16:1447
40. Kovacs AJ (1963) Fortschr Hochpolym Forsch 3:394
41. Brather A, Link G, Luchschneider R (1980) Colloid Polym Sci 258:1307
42. Broens O, Müller FH (1955) Kolloid Z Z Polym 140:121
43. Saito S, Nakajima T (1959) J Appl Polym Sci 2:93
44. Matsuoka S, Williams G, Johnson GE, Anderson EW, Furukawa T (1985) Macromolecules 18:2652
45. Matsuoka S (1986) J Rheol 30:869
46. Koppelmann J, Leder H, Royer F (1979) Colloid Polym Sci 257:673
47. Schlosser E (1982) Polymer Bull 8:461
48. Weitz A, Wunderlich B (1974) J Polym Sci, Polym Phys Ed 12:2473
49. Wetton RE, Moneypenny HG (1975) British Polymer J 7:51
50. Brown IG, Wetton RE (1978) Polymer 19:659
51. Kogovski GJ, Filisko FE (1986) Macromolecules 19:829

Received April 21, 1987;
accepted February 12, 1988

Authors' address:

Dr. B. Stoll
Universität Ulm
Abt. Angewandte Physik
Oberer Eselsberg
D-7900 Ulm, F.R.G.

Progress in Colloid & Polymer Science Progr Colloid Polym Sci 78:54–58 (1988)

Modified epoxies as studied by ultrasonic methods

I. Alig[1]), M Fedtke[2]), K.-G. Häusler[2]), W. Tänzer[2]), and S. Wartewig[1])

Technical University "Carl Schorlemmer" Leuna-Merseburg,
[1]) Department of Physics and
[2]) Department of Chemistry, Merseburg, GDR

Abstract: The cure of modified epoxy resins was monitored by taking ultrasonic measurements. From the time dependence of acoustic parameters it was possible to elucidate the different stages of curing, the gel point, the dynamic glass transition, and the end of the crosslinking reactions. The chemical composition of the epoxy systems was varied in a systematic manner and its influence on the ultrasonic properties was studied. The results show that ultrasonic methods serve as a useful tool for the characterization of curing reactions and highly crosslinked networks.

Key words: Ultrasonic velocity and attenuation, epoxy, crosslinks, moduli, curing.

1. Introduction

The increasing technological importance of modified epoxies demands comprehensive physical and chemical characterization of the curing process and the final network itself. The properties of the fully cured resins depend primarily on their chemical composition and degree of crosslinking as well as on the details of network structure. Highly crosslinked products cannot be easily characterized by conventional methods. Therefore, techniques for the characterization of these network systems need further development.

During the curing reaction, the epoxy system transforms from the liquid state to a gel and then vitrifies to the glassy state. This transformation is accompanied by marked changes in the viscoelastic properties of the material which are related to the molecular structure of the network. Therefore, ultrasonic investigations during the cure may give information about the extent of the curing reaction and the mechanical properties of the material.

From literature, the applicability of ultrasonic methods for monitoring the crosslinking of various resins has been known for a long time [1–10]. The first such study seams to be that of Sofer and Hauser, who followed the change in ultrasonic velocity and attenuation during the cure progress [1, 2].

Ultrasonic investigation is now well established in polymer research as far as to study the dynamic properties of polymer molecules in both liquid and solid state [11, 12].

The aim of the present paper is to monitor changes in the longitudinal sound velocity and absorption during the cure of modified epoxy resins whose chemical composition has been varied in a systematic manner. In this way it is possible to obtain information about structural changes during the crosslinking. Furthermore, we will illustrate that ultrasonic methods serve as a useful tool for the characterization of highly crosslinked networks.

2. General remarks

The propagation properties of longitudinal ultrasonic waves in a viscoelastic medium are usually described in terms of the velocity c and absorption coefficient α which are related to the elastic moduli and the dynamic viscosities of the medium

$$c^2 = \frac{M'}{\varrho} = \frac{K' + \frac{4}{3}G'}{\varrho} \tag{1}$$

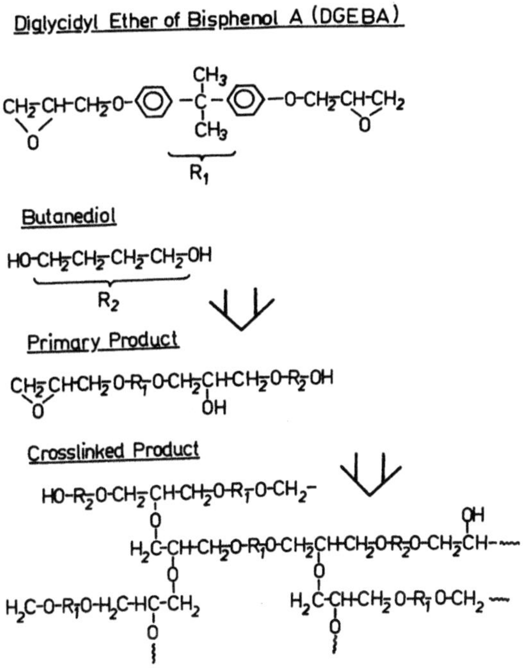

Fig. 1. Starting materials and expected chemical structures in dependence on the synthesis routes for DGEBA/butane – 1,4-diol

$$\alpha = \frac{\omega}{2\varrho c^3} M'' = \frac{\omega}{2\varrho c^3}\left(K'' + \frac{4}{3} G''\right) = \frac{\omega}{2\varrho c^3}\left(\eta'_v + \frac{4}{3} \eta'_s\right) \tag{2}$$

where $M^* = M' + iM''$ is the complex longitudinal wave modulus, K' and K'', as well as G' and G'' are the real and imaginary parts of the bulk and shear modulus, respectively; η'_s and η'_v are the dynamic shear and the dynamic volume or bulk viscosities, respectively, ϱ is the density, and ω is the angular frequency. We assume that additional dissipation mechanisms due to scattering and transport of heat can be neglected for the systems studied. It should be noted that in the case of longitudinal waves we always measure a combination of shear and bulk properties, which often renders the interpretation more difficult.

3. Experimental

Synthesis of modified epoxies

The modification of the most commonly used expoxy diglycidyl ether of bisphenol A (DGEBA) has been carried out by crosslinking DGEBA in the presence of i) various diols and ii) bifunctional carboxyl-terminated oligostyrene (COS) with different molar mass. For the DGEBA/butane-1,4-diol system the starting materials and the expected chemical structures in dependence on the synthesis routes are schematically shown in Fig. 1. The full-synthesis schema is described elsewhere [13,14]. In order to monitor the curing process initial products of the systems

DGEBA: butane-1,4-diol : $Mg(ClO_4)_2$ = 1 : 0.25 : 0.03 and

DGEBA: but-2-ene-1,4-diol : $Mg(ClO_4)_2$ = 1 : 0.25 : 0.03

were investigated.

Ultrasonic method

The sound velocity of the cured samples was determined by using the pulse transmission technique at a frequency of 1 MHz or 12 MHz [15]. Velocities were measured with a precision of Δc = 10 m/s. The sample thickness varied between 4 and 10 mm. A modification of the pulse transmission technique described in [16] was used in order to monitor the curing. In this way it was possible to measure the velocity and the relative attenuation $\Delta\alpha \cdot x$ simultaneously at a frequency of 1 MHz. The relative attenuation is defined by $\Delta\alpha \cdot x = \ln(A_0/A)$, where A_0 and A are the pulse amplitudes at the times t_0 and t. The thickness of the sample is designated by x. With this equipment we were able to study a sample amount of 3 ml in the temperature range – 20 °C–150 °C. The temperature within the sample cell was controlled with a precision of ± 0.3 K.

4. Results and Discussion

4.1 Curing reactions

For the cure of DGEBA/but-2-ene-1,4-diol (molar ratio 1 : 0.25) in the presence of 3 mol % $Mg(ClO_4)_2$ as catalyst at a curing temperature of 100 °C, the variation of the sound velocity c and the relative change of the absorption $\Delta\alpha \cdot x$ as function of the curing time is shown in Fig. 2. Obviously, we can distinguish different stages of curing. In the pre-gel stage there is no significant change in the acoustic properties. After the gel point, i.e., in the post-gel stage we observed a distinct increase in the velocity and the absorption, which is the result of the network formation. Owing to the solidification of the sample, there is a shift in the position of the glass transition to lower temperatures. Consequently, the dynamic glass transition can pass through the measuring frequency according to the chosen curing temperature. The corresponding WLF-curves are schematically plotted in Fig. 3. It is well known that the dynamic glass transition causes a maximum of the ultrasonic absorption and a dispersion of the velocity. Therefore we can identify the dynamic glass transition T_{dyn} in Fig. 2. In the course of the crosslinking process, the acoustic parameters finally approach the values of the fully cured network.

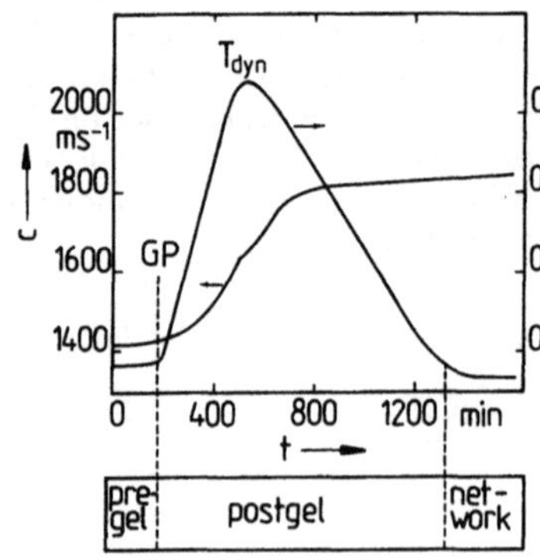

Fig. 2. Ultrasonic velocity c and relative ultrasonic attenuation $\Delta\alpha \cdot x$ in dependence on the curring time for DGEBA/but-2-ene-1,4-diol (molar ratio 1:0.25) in the presence of 3 mol % $Mg(ClO_4)_2$. Measuring frequency 1 MHz, curing temperature 100 °C, GP gel point, T_{dyn} dynamic glass transition

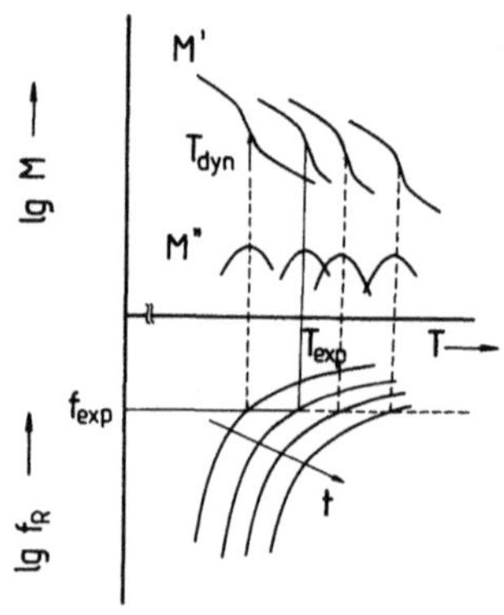

Fig. 3. Schematic of the shift of the dynamic glass transition (WLF-curve) during the cure process; T_{exp} curing temperature, f_{exp} measuring frequency, and f_R relaxation frequency

A detailed interpretation of the plots shown in Fig. 2 is rather complicated. Immediately after the gel point the system can be considered as a swollen network in which the portion of the network increases during curing, the sol component decreases, and the molar mass of the chain between the network nodes decreases. Therefore, we have to discuss the dynamic shear and bulk properties (G^*, K^* or η_s^*, η_v^*) in terms of crosslinked systems where the influence of solvent should also be taken into consideration. After the formation of the network the moduli reflect the rubber elastic behavior. In the range of relatively low crosslinking densities the shear modulus can be described by theories of rubber elasticity (see [17, 18, 19] and references therein). From experimental and theoretical points of view there has been little research carried out on bulk modulus and highly crosslinked systems. It should be noted here that all motional processes (network modes, rubber elastic behavior, local motions, etc.) shift their frequency position during the crosslinking reaction.

In the course of curing the motional modes between the network nodes become effective and the freezing-in of the motions at the dynamic glass transition can shift through the measuring frequency. In an advanced stage of curing, secondary processes can also pass through the measuring conditions. Finally, in the highly crosslinked epoxy resins the acoustic properties reflect the moduli in the glassy state and it should be possible to consider these moduli as superpositon of rubber elastic moduli and glass moduli of highly crosslinked systems. As in the case of the rubber plateau modulus of highly crosslinked systems there is, so far, no molecular description of the glass transition. Furthermore, the relation of shear and bulk properties for the different motions is unknown.

However, qualitative conclusions can be drawn about curing experiments which have been differently performed. Figures 4 and 5 show the results for the curing of DGEBA in the presence of butane-1,4-diol and but-2-ene-1,4-diol under the same conditions (molar ratio 1 : 0.25, 3 mol % $Mg(ClO_4)_2$, curing temperature 100 °C). For the two systems distinct differences in the behavior of the sound velocity and absorption are evident. Density changes were only 1 % or 2 %. The higher value of the sound velocity and the lower sound absorption for cured DGEBA/but-2-ene-1,4-diol indicate that a harder network is formed due to the hindered molecular mobility caused by the double

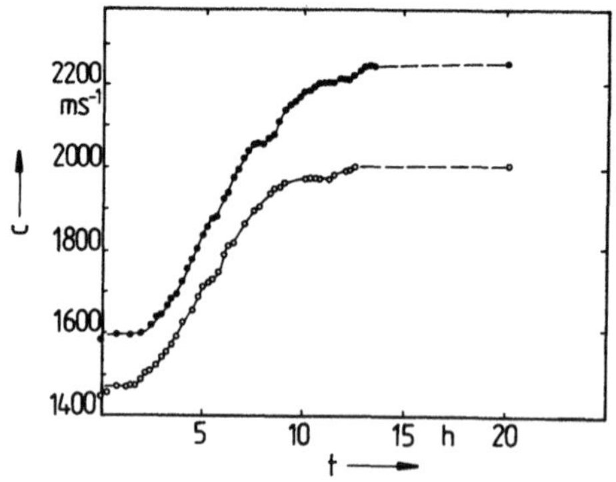

Fig. 4. Ultrasonic velocity c measured at 1 MHz as a function of curing time t for DGEBA/but-2-ene-1,4-diol (●) and DGEBA/butane-1,4-diol (○), molar ratio 1 : 0.25, 3 mol % $Mg(ClO_4)_2$, and curing temperature 100 °C

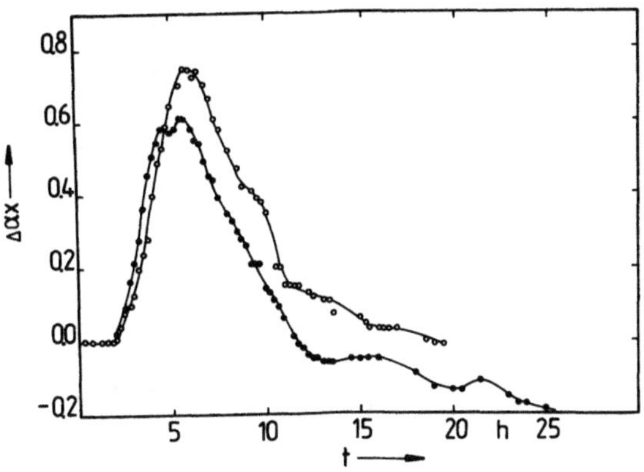

Fig. 5. Relative ultrasonic attenuation $\Delta\alpha \cdot x$ at 1 MHz as a function of curing time t for the systems explained in Fig. 4

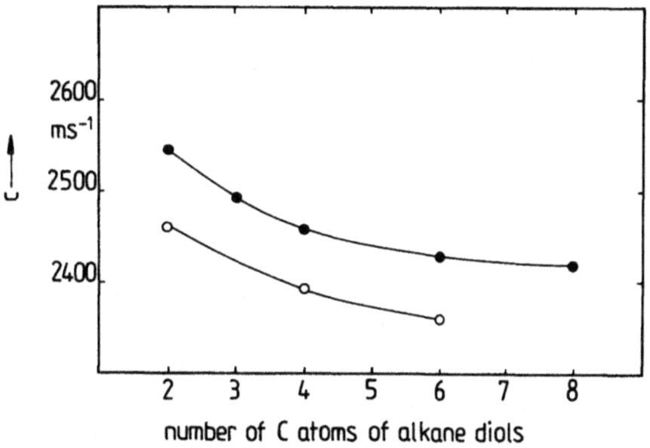

Fig. 6. Ultrasonic velocity c as a function of the number of C atoms of alkane diols. Fully cured system studied at 12 MHz and room temperature; Catalyst $Mg(ClO_4)_2$ (●) and N,N-dimethyl benzylamine (○), curing temperature 130 °C

bond. The irregularities of the curves should be related to the "passing through" of the various molecular mechanisms during curing as discussed above. On the other hand, it is remarkable that for both systems the changes in the absorption are nearly the same at the beginning of the cure reaction, e.g., the gel points occur after the same time interval. This indicates that the start phase of the network formation is almost identical for the two systems.

4.2. Characterization of modified epoxy resins

If the temperature of the dynamic glass transition at the sound frequency applied is higher than the measuring temperature, then the moduli of the glassy state should mainly determine the acoustic data. As shown above, the dynamic glass transition (at 1 MHz) of the modified epoxy resins studied occurs at about 100 °C. Therefore, we can deduce the longitudinal wave modulus M' of the glassy state from acoustic experiments at room temperature. Since a molecular theory of the glassy state does not exist, we cannot elucidate the corresponding molecular interaction.

Fortunately, an empirical dependence of the longitudinal sound velocity on parameters of the synthesis could be verified. The velocity c as a function of the number of C atoms of the alkane diols is shown in Fig. 6. It is evident that the increasing chain length causes a decrease in c. Since the diol acts as a chain lengthening agent, we can qualitatively understand that the increasing number of C atoms leads to a "softening" of the crosslinked network. Figure 6 reveals another interesting effect: the relative change in the sound velocity becomes smaller with increasing numbers of C atoms. Therefore, we can assume that the number of C atoms exerts no influence on the velocity if this number is larger than eight. From this fact we can conclude that the molecular motions responsible for the moduli in the glassy state are relatively localized. It should be mentioned that the dependence of the glass transition temperature on the numbers of C atoms shows a similar behavior (see Fig. 7).

The comparison of the experimental results for two samples which have been crosslinked with different catalysts ($Mg(ClO_4)_2$ and N,N-dimethyl benzylamine) illustrates the higher sound velocity for the network cured in the presence of $Mg(ClO_4)_2$ (see Fig. 7). As investigations on monofunctional glycidyl ether have shown, in the case of amine catalyst, the selectivity ratio of primary to secondary hydroxyl groups is

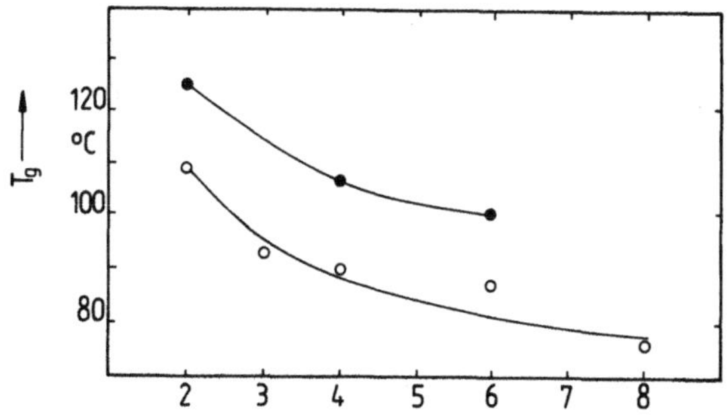

Fig. 7. Glass transition temperature T_g as a function of C atoms of alkane diols. T_g were determined using a differential scanning calorimeter. Epoxy systems studied explained in Fig. 6

10 : 1. In other words, long chains with a low branching degree will predominantely appear at the beginning of the cure. Then these long chains will crosslink at the advanced stage of reaction. In the case of DGEBA cured with $Mg(ClO_4)_2$ as catalyst (selectivity ratio 3 : 1), branching occurs at the beginning and the network formation starts very early. These ultrasonic investigations verify that different crosslinking regimes also produce different densities of crosslinking. In the case of DGEBA/COS, the crosslinking density of the sample was varied by incorporation of carboxyl-terminated oligostyrenes with different molar mass $\bar{M}_n$ [20]. Figure 8 demonstrates the expected dependence of the longitudinal wave modulus M' on the molar mass between the junction points. The decrease in M' with increasing $\bar{M}_n$ can be described by the empirical equation

$$M' = 3 \exp(-9.7 \times 10^{-3} \bar{M}_n) + 5.4$$

in the region $\bar{M}_n = 0 \ldots 2\,000$ g/mol, where M' and $\bar{M}_n$ are given in *GPa* and g/mol, respectively. The reference sample of pure DGEBA ($\bar{M}_n = 0$) may be considered as strongly crosslinked. The transition to the range of moderate crosslinking results in a smaller dependence of the modulus M' on the structural parameter $\bar{M}_n$.

5. Conclusion

The curing of epoxy resins modified by diols and carboxylterminated oligostyrene was monitored by longitudinal ultrasonic velocity and attenuation measurements. From the time dependence of the acoustic parameters it was possible to determine the gel point, the dynamic glass transition, and the end of the crosslinking reaction for the different resins. The results show that the formation and structure of the network depend on the chain length and the structure of the crosslinking agents as well as on the catalyst used.

The experimentally determined ultrasonic velocities of the fully cured systems reflect the properties of the corresponding moduli of the glassy state. At present, the molecular relations between the parameter of the network and the moduli of the glassy state cannot be satisfactorily explained.

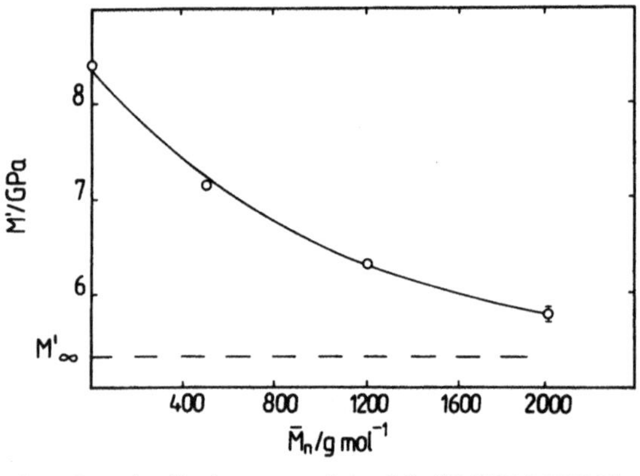

Fig. 8. Longitudinal wave modulus M' of DGEBA/COS in dependence on the molecular mass $\bar{M}_n$ of COS

References

1. Sofer GA, Hauser EA (1952) J Polym Sci 8:611
2. Sofer GA, Dietz AGH, Hauser EA (1953) Ind and Engn Chem 45:2743
3. Ranachowskij J, Łaš T (1965) Postepy Techniki Wysokich Napięć 4:365, Warsaw
4. Papadakis EP (1974) J Appl Phys 45:1218
5. Shchurow AF, Gubkina NL (1974) Fr Khim Tekhnol 2:43
6. Lindrose AM (1978) Exper Mech 18:227
7. Baeri JC, Courdille JM, Dumas J, Rajaonarison R (1980) J Physique-Lettres 41:L369
8. Taki S, Furuta Y, Takemura T (1981) Rev Sci Instr 52:1388
9. Krylovich VJ, Ermolenko IN, Alexenko VI (1981) Zavod Lab 50:41
10. Rokhlin SI, Lewis DK, Graff KF, Adler L (1986) J Acoust Soc Amer 79:1786
11. Bailey RT, North AM, Pethrick RA (1980) Molecular Motion in High polymers. Oxford University Press, Oxford
12. Pethrick RA (1983) Progr Polym Sci 9:197
13. Härtel E, Fedtke M, Plaste und Kautschuk, in press
14. Tänzer W, Fiedler H, Fedtke M (1986) Acta Polymerica 37:70
15. Meischner Ch, Hauptmann P (1985) Exp Technik Physik 33:175
16. Hauptmann P, Alig I, Wartewig S (1984) Wiss Zeit TH Leuna-Merseburg 26:499
17. Burchard W (1985) Ber Bunsenges Phys Chem 89:1154
18. Valles EM, Macosco CW (1976) Macromol 9:206 and (1979) Macromol 12:521, 673
19. Edwards SF (1967) Proc Roy Soc (London) 9:9
20. Häusler KG, Hauptmann P, Meischner Ch, Fedtke M, Härtel E, Wartewig S (1987) Polymer Comm 28:154

Received June 25, 1987;
accepted April 21, 1988

Authors' address:

Prof. Dr. S. Wartewig
Technische Hochschule "Carl Schorlemmer"
Leuna-Merseburg
Sektion Physik
DDR-4200 Merseburg, G.D.R.

Progress in Colloid & Polymer Science Progr Colloid Polym Sci 78:59–62 (1988)

Constraint release effects in polymer networks viscoelastic analysis

G. Heinrich[1]) and A. Havránek[2])

[1]) Technische Hochschule "Carl Schorlemmer" Leuna-Merseburg, Department of Physics, Merseburg, G.D.R.
[2]) Charles University, Department of Polymer Physics, Prague, Czechoslovakia

Abstract: The expression for the relative change in the network shear modulus during time (relaxation process) has been calculated using the concept of topological constraint release effects. The length distribution of the chains between two neighbouring cross-link points is taken into account in the calculation. The expression obtained is compared to theoretical results gained from other concepts and with some experimental results.

Key words: Viscoelasticity, network, constraint release effect, entanglement, cross-link density.

Introduction

The viscoelastic behaviour in the main transition zone of different types of polymer networks is discussed in a separate paper [1]; in this the short time relaxation process is attributed to the glass-rubber transition, T_g. Slow relaxation processes in polymer networks are often attributed to the relaxation of dangling chains [2–4]. We assume that these relaxation processes may also be connected with the topological constraint release effects of network strands, both ends of which are fixed at cross-linkage points [5].

In this paper we present some theoretical considerations concerning the relaxation strength, $\Delta(1)$, and the characteristic relaxation time, $\tau'_{m,2}$, of the proposed model for slow relaxation process in polymer networks. In particular, we propose an explanation of the relation

$$\tau'_{m,2} \sim \nu^{-\varphi} \qquad (1)$$

which reflects the dependence of the time position of the slow relaxation process relative to T_g, on cross-link density ν [1]. Values of φ varying from 3 to 12 were obtained from viscoelastic measurements of various polymer networks [1, 6].

Theory

The cooperative relaxation results from rearrangements of the cross-linkage points positions after deformation. This occurs in such a way that the overall distribution of junction points remain homogeneous and leads to a smaller deformation, at the molecular level, than that predicted from the phantom models of rubber elasticity [9]. As elementary steps, we may assume that the fluctuating network chains and junctions push the surrounding polymer material back and forth; when such a wobble motion is averaged over a certain time scale the network mesh rotates about its mean position to fulfil the above mentioned condition of homogenity. This process leads to a decrease of the deformation dependence of the topological constraints and depends on the degree of network chain interpenetration [5, 9]. The deformation dependence of the topological constraints in slightly cross-linked networks has been described by the relation [5]

$$d_\mu = d_o\, \alpha_\mu^{1/2}.$$

where d_o denotes the undeformed configurational tube radius and d_μ ($\mu = x,y,z$) denotes its deformed dimensions. The microscopic network chain deformation

$$\alpha_\mu = \lambda_\mu^{\beta_e}$$

connects the macroscopic deformation of the sample λ_μ, with the equilibrium constraint release parameter β_e [5]. This parameter characterizes the final (equilibrium) state of the relaxed microscopic network chain deformation [5]. The value of β_e depends on the amount of solvent, sol fraction, network defects and cross-link density. The β_e parameter was originally introduced [5, 10, 11] to explain the swelling and small angle neutron scattering properties of polymer networks.

The time dependence of the constrained release parameter is assumed to follow a fractional-exponential law:

$$\beta^2(t) = \beta_e^2 + (1 - \beta_e^2)\, e^{-(t/T_c)^\alpha} \tag{2}$$

where T_c is the characteristic time of the constraint release process of a network chain. We propose that

$$T_c = \tau\, (N_c/N_e)^\varphi\, e^{\varkappa(N_c/N_e)}, \tag{3}$$

where τ corresponds to the characteristic relaxation time of the fast relaxation process, N_e is the number of segments between two succesive entanglement points, N_c denotes the number of segments between two cross-linkage points, and α, $\varkappa$, φ are material constants. The parameter φ depends on the network functionality. It will be shown that φ here can be identified with φ from Eq. (1).

Although we have no microscopic theory of the relaxation of obstacle-enclosing but ramified cyclic network meshes, we assume the knot rearrangement process (in a very crude approximation) to be a quasi-free diffusion of branched f-star molecules (junction + f network strands) over a distance d_o. We have used the exponential law for the characteristic relaxation time [12] and have modified the pre-exponential factor by introducing φ to express the higher cooperativity of networks (Eq. (3)).

We now introduce the relative change of modulus (cf. [1]) as

$$\Delta'(t) \equiv \frac{G(t) - G_e}{G_e} = \frac{G_N}{G_e}\,\frac{1-\beta_e^2}{\beta_e^2}\, e^{-(t/T_c)^\alpha} \tag{4}$$

where $G_e = G_c + G_N$ is the equilibrium value of the total shear modulus; G_c is the shear modulus associated with purely chemical cross-links and G_N represents the contribution of the local topological constraints. This topological modulus may be expressed as $G_N = a'\, G_N^o\, \beta_e^2$. The order of magnitude of a' is one, and G_N^o denotes the plateau modulus of the corresponding uncrosslinked sample [5].

The N_c-dependence of the right hand side of Eq. (4) is mainly attributable to T_c in the exponential term. Therefore, when summing contributions to the relaxation modulus from a distribution of network chain lengths, we replace, for simplicity, N_c by the mean value $\bar{N}_c$ in the pre-exponential factor of Eq. (4). This gives (see Appendix)

$$\sum_{N_c=0}^{\infty} P(N_c)\, e^{-(t/T_c)^\alpha} \approx q \exp\left(-\mu(t/\tau_c)^\alpha\right) \tag{5}$$

with $\mu = e^{-\alpha\varkappa} = \text{const.}$

and

$$\tau_c = \tau\, (\bar{N}_c/N_e)^\varphi \sim \nu^{-\varphi}. \tag{6}$$

The $P(N_c)$ is the probability of having a network chain of N_c segments. This probability has been estimated for randomly cross-linked system with a cross-link density ν as (e.g. [3])

$$P(N_c) = q\,(1-q)^{(N_c-1)}, \tag{7}$$

where

$$q = \nu/n_s \sim \bar{N}_c^{-1}$$

and n_s is the number of chain segments per unit volume. Using Eq. (5) we may then write the expression for the relaxation strength

$$\Delta(t) = \sum_{N_c=0}^{\infty} P(N_c)\Delta'(t) \approx \gamma_1 q \exp(-\mu(t/\tau_c)^\alpha) \tag{8}$$

with

$$\gamma_1 = \frac{G_N}{G_e}\,\frac{1-\beta_e^2}{\beta_e^2}. \tag{9}$$

Discussion and conclusions

Equation (8) has a similar fractional-exponential law form to that predicted by Gaylord et al. [4], who assumed that the stress relaxation behaviour originates in the hindered diffusion of dangling ends in the presence of entanglements; the diffusion is described within the continuous-time random walk concept. Therefore, the effect of dangling chains, which add to the relaxation properties of networks, can hardly be distinguished from the constraint release effect of network strands. Gaylord et al. have shown that Eq. (8) fits the long-time behaviour of stress-relaxation data reasonably well. The fit is practically indistinguishable from that of a fractional-power law ($\Delta \sim t^{-\alpha}$) [4].

The junction rearrangement process depends on the degree of network chain interpenetration [9]. Therefore we propose, in the case of completely random cross-linked networks [5], the relation

$$\beta_e \simeq (n_F/f)^{-1} \sim \bar{N}_c^{-1/2}$$

where n_F denotes the total number of cross-links within a Flory-coil (see e.g. [5]) and f is the functionality. We may now conclude that with increasing density of cross-links the interpenetration effect of the vulcanized networks is reduced ($\beta_e \to 1$) and the value of γ_1 decreases. Such behaviour has been observed experimentally [13].

In the case of networks of small polyfunctional units (e.g. the POPT-MDI system investigated in [8]) the discussion changes somewhat. Here the corresponding structure of the unswollen network sample is not characterised by a strong overlap of the network meshes [14]. For this case we propose the relation $\beta_e^2 \simeq T_e$ [5], where T_e denotes the trapping factor [15], with the limiting values zero at the gel point and unity for a perfect network without defects. This relation together with Eq. (8) show that in POPT-MDI networks the slow relaxation process vanishes for the case where the molar ratio of the reactive groups (and, therefore, T_e) is approximately equal to one [8,16].

Finally, we notice that our results agree qualitatively with some conclusions drawn from computer simulations of the dynamics of end-linked random networks by Neuburger and Eichinger [17]. They concluded that there are motions in polymer networks which have relaxation times longer than or comparable to those directly attributable to the motion of dangling chains.

Appendix

We have to calculate the following sum

$$\sum_{N_c=1}^{\infty} P(N_c) \exp[(-t/T_c)^{\alpha}]$$

$$\approx \sum_{N_c=1}^{\infty} P(N_c) \exp(-a x^{N_c}) \qquad \text{(A 1)}$$

with

$$a = [(t/\tau)(N_e/\bar{N}_c)^{\varphi}]^{\alpha} \text{ and } x = \exp(-\alpha x/N_e) < 1.$$

The probability $P(N_c)$ is given in Eq. (7). The right hand side of Eq. (A 1) can be rewritten as

$$S = \sum_{i=0}^{\infty} (-1)^i a^i/i! \sum_{N_c=1}^{\infty} P(N_c)(x^i)^{N_c} \qquad \text{(A 2)}$$

For the second sum it follows

$$S' = \sum_{N_c=1}^{\infty} q(1-q)^{(N_c-1)} (x^i)^{N_c}$$

$$= \frac{q}{1-q} \sum_{N_c}^{\infty} [x^i(1-q)]^{N_c}. \qquad \text{(A 3)}$$

We define the quantity q' through the relation $(1-q)x^i = 1-q'$. Then we get from Eq. (A 3)

$$S' = \frac{q}{1-q} \frac{1-q'}{q'} \sum_{N_c}^{\infty} q'(1-q')^{(N_c-1)} = x^i q/q'.$$

For further simplification we assume that $q' \approx 1$ for all i. That is, of course, a good approximation only in the case if $i > 1$ ($x < 1$). Then Eq. (A 2) gives

$$S \approx q\, e^{-ax}. \qquad \text{(A 4)}$$

Eq. (5) of the text is obtained from Eq. (A 4).

References

1. Havránek A, Heinrich G (1988) Acta Polymerica 39:540
2. Ferry JD (1980) Viscoelastic Properties of Polymers. 3rd ed Wiley, New York
3. Curro JG, Pincus P (1983) Macromolecules 16:559
4. Gaylord RJ, Joss B, Bendler JT, Di Marzio EA (1985) Brit Polym J 17:126
5. Heinrich G, Straube E, Helmis G (1988) Adv Polym Sci 85:33
6. Havránek A (1988) Rheol Acta Suppl (Progress and Trends in Rheology II) 26:202
7. Havránek A (1982) Polym Bull 8:133
8. Havránek A, Ilavsky M, Nedbal J, Böhm M, v. Soden W, Stoll B (1987) Coll and Polym Sci 265:8
9. Bastide J, Picot C, Candau S (1981) J Macromol Sci, Phys B 19:13
10. Helmis G, Heinrich G, Straube E (1984) Wissenschaftl Zeitschr TH Leuna-Merseburg 26:461
11. Heinrich G, Straube E (1987) Polym Bull 17:247
12. Klein J (1986) Macromolecules 19:105
13. Thirion P, Chasset R (1967) Chimie et Industrie – Genie Chimique 97(5):617

14. Daoud M, Bouchard E, Jannik G (1986) Macromolecules 19:1955
15. Langley NR (1968) Macromolecules 1:348
16. Ilavský M, Dušek K (1983) Polymer 24:981
17. Neuburger NA, Eichinger BE (1985) J Chem Phys 83:884

Received April 24, 1987;
accepted April 21, 1988

Authors' address:

G. Heinrich
Technische Hochschule "Carl Schorlemmer"
Leuna-Merseburg
Department of Physics
DDR-4200 Merseburg, G.D.R.

Entanglement and reversible gelation for polymers of different architectures

W. Burchard

Institute of Macromolecular Chemistry, University of Freiburg, Freiburg, F.R.G.

Abstract: Entanglement has been studied in the past mainly with flexible linear chains. Early predictions were based upon scaling theory which gave only data on exponents of power laws but no prefactors. Tests were strongly inhibited by ambiguities in the definition of the overlap concentration c^*. The present study uses the thermodynamically well-defined scaling parameter $X = A_2 M_w c$. Recent renormalization group theories succeeded in the derivation of an analytic expression for the osmotic modulus in terms of X. The osmotic modulus was measured by static light scattering at the scattering angle zero. Experimental data from polystyrene chains of different molecular weights and from many other linear polymers demonstrate a good agreement with theory up to values of $X =$ 3–5. Beyond that point the osmotic modulus increases stronger than predicted. The final exponent is 1.40 ± 0.03 instead of the predicted 1.25. Good agreement with the theoretical prediction by Carnahan and Starling is also found for impenetrable particles (spheres). As a third architecture the experimental curve for stiff chain molecules is shown for which no theory is available to date. The results can, however, be interpreted qualitatively in a consistent manner.

Many polymers show at moderately high concentrations pronounced deviations from the pure entanglement behavior. Three effects are observed: 1) a strong low-angle excess scattering is found; 2) the osmotic modulus shows a turnover and decreases again; and 3) a pronounced slow mode of motion becomes apparent. The three effects increase strongly with concentration. Finally, behavior is observed which indicates reversible gelation.

Key words: Polymer solution, entanglement, gelation, light scattering.

1. Introduction

A solution is called semidilute if the polymer concentration is larger than the overlap concentration c^*, but the concentration should, nevertheless, be fairly low, i.e., less than 10% by weight [1]. Around c^* a cross-over from the typical dilute to semidilute behaviour is expected. Unfortunately c^* cannot be defined clearly and at present mainly the following three definitions are in use

$$c^*_\eta = 1/[\eta] = M/(\phi R_g{}^3) \tag{1a}$$

$$c^*_{Rg} = M/(N_A(4\pi/3)R_g^3) \tag{1b}$$

$$c^*_{Rh} = M/(N_A(4\pi/3)R_h^3)\,. \tag{1c}$$

In these equations $[\eta]$ is the intrinsic viscosity, $R_g = \langle S^2 \rangle^{1/2}$ the radius of gyration, ϕ the prefactor in the Fox-Flory equation for the intrinsic viscosity, N_A is Avogadro's number, and R_h the hydrodynamic radius of the macromolecule, which is defined via the Stokes-Einstein relationship by the translational diffusion coefficient D

$$R_h = kT/(6\pi\eta_o D)\,. \tag{2}$$

The three definitions in Eqs. (1a), (1b), and (1c) differ in their numerical values but not in their molecular weight dependences.

Theoreticians are often satisfied by deriving asymptotic relationships which in recent years were obtained mostly from scaling techniques. Scaling, however, gives information only on asymptotic behavior and fails to predict the corresponding prefactors. Experimentalists, on the other hand, need to know the exact position for the cross-over from dilute to semidilute behavior and thus require a parameter which is clearly defined for all molecular architectures and one that can be easily measured.

In the following the dimensionless parameter

$$X = A_2 M_w c \tag{3}$$

is used which meets the requirements of theory and its experimental verification [2–6]. This may be shown with the osmotic compressibility for a system of hard spheres and of linear flexible chains.

2. Theoretical relationships for semidilute solutions

The osmotic modulus $(\partial\pi/\partial c)$ is obtained from static light scattering measurements at zero scattering angle [7] according to the equation

$$Kc/R_{\theta=0} = (1/RT)(\partial\pi/\partial c) \tag{4}$$

with K an optical contrast factor, c the weight concentration, and Rayleigh ratio

$$R_\theta = (i(\theta)/I)\, r^2 \tag{5}$$

$i(\theta)$ is the scattering intensity at the scattering angle θ, I the primary beam intensity, and r the distance of a detector from the center of the scattering volume.

For small concentration (dilute solutions) the usual virial expansion can be applied to the osmotic pressure, which gives for the osmotic modulus

$$(1/RT)(\partial\pi/\partial c) = (1/M_w)(1 + 2X + 3a_2X^2 + 4a_3X^3 + \ldots) = 1/M_{app} \tag{6}$$

where $X = A_2M_wc$. For *hard spheres* the coefficients a_i are known exactly up to a_5 [8]. For other molecule architectures these coefficients are in general not known, and for flexible chains only a_2 has been calculated [9].

A virial expansion of the kind of Eq. (6) is useful up to values of $X \cong 0.6$; above this value, i.e., at higher concentrations, the series no longer converges. Several approximate analytic theories have been derived in the meantime which cover the whole concentration region that can be physically realized with hard spheres. Carnahan and Starling [10] derived a simple analytic expression and obtained almost complete agreement with simulation data. The corresponding reduced osmotic modulus is given by

$$(M_w/RT)(\partial\pi/\partial c) = M_w/M_{app} = \frac{1 + 4y(1 + y - y^2) + y^4}{(1-y)} \tag{7}$$

with

$$y = X/4 = N_A(V_m/M_w)c \tag{8}$$

where V_m is the volume of a sphere and M_w its molar mass. Expansion of Eq. (7) in terms of X produces the correct virial efficients to a high accuracy. Eq. (4) indicates an increase of the reduced osmotic modulus $(M_w/RT)(\partial\pi/\partial c)$ beyond all limits if $y = 1 = 4X$ is approached. Near this point we have a dense packing of spheres, and an infinitely large osmotic change in the osmotic pressure is needed to compress the system further.

The curve for the hard spheres is shown in Fig. 1. This Fig. also contains the corresponding curves for a flexible linear chain and a 12-arm star-branched macromolecule with flexible arms. These equation have been derived from the space renormalization

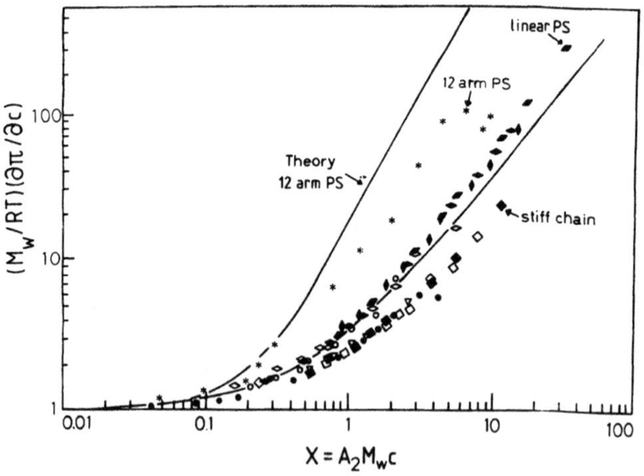

Fig. 1. Reduced osmotic modulus $(M_w/RT)(\partial\pi/\partial c) = Kc/R_{\theta=0}$ as a function of the parameters $X = A_2M_wc$ for different macromolecular architectures. Full lines correspond to theories for flexible linear chains by Ohta and Oono et al. [2], and for a 12-arm star-molecule by Cherayil et al. [4], and for hard spheres by Carnahan and Starling [10]. The symbols indicate measurement for *PS* in toluene, 12-arm *PS*-star, and stiff Xanthan chains in 0.1*N* NaCl aqueous solution

group theory by Ohta and Oono [2] and by Freed and his coworkers [3,4], and are given by the equation

$$(M_w/RT)(\partial\pi/\partial c) = 1 + X_{\exp}[-I/32X^2]\,2 + I/16X^2 - DI/32X^2$$

$$I = \int_0^\infty dzz\,[\ln(1+V) - V/(1+V) - (1/2)V^2]$$

$$DI = \int_0^\infty dzz\,[V/(1+V) - V/(1+V)^2 - V^2]$$

$$V = 4Xg(z)$$

$$g(z) = (1/z) + (1/z^2)f[\exp(-z/f) - 1] \times \{1 + (1/2)(f-1)[\exp(-z/f) - 1]\}\,. \quad (9)$$

This equation gives the asymptotic scaling laws as derived previously by the Cloizeaux [11] and de Gennes [1] and by Daoud et al. [12]. In addition, the prefactors were obtained and the whole cross-over region from dilute to semidilute behavior is covered.

This success of renormalization theory results from the attempt to extract the "long wave" character of a chain of finite length. This long wave behavior is expected to show universal character as all details of a micro-structure is thrown out by a special regularization technique of the excluded volume and by a well-defined renormalization transformation [13].

The renormalization technique has been applied also to dynamic properties, however, with less success. In the limit of good solvents and large hydrodynamic interaction an equation was derived by Oono et al. [14] which covers the whole transition region from dilute to semidilute behavior and gives the familiar asymptotic exponential behavior for the mutual diffusion coefficient, $D_c/D_o \sim X^{3/4}$, derived first by de Gennes [1]. For small X the Oono et al. theory gives

$$D_c = D_o(1 + KX + \ldots) \quad (10)$$

with

$$K = 1.348\,. \quad (11)$$

Equations (10) and (11) apply so far only to linear chains. A similar relationship was derived for spheres from elementary hydrodynamics which takes only a two-body hydrodynamic interaction into account. The corresponding constant in Eq. (10) is given as [16,17]

$$K_{\text{sphere}} = 0.33\,. \quad (12)$$

Equations for star-branched macromolecules and for rings are at present not known.

3. Results and discussion

3.1 Osmotic modulus

Static LS from linear PS molecules in toluene gave one common curve for the various molecular weights in the range of $M_w = 50 \times 10^3$ to $8\,000 \times 10^3$, if the reduced osmotic modulus $(M_w/RT)(\partial\pi/\partial c) = M_wKc/R_{\theta=0}$ is plotted against $X = A_2M_wc$. This curve is shown together with the 12-arm star-branched macromolecules in Fig 1. One notices a very satisfactory agreement for linear PS with theory up to values of $X \simeq 3$. At larger X the experimental curve increases stronger than predicted, but the universal character, i.e., independence of the chain length, is still preserved. The same curve is obtained also in marginal solvents and for a number of other linear chains (Polyvinylcaprolactam in water [18], Cellulose-tri-carbanilate in dioxane [19], and Polybutadiene in dichloroethane [20]), however, only up to a certain X-value. Beyond these values a strong deviation occurs which will be discussed below.

Figure 1 also contains experimental results from native and modified Xanthans in water of 0.1 N NaCl salt content [6]. These polysaccharides belong to the stiffest chains we know of and consist of, in spite of the high molecular weight, only 3.5 – 6 Kuhn segments. Evidently, rod-like chains produce a significantly lower repulsion when forced into a high concentration.

Randomly branched chains [21] and star-branched chains with a small cross-linked microgel as a star-center and many attached arms [22], show opposite behavior. This feature is qualitatively easily understood. A randomly branched macromolecule, and to a greater extent a cross-linked microgel, has little opportunities for chain interpenetration and will soon show behavior of a soft sphere that can be deformed. Thus, these molecules should lie in between the curves for hard spheres and flexible linear chains. Star molecules can be expected to interpenetrate to a considerably larger extent as long as the number of arms is smaller than 10. If there are more arms attached to a microgel, the center is densely packed, and again the behavior will

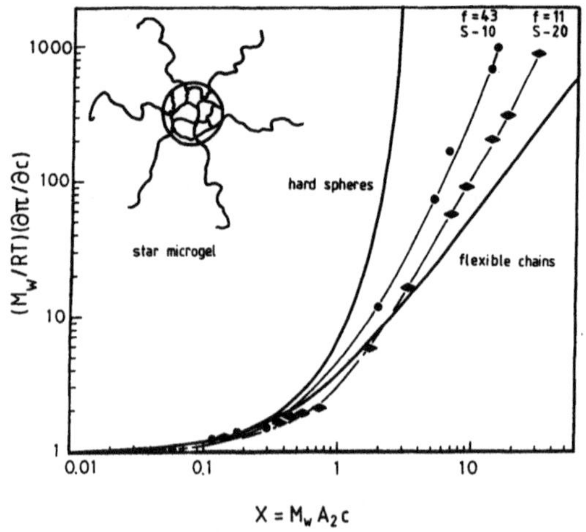

Fig. 2. Measurements for osmotic compressibility for star-microgel with 43 and 11 arms, respectively, in comparison with theoretical curve for hard spheres and flexible chains. The structuree of the star-microgel is sketched in the insert

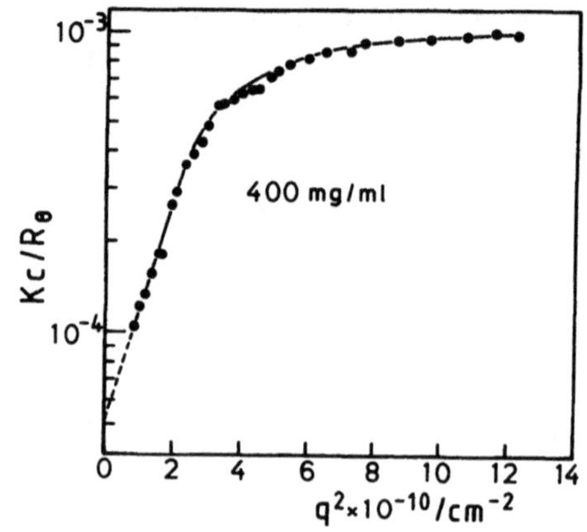

Fig. 4. Static *LS* from the *S*-10 PMMA star-microgel in *THF* at $c = 400$ mg/ml

be more that of a hard sphere. For the star-microgels this behavior is found at large X (see Fig. 2), but at small X the curve agrees more with that for linear chains than with the compressibility curve for regular stars. At present we have no explanation for this behavior. Possibly the polydispersity of the microgel-center, which implies a distribution in the number of arms, is responsible for this feature.

Summarizing the theories and experimental findings one can state: i) measurements of the osmotic compressibility gives a good means to differentiate among various macromolecular or colloidal architectures at rather high concentrations; ii) the region of validity of theories for semidilute solutions appears to be much more extended than was hitherto expected, and iii) the parameter $X = A_2 M_w c$ is a very useful quantity for the various architectures.

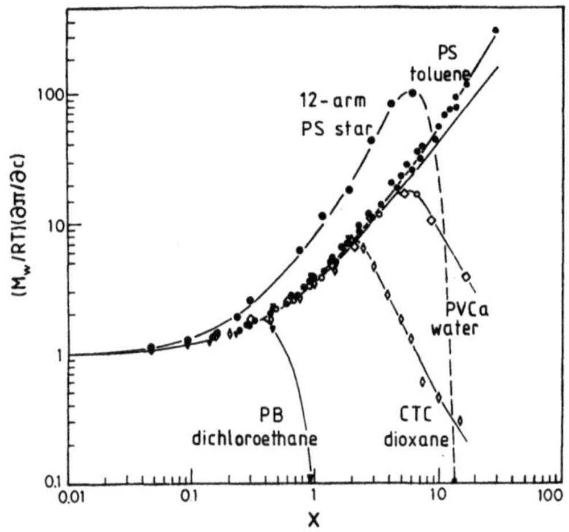

Fig. 3. Experimental data for the osmotic modulus of different linear chains and a 12-arm star PS in good solvents. PB = Polybutadiene, CTC = Cellulose-tri-carbanilate, PVCa = Polyvinylcaprolactam

3.2 Association and gelation

As was already mentioned, in most realistic systems the modulus curve can be realized only up to a certain X-value. Beyond this point a sharp turn-over occurs, (Fig. 3). Simultaneously a strong excess low angle scattering is observed, (Fig. 4). There exists a general agreement, at least among experimentalists, that this excess low angle light scattering results from very large objects [23, 24]. These grow very quickly in size as the concentration is increased, and these objects disappear again on dilution.

The turn-over in the reduced modulus is in accordance with this interpretation, which can be recognized when writing

$$(M/RT)(\partial\pi/\partial c) = M_w/M_{app}(c) \qquad (13)$$

where M_w is the molecular weight at infinite dilution. As long as only non-attracting objects are present, $M_{app}(c)$ decreases with c and $M_w/M_{app}(c)$ increases. Thus a turn-over of $M_w/M_{app}(c)$ to lower values can be only produced by an increase of $M_{app}(c)$ rather than by a decrease as a result of the repulsing force. Since this turn-over is reversible and disappears on dilution we come to the conclusions that specific attractive

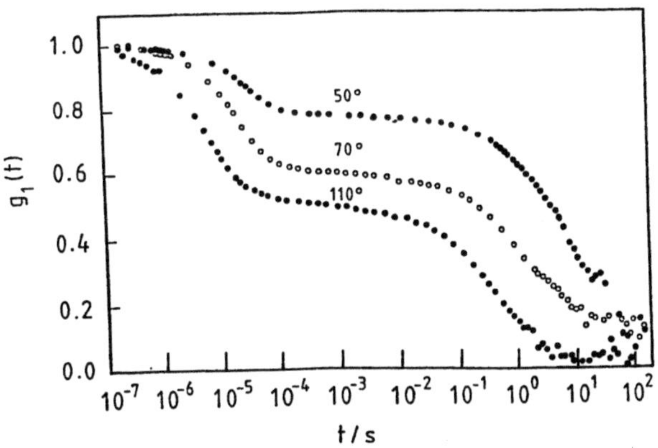

Fig. 5. Time correlation function $g_1(t)$ of dynamic *LS* for the *S*-10 star-microgel at $c = 300$ mg/ml at three scattering angles of 50, 70 and 110 degrees

interactions become effective if the segments come close enough to each other and the overall concentration is sufficiently large. According to the law of mass action, association will take place and should eventually lead to reversible gelation.

The drastic growth of clusters is particularly well seen in the time correlation function (TCF) of the dynamic light scattering. Figure 5 shows as an example the TCF of a star-microgel with 43 arms at a concentration of about 30%. The plot shown here differs from the commonly used plot of ln $g_1(t)$ versus t, in that the *linear* TCF $g_1(t)$ is shown as a function of the logarithmic delay time. A single exponential decay shows in this plot a sigmoidal shape. One clearly recognizes two relaxation processes which differ by approximately four decades in time. The fast process can be assigned to the irregular vibration of the "gel-mode", and the slow process corresponds to a translational diffusion of very large clusters superimposed by many internal relaxation modes of motion. The pure translational diffusion of the center of mass of these clusters is obtained after extrapolation of $g_1(t)$ to zero scattering angle.

A plot of these data against concentration allowed the determination of critical c_{gel} where 100% slow modes of motion are obtained. This point was taken as the gel point.

This approximate proceedure is confirmed by the observation that at this concentration the cell can be turned upside down without observing a flow within 10 min. Usually this tilting test is taken as indication for gelation.

Acknowledgement

This contribution is based on experimental studies performed by several of my coworkers. I am grateful for their confidential cooperation and wish to thank in particular, Dorothea Friebe for her work with linear PS, Dr. Klaus Huber for the data obtained with the 12-arm star-branched PS, Peter Lang for those with the PMMA-star-microgels, Elke Wachenfeld for the epoxy data, Markus Wenzel for measurements with cellulose-tri-carbanilate, Michael Eisele for those with poly(vinyl-caprolactam) and Dr. Raimund Stadler for the data from poly-(butadiene). Most of the data have not yet been published. The auther is indebted also to Professor W. H. Stockmayer, Dartmouth College, Hanover, and Professor K. F. Freed, University of Chicago, for helpful discussions.

The project was kindly supported by the Deutsche Forschungsgemeinschaft within the scheme SFB 60.

References

1. De Gennes P-G (1979) Scaling Concepts in Polymer Physics. Cornell University Press, NY
2. Ohta T, Oono Y (1982) Physics Letters 89:460
3. Freed KF (1983) J Chem Phys 79:6357
4. Cherayil SJ, Bawendi MG, Miyake A, Freed KF (1980) Macromolecules 20:2770
5. Wiltzius P, Haller HR, Cannell DS, Schaefer DW (1983) Phys Rev Letters 51:1183
6. Coviello T, Burchard W, Dentini M, Crescenzi V (1987) Macromolecules 20, in press
7. Debye P (1947) J Colloid Phys Chem 51:18
8. McQuarrie DA (1973) Statistical Mechanics. Harper & Row, New York
9. Carpenter DK (1970) Enzyclopedia of Polymer Science and Technology. Wiley New York, 12:626
10. Carnahan NF, Starling KE (1969) J Chem Phys 51:635
11. des Cloizeaux J (1982) J Phys 36:281
12. Daoud M, Cotton JP (1982) J Phys 43:531
13. Miyake A, Freed KF (1983) Macromolecules 16:1228
14. Oono M, Baldwin PR, Ohta T (1984) Phys Rev Letters 53:2149
15. Wiltzius P, Haller HR, Cannell DS, Schaefer DW (1984) Phys Rev Letters 53:834
16. Batchelor GKJ (1976) J Fluid Mech 52:245
17. Hess W, Klein R (1976) Physica A 85:509
18. Eisele M, Burchard W, Manuscript in preparation
19. Wenzel M, Burchard W, Schätzel K (1986) Polymer 27:195
20. Stadler R, Burchard W, Manuscript in preparation
21. Wachenfeld-Eisele E, Burchard W (1987) Proceedings of the 8th Networks Meeting, Ed O Kramer, Copenhagen, Elsevier
22. Lang P (1987) Diploma Thesis Freiburg
23. Koberstein JT, Picot C, Benoit H (1985) Polymer 28:673
24. Gan JYS, Francoise J, Guenet H-M (1986) Macromolecules 19:173

Received April 24, 1987;
accepted April 21, 1988

Author's address:

W. Burchard
Institut für Makromolekulare Chemie
Universität Freiburg
Stefan-Meier-Str. 31
D-7800 Freiburg, F.R.G.

Progress in Colloid & Polymer Science Progr Colloid Polym Sci 78:68–71 (1988)

Dynamic light scattering measurements on semidilute solutions of polystyrene in ethyl acetate

P. Štěpánek[1]), J. Jakeš[1]), and W. Brown[2])

[1]) Institute of Macromolecular Chemistry, Czechoslovak Academy of Sciences, Prague, Czechoslovakia
[2]) Institute of Physical Chemistry, University of Uppsala, Uppsala, Sweden

Abstract: The dynamic behavior of semidilute solutions of polystyrene in ethyl acetate is studied by quasi-elastic light scattering as a function of concentration. The complex nature of autocorrelation functions is established and the corresponding spectra of relaxation times are obtained using a suitable Laplace transformation technique. Several components are identified in the spectra and their properties are discussed as a function of scattering vector and concentration.

Key words: Semidilute solutions, light scattering, Laplace transformation, spectra of relaxation times.

Introduction

Recent communications [1–5] have described the complex character of the autocorrelation functions obtained in quasi-elastic light scattering (QELS) measurements on high molecular weight polystyrenes (PS) in marginal and *theta* solvents. With PS in ethyl acetate (marginal solvent) the quality of the solvent may be progressively reduced toward the *theta* point, either by addition of a non-solvent, or by changing the temperature. For example, addition of the isorefractive non-solvent ethanol [3] corresponds to θ-conditions at about 10 % ethanol in ethyl acetate, and $\theta \cong -44$ °C in pure solvent which is still an experimentally-accessible [5, 6] temperature. In the studies referred to, the correlation functions were analyzed by various methods: multi-exponential fit, Contin [7], and PES [8] (positive exponential sum). The presence of three relaxational modes was demonstrated in most cases (depending on angle and temperature), of which the fastest corresponds to the dynamic behavior of the transient gel. A recent report [5] dealing with a single semidilute concentration as a function of temperature showed that the relaxational rates Γ of the modes were approximately K^2-dependent (K is the scattering vector), thus diffusive, since then $\Gamma = DK^2$, D being the diffusion coefficient [4]. The present communication is directed to an examination of the concentration dependence characterizing each of the contributory modes at ambient (26 °C) temperature.

Experimental

Polystyrene ($\bar{M}_w = 2.95 \times 10^6$, $\bar{M}_w/\bar{M}_n = 1.05$) was from Toya Soda Ltd., Tokyo. Ethyl acetate was of chromatography grade. Parallel series of measurements were made, as previously, [5] in Prague and Uppsala on the same solutions. Use of the Contin programme [7] for analysis of correlation functions $G(t)$ gave superimposable decay time spectra $A(\tau)$, when the curves (as in Fig. 1) were obtained, either by computer-splicing an appropriate number of runs at increasing sample times, or, alternatively, taken up with a multi-tau autocorrelator [9] with 23 simultaneous sampling times typically covering the lagtime range 1 µs to 1 min (see [5] for further details).

Results and discussion

Figure 1a shows typical correlation functions obtained at dilute and semidilute concentrations. Distinct changes in the shape of the functions with increasing concentration are apparent. Figure 1 b depicts the results of the Contin inversion of the curves in 1a. These results confirm the previously made observa-

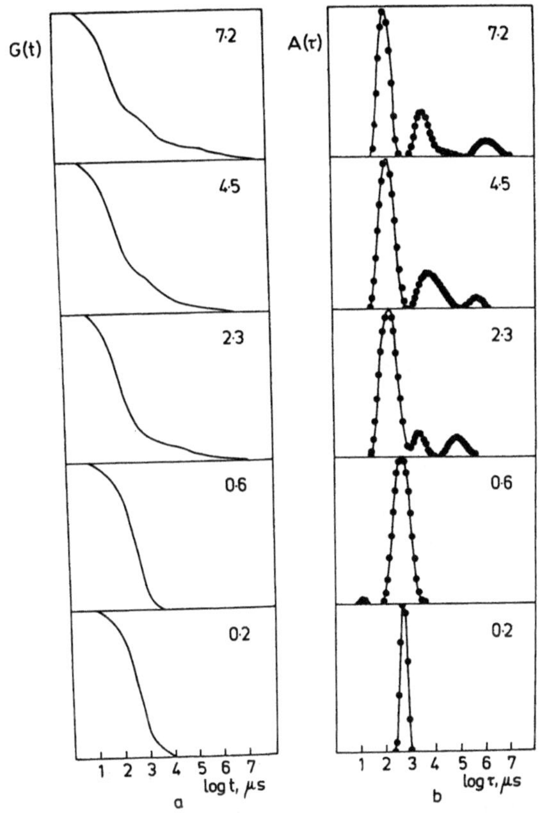

Fig. 1. a) Correlation functions ($G(t)$ vs. time $t/\mu s$) at concentrations as shown ($C \times 10^2/\text{gml}^{-1}$) for PS ($\bar{M}_w = 2.95 \times 10^6$) in ethyl acetate at 26 °C and angle $\theta = 60°$; b) The corresponding decay time spectra obtained using Contin [7]. The plots show the relative amplitude ($A(\tau)$) vs. decay time, $\tau/\mu s$

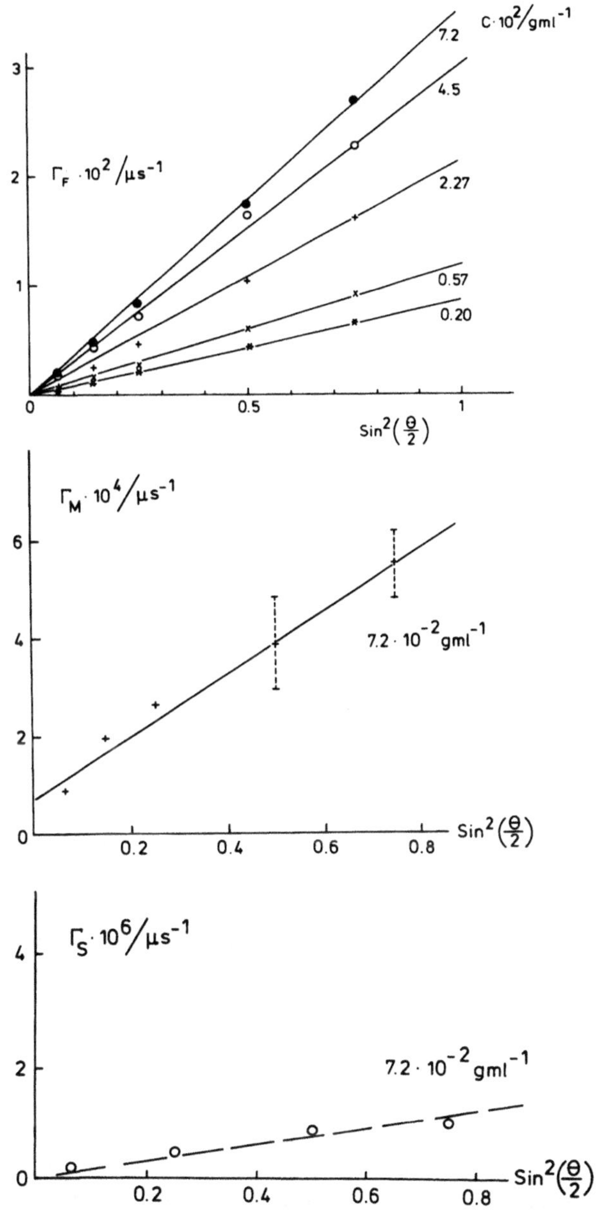

Fig. 2. Relaxation frequency ($\Gamma/\mu s^{-1}$) as a function of $\sin^2(\theta/2)$ at different concentrations for the three modes (fast, middle and slow) depicted in Fig. 1b

tion [5] that three different decay processes exist, widely spaced in decay time, and are each required to adequately describe the data at semidilute concentrations. Below the crossover concentration, the correlation functions may be approximated by a single exponential.

Figure 2 shows the angular dependences of the relaxation rate (Γ) as evaluated from the moments of the respective Contin inversion peaks. The fastest mode is clearly diffusive $\left(\Gamma \sim K^2 \sim \sin^2\left(\frac{\theta}{2}\right)\right)$ and is considered to represent the dynamic behavior of the transient gel. The middle mode shows a finite intercept in the Γ vs. $K^2 \sin^2 \frac{\theta}{2}$ plot, which may possibly derive from a contribution from a structural relaxation (K-independent). The slow mode is dependent on $K^2 \sim \sin^2 \frac{\theta}{2}$ within experimental error; owing to the wide spacing in time scale it was found possible to improve the resolution of this mode by approximating the decay data by a single exponential [5]. It is noted that the presence of three relaxational processes is not accounted for by current theory of semidilute polymer solutions [10]. However, the concentration dependences of the individual decay rates, as illustrated in Fig. 3, suggest the following speculative interpretation: 1) The fastest component ($\Gamma_F \sim C^{0.43}$) is characterized by an expo-

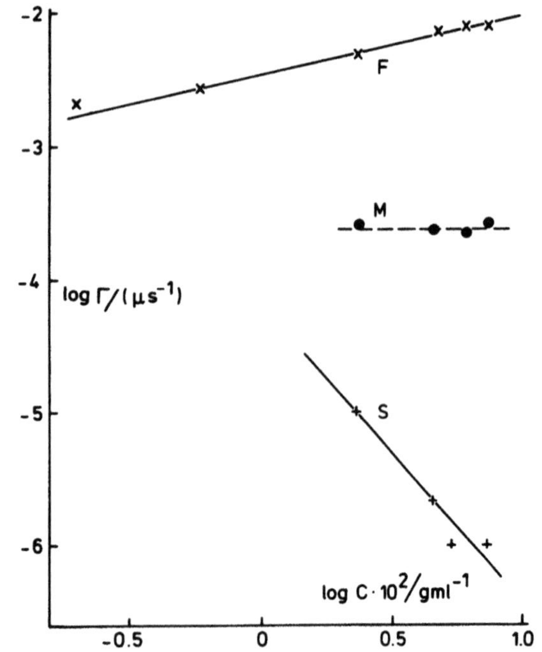

Fig. 3. Log-Log plots of the relaxation rate (Γ) versus concentration for the fast (F), middle (M) and slow (S) modes. Exponents: $\Gamma_F \sim C^{0.43}$; $\Gamma_M \sim C^0$; $\Gamma_S \sim C^{-2.2}$

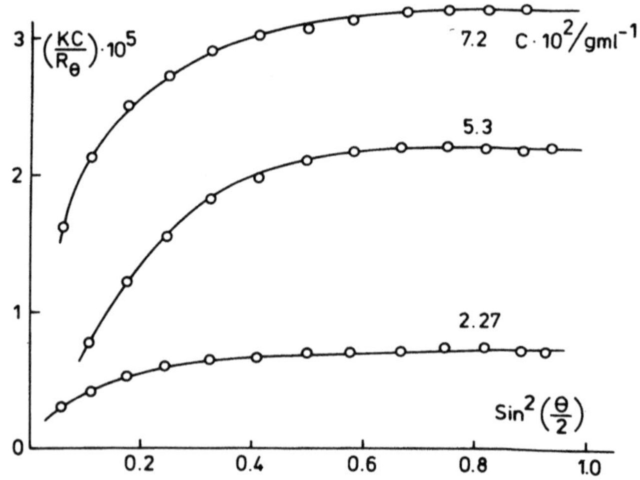

Fig. 4. Intensity light scattering data. (KC/R_θ) vs $\sin^2(\theta/2)$ at different PS concentrations in ethyl acetate at 26 °C

nent which agrees with previous findings for the gel mode [1–3, 11] in semidilute, marginal solvents; 2) The approximately concentration-independent middle mode may be assigned to translational (Stokes-Einstein) motions of the whole coil in the (imperfect) gel network. This conclusion is supported by data for the concentration dependence at a series of temperatures down to the *theta* temperature in ethyl acetate [12] which show that the concentration exponent for the middle mode changes with temperature in a characteristic manner: it exhibits the change expected as the solvent quality decreases, as indicated by a reduction in the second virial coefficient A_2; 3) The data for the slowest mode is described by $\Gamma_S \sim C^{-2.2}$, which is consistent with the reptation in a network swollen by a marginal solvent (exponent $= -1.75$ (good solvent) and $= -3$ (*theta* solvent) according to scaling theory [10]). We note that the decay rate of the slowest mode approximately matches the exponent of -3 at -40 °C (approximamtely *theta* conditions).

Deschamps and Léger [13] recently reported self-diffusion coefficients for single chains in a *theta* solvent. Comparison with their data suggests that the present slowest mode is about two orders of magnitude slower than that corresponding to single chain motion. This indicates the presence of locally clustered chains. A similar phenomenon was established in semidilute solutions of polyethylene oxide [14] by comparison of QELS and pulsed-field-gradient NMR data.

This interpretation is reinforced by data from intensity light scattering measurements (Fig. 4). The reduced intensity function diminishes at low angles suggesting the presence of supermolecular particles (clusters) of a size about 1700 Å at $C = 0.072$ gml^{-1}.

We can exclude dust as the origin of this effect by the method of preparation of the solutions and measurements [5] at +70 °C which did not reveal this trend.

Figure 5 shows the relative intensities of the three modes as a function of both angle and concentration. At low concentration the gel mode is the main feature, while the middle mode becomes the most accentuated towards the highest concentration. It was previously noted [5] that when the total intensity is apportioned between the modes according to the weighted average, the middle mode is dominant in the vicinity of the *theta* temperature.

Conclusions

The presence of three relaxational modes whose decay rates are approximately K^2-dependent, is established in semidilute solutions of polystyrene in ethyl acetate. The following speculative interpretation is advanced: 1) the fastest mode corresponds to motions characterizing the transient gel; 2) the middle mode reflects the translational (Stokes-Einstein) movement of the whole coil in the imperfect network; 3) the slo-

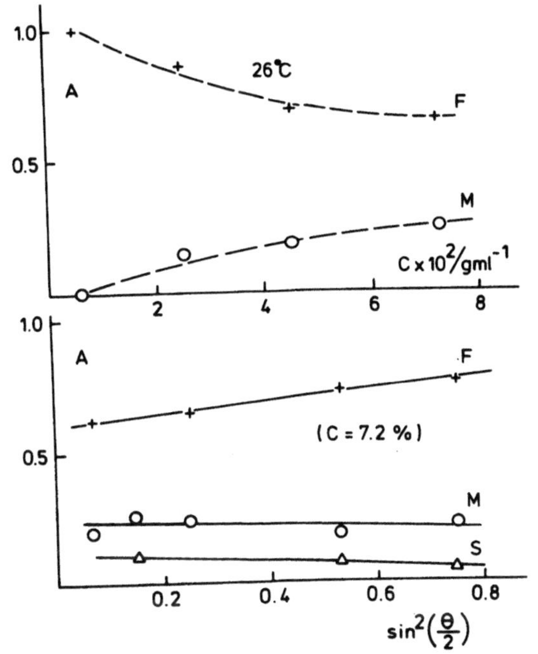

Fig. 5. Relative intensities (A_i) for the three modes derived from the moments of the Contin peaks shown as a function of concentration and angle

west mode describes the motions of clusters of chains in the network.

Further work with other molecular weight fractions would assist in establishing the assignments made. A parallel study of PS in cyclopentane [15] supports the above interpretation in terms of three components.

References

1. Brown W, Johnsen RM (1985) Macromolecules 18:379
2. Brown W (1986) Macromolecules 19:387, 1083
3. Brown W, Johnsen RM (1986) Macromolecules 19:2002
4. Štěpánek P, Koňák Č, Jakeš J (1986) Polymer Bull 16:67
5. Štěpánek P, Jakeš J, Koňák Č, Johnsen RM, Brown W (1987) Polymer Bull 18:175
6. Saeki S, Konno S, Kuwahara N, Nakata M, Kaneko M (1974) Macromolecules 7:521
7. Provencher SW (1979) Makromol Chem 180:201
8. Jakeš J (1988) Czech J Phys, in press
9. manufactured by ALV Langen Co. Ltd., FRG
10. de Gennes PG (1979) Scaling Concepts in Polymer Physics. Cornell University Press, London
11. Schaeffer DW, Joanny JF, Pincus P (1980) Macromolecules 13:1280
12. Brown W, Štěpánek P, Jakeš J (1988) Macromolecules 21:1791
13. Deschamps H, Léger L (1986) Macromolecules 19:2760
14. Brown W (1984) Macromolecules 17:66
15. Brown W, Štěpánek P, Macromolecules, in press

Received April 24, 1987;
accepted April 21, 1988

Authors' address:

P. Štěpánek
Institute of Macromolecular Chemistry
Czechoslovak Academy of Sciences
162 06 Prague 6, Czechoslovakia

Progress in Colloid & Polymer Science Progr Colloid Polym Sci 78:72–74 (1988)

Quasielastic light scattering from solutions of branched epoxy resins below the gel point

P. Štěpánek, J. Jakeš, Č. Koňák, and K. Dušek

Institute of Macromolecular Chemistry, Czechoslovak Academy of Sciences, Prague, Czechoslovakia

Abstract: Quasielastic light scattering was used to study samples of epoxy resins prepared by copolymerization in bulk of diglycidylether of Bisphenol A with Jeffamine D 400. First cumulants of the autocorrelation functions obtained with dilute solutions of these samples were determined, as well as distributions of sizes of the clusters below the gel point.

Key words: Branched epoxy resins, quasielastic light scattering, distribution of sizes, fractal dimension.

Introduction

The synthesis of a great number of branched or crosslinked polymeric materials (epoxides, polyurethanes, etc.) proceeds by random branching reactions [1]. At conversions corresponding to the vicinity of the gel point highly branched or even crosslinked particles appear which have a broad distribution of molecular weights [2]. In the vicinity of the gel point, physical properties of the system are described by scaling laws [3] that have as a variable the relative distance ε to the gel point, $\varepsilon = 1 - p/p_c$, where p is the degree of conversion and p_c is the critical conversion at the gel point. The main theories describing the gelation process are the classic theory of gelation [1, 4], the percolation theory [3, 5], and the theory of branching processes [2, 6].

With respect to quantities that can be determined by quasielastic light scattering, these theories predict [2] the z-average diffusion coefficient D_z

$$D_z \sim \varepsilon^{\alpha} . \tag{1}$$

where $\alpha = 0.609$ in the percolation theory and $\alpha = 1/2$ in the classic theory of gelation. At the present time, no prediction exists for the distribution of hydrodynamic radii of the clusters below the gel point.

The objective of this contribution is to determine the exponent α in relation (1) to the diffusion coefficient D_z and the distribution of hydrodynamic radii of clusters for samples of varying starting molar ratios r_A of amine and epoxy groups, dissolved in dimethylformamide.

Experimental

Materials used: poly(oypropylene) polyamine (Jeffamine® D 400); number average molecular weight $M_n = 397$; functionality $f_n = 3.75$. Diglycidylether of Bisphenol A (CIBA-Geigy) with $M_n = 350$ and $f_n = 2.0$. Dimethylformamide was of analytical grade purity, dried, and redistilled.

Sample preparation: diglycidylether of Bisphenol A was dissolved in Jeffamine at 50 °C, stirred for 5 min, and degassed at room temperature. To ensure a nearly complete conversion of the epoxy groups, the samples were allowed to react for 72 h at 100 °C in a nitrogen atmosphere. For experiments in the vicinity of the gel point samples with an excess of amine groups were prepared which had the initial molar ratio $r_A = 2[NH_2]/[E]$ in the range 2.35–2.45. The critical molar ratio was determined by testing the solubility of the samples by light scattering. The presence of a very small macroscopic quantity of gel particles was indicated by an obvious turbidity of the sample. This determination of the gel point led to the critical value r_{Ac} at the gel point $r_{Ac} = 2.395 \pm 0.002$.

Quasielastic light scattering: autocorrelation functions were measured on a homodyne photon correlation spectrometer equipped with a HeNe laser and a 96-channel digital correlator. The instrument has been fully described elsewhere [7]. In the case of highly polydisperse samples, multi-tau autocorrelation functions were measured, the correlator being operated with three different sampling times tau simultaneously.

Data analysis: single-tau autocorrelation functions were analyzed using the method of cumulants [8]; the value D_z was determined from the first cumulant of the autocorrelation function Γ by an extrapolation of Γ/K^2, where K is the scattering vector, to zero concentration and zero scattering vector. The multi-tau autocorrelation functions were analyzed first by the constrained regularization method of Contin [9] which yields a distribution $A(\tau)$ of relaxation times τ which may be further converted to a distribution of hydrodynamic radii R_H via the Stokes-Einstein relation $R_H = \tau K^2 \frac{kT}{6\pi\eta}$, ($k$ is the Boltzmann constant, T is the absolute temperature, and η is viscosity of the solvent). Results of Contin's calculations showed distributions with a weak side peak or a shoulder on the low τ side. In view of the recent testing [10] of Contin that showed this might be an artifact, it was decided to fit the experimental multi-tau correlation functions by a modified Schulz-Zimm distribution of τ,

$$A(\tau) = \Gamma^{-1}(n)\, x^{n-1}\, \tau_0^{-n} \exp(-x/\tau_0)\, t_0^2\, (t_0 - \tau)^{-2} \quad \text{for } \tau < t_0,$$

$$A(\tau) = 0 \quad \text{for } \tau \geqq t_0, \tag{2}$$

where $x = t_0\tau/(t_0 - \tau)$ and n (half integer), τ_0, and t_0 are adjustable parameters (Γ is the gamma function). The change from variable τ to x in the Schulz-Zimm distribution introduced another adjustable parameter t_0 while preserving the closed-form integrability of the distribution. This particular form was chosen on the basis of the Contin results; it was found that for samples studied in this contribution this function fits the experimental data with no systematic deviations (random distribution of residuals).

Results and discussion

The dependence of the diffusion coefficient as a function of the distance to the gel point is presented in log-log coordinates in Fig. 1. A straight line in this representation means that the relationship (1) between D_z and ε is described by a power law (cf. Eq. (1)) and the slope of this line gives the value of the exponent $\alpha = 0.49 \pm 0.03$. This finding indicates that for the present system gelation is well described by the classical theory.

Distributions of hydrodynamic radii $A(R_H)$ for samples having different distances ε to the gel point are shown in Fig. 2. These were obtained by extrapolation of the $A(\tau)$ distribution functions to zero angle and zero concentration. Relative half-widths w of the $A(R_H)$ distribution were calculated as the ratio of the half-width at half-height to the position of the maximum of the distribution and are plotted in Fig. 3. Contrary to expectation [11], w decreases instead of

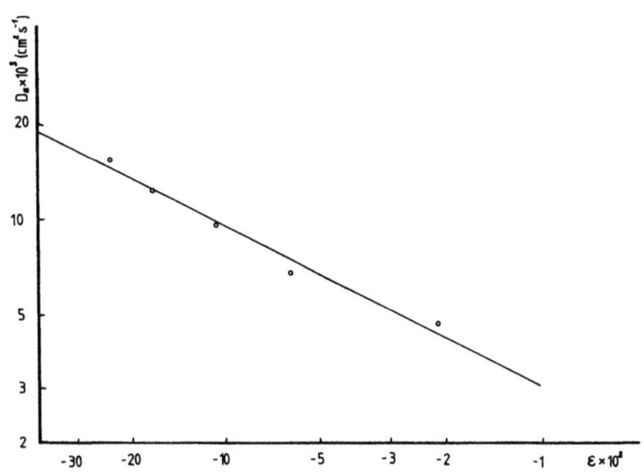

Fig. 1. Dependence of the z-average diffusion coefficient D_z on the relative distance ε to the gel point

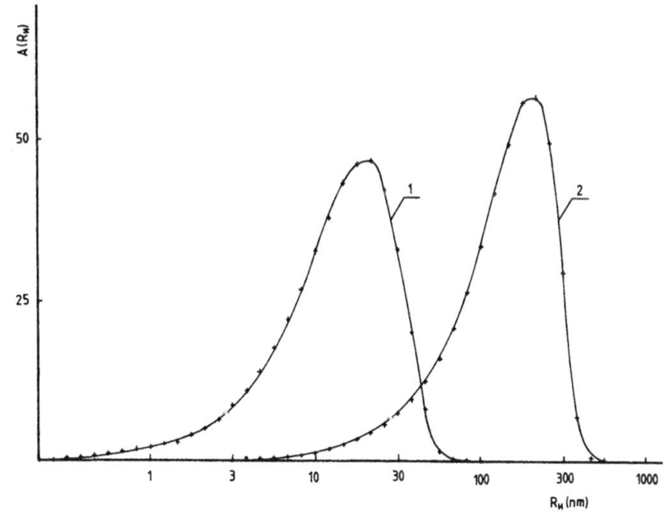

Fig. 2. Distribution function of hydrodynamic radius R_H for sample: (1) with largest distance to the gel point $\varepsilon = -2.27 \times 10^{-2}$; (2) with the smallest distance to the gel point $\varepsilon = -2.05 \times 10^{-3}$

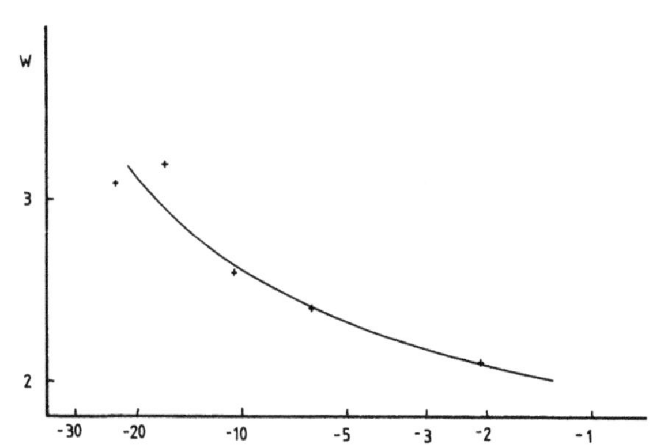

Fig. 3. Dependence of the relative half-width w of the distributions of hydrodynamic radius on the relative distance ε to the gel point

Table 1. Values of the fractal dimension for different structures

System	D_f
Linear polymer, good solvent	1.67
Linear polymer, theta solvent	2
Branched polymer, good solvent	2
Branched polymer, theta solvent	2.29
Diffusion limited aggregation	2.5
Solid sphere	3

increasing as the gel point is approached. The current theories, however, predict an increase of polydispersity in molecular weight M and, as already mentioned, no prediction is available for the distribution of hydrodynamic radii. The molecular weight M of a polymer is linked to its radius R by

$$M \sim R^{D_f}, \tag{3}$$

where D_f is the fractal dimension. If we accept the fact that the relative half-width of the distribution of molecular weight should increase and take into account the observation that the relative half-width of the distribution of hydrodynamic radii decreases as the gel point is approached, we are led to the conclusion that the fractal dimension D_f of the polymer [12] increases as the reaction proceeds. Exact values of D_f are not known; but typical values for some possible structures are recalled in Table 1. The polymers become more compact in the vicinity of the gel point, not just larger in size and in molecular weight. Comparison of a distribution of R_H obtained from quasielastic light scattering, as described here, with the distribution of molecular weight obtained, e.g., from GPC combined with low angle light scattering, should provide valuable information on quantitative changes in D_f, and thus in the structure of polymers during the crosslinking reaction.

Acknowledgement

Stimulating discussions with Prof. W. Burchard are gratefully acknowledged.

References

1. Flory PJ (1953) Principles of Polymer Chemistry. Cornell University Press, Ithaca, New York
2. Burchard W (1983) Adv Polym Sci 48:1
3. de Gennes PG (1979) Scaling Concepts in Polymer Physics. Cornell University Press, Ithaca, New York
4. Stockmayer WH (1943) J Chem Phys 11:45
5. Stauffer D, Coniglio A, Adam M (1982) Adv Polym Sci 44:103
6. Dušek K (1986) Adv Polym Sci 78:1
7. Koňák Č, Štěpánek P, Sedláček B (1984) Czech J Phys A 34:497
8. Koppel DE (1972) J Chem Phys 57:4814
9. Provencher SW (1979) Makromol Chem 180:201
10. Jakeš J (1988) Czech J Phys, in press
11. Mikeš J, Dušek K (1982) Macromolecules 15:93
12. Mandelbrot BB (1982) The Fractal Geometry of Nature. Freeman WH and Co San Francisco

Received June 25, 1987;
accepted April 21, 1988

Authors' address:

P. Štěpánek
Institute of Macromolecular Chemistry
Czechoslovak Academy of Sciences
Heyrorského no. 2
162 06 Praha, CSSR

ESR on polymer semiconductors of poly(tetrathiafulvalene)

H.-K. Roth[1]), H. Gruber[1]), E. Fanghänel[2]), and Trinh vu Quang[2])

[1]) Department of Natural Sciences, Leipzig University of Technology, G.D.R.
[2]) Department of Chemistry, Merseburg Technical University "Carl Schorlemmer", G.D.R.

Abstract: Results of ESR and conductivity measurements on undoped and iodine doped poly(tetrathiafulvalenes) are reported. The ESR properties of phenylene-bridged polymeric tetrathiafulvalenes which are substituted by CH_3- or C_2H_5- groups and poly(tetrahydroanthratetrathiafulvalene) are described. The maximum spin concentration in the doped speciality polymers is about 10^{20} spins per gram. The conductivity of the variously doped samples lies in the area of 10^{-10} to 10^{-4} Scm^{-1} at room temperature and increases with temperature, as is typical for semiconductors.

Key words: Conducting polymers, ESR, doping with iodine, poly(tetrathiafulvalenes).

1. Introduction

The conductivities of the semiconductive or metallic regime attained by doping organic substances with electron acceptors or donors have produced wide interest, for both the practical use of new electrical and electronic materials and the basic physical understanding of their peculiar conduction mechanism [1–4]. In recent years, the electron donor tetrathiafulvalene (TTF) and its derivatives have been the subject of chemical and physical studies, due to the fact that many compounds of this group can form electrically conducting charge-transfer (CT) salts. This peculiarity is dependent upon the relative ease with which the TTF can be oxidized to obtain the radical cation or the dication.

The best conductivity of CT complexes, consisting of molecules with plane structural elements, can be observed when the donor and the acceptor molecules (i.e. tetracyanoquinodimethane) form separate stacks and when only an incomplete charge transfer is realized. The conductivity of low molecular CT complexes with TTF is about 600 Scm^{-1} at room temperature [5–7]. The high electrical conductivity of tetrathiafulvalene charge-transfer complexes has stimulated the synthesis of polymers with TTF units [8–17].

The formation of new polymeric semiconductors which are also stable against humidity and oxygen has been the aim of our recent research work.

The change-over from a "monomer" TTF CT complex to a macromolecular material having sufficient conductivity is a complicated process because the formation of stacks in polymers is extremely difficult. The loss of conductivity caused by these ordering problems may be compensated partly by the transport of the charge carriers along the individual polymer chains. An extended conjugated and planar π-electron system which can be doped by acceptor (i.e. iodine, bromine) or donor molecules is regarded as a necessary condition for electrical conductivity.

Therefore, we have synthesised various poly(tetrathiafulvalenes) (PTTF), in which the TTF units are bridged by 1,4-phenylene groups. The planarity of the PTTF is influenced by the substituents connected with the TTF unit. One can expect that the conjugative interaction between the TTF units via the phenylene groups is not very strong. For these reasons a conductivity can be expected which is not that of high conducting polymers, like doped polyacetylene or of a CT-complex like TTF/TCNQ, but in the area of semiconductors [18].

ESR studies, on PTTF which is undoped or doped with iodine, have been carried out in order to search for relations between the spin concentration and the conductivity. Characterisation of the paramagnetic species and obtaining information on the doping mechanism in polymers of various degrees of planarity were further aims of the ESR investigations.

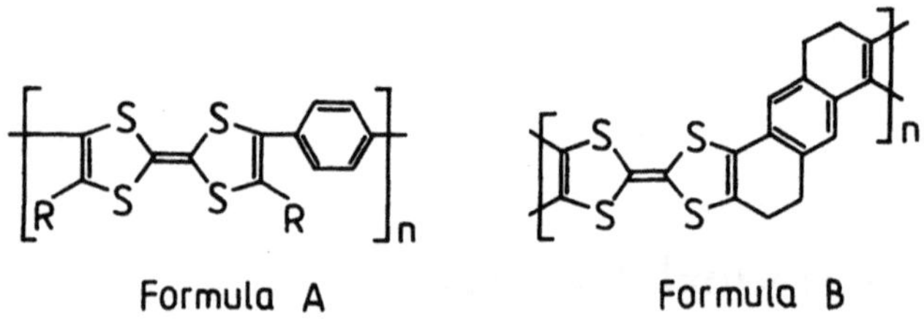

Fig. 1. Chemical formulae of the compounds A, B. (A): 1,4 phenylene-bridged poly(tetrathiafulvalene), PTTF (B): Poly(1,2,5,6,tetrahydroanthratetrathiafulvalene), PTHA-TTF

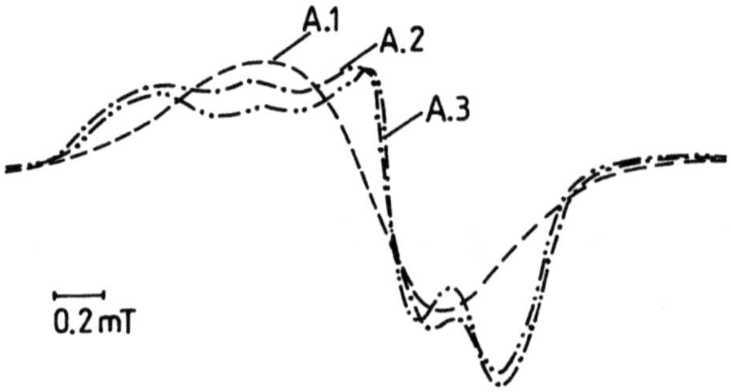

Fig. 2. ESR properties of undoped poly(tetrathiafulvalenes), compounds A.1; A.2; A.3

Table 1.

Compound	General chemical formula	R	m	n
A.1	A	$-H$	7150	25
A.2	A	$-CH_3$	7660	25
A.3	A	$-C_2H_5$	10900	32
B.1	B		6810	20

2. Experimental

The general formulae of the investigated compounds are presented in Fig. 1. The substituents R, the average molecular weights m and the average number of the repeatable units n are listed in Table 1.

The polymers A.1 and A.2 were synthesized according to Ref. [10] from the corresponding 1,4-phenylene-di(1,3-dithiolium-perchlorates) by deprotonation with tertiary aliphatic amines. The new polymer A.3 was obtained in a similar manner.

To synthesize B.1, the 4,5,10,11-tetrahydroanthra[1,2-d; 5,6-d']bis(1,3-dithioliumperchlorate) was used [19]. The details of the synthese of the compounds A.3 and B.1 will be published elsewhere.

The molecular structure is given by the synthesis and is in accordance with the elemental analysis. All studied polymers are red to brown powders. They are insoluble in normal organic solvents, but soluble in trifluoracetic acid. The samples were doped by iodine in chloroform, using different concentrations where the iodine content was referred to the TTF-unit. The iodine content of the samples was determined by elemental analyses. After doping, the colour of the polymers was black. The ESR investigations were carried out in contact with air using an ESR 231 spectrometer (from ZWGB Berlin, G.D.R.), which was coupled on-line with a microcomputer. For concentration measurements a DPPH standard was used. The magnetic induction was measured with the MJ-110R-NMR magnetometer. DC conductivity was measured by means of the four-probe technique on pellets from the powder.

3. Results and discussion

3.1. ESR on undoped material

The ESR spectra of PTTF do not show a hyperfine splitting, but consist of a single line or of a superposition of single lines. In the case of the undoped PTTF with $-CH_3$ and $-C_2H_5$ substituents (samples A.2 and A.3), the predominant ESR signal is asymmetrical, whereas with the unsubstituted PTTF (sample A.1) and with PTHA-TTF (sample B.1) the predominant signal is symmetrical, as is the case with all highly doped PTTF (see Fig. 2).

The asymmetric spectra are probably due to peroxy radicals. Our spectra show a strong similarity to the spectra obtained, for example, from irradiated amorphous polyvinylchloride or semicrystalline polyethylene, if the samples are in contact with oxygen before or during the irradiation [20]. This is one reason why we suppose that the PTTF with $-CH_3$ or $-C_2H_5$ groups have partly reacted with oxygen to peroxy radicals.

All samples of "monomer" TTF with various substituents (including $R = -H$), which are stored in air, show an asymmetric ESR signal. By investigation of Carlsen [21] and Lakshmikantham [22] it is known that crystallites of "monomer" TTF are oxydable on their surfaces by exposure to air. These thin layers of TTF-S-oxides can be formed via the formation of peroxy radicals from the structure I or via biradicals of the type II.

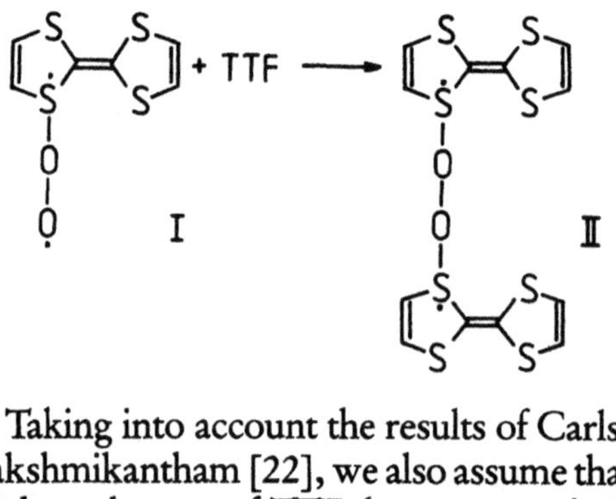

Taking into account the results of Carlsen [21] and Lakshmikantham [22], we also assume that in the case of the polymers of TTF the asymmetric ESR spectra are mainly due to peroxy radicals of type I generated by the reaction of triplet oxygen with sulphur.

In the case of PTTF with $R = -H$ and also on PTHA-TTF we do not observe asymmetric peroxy spectra but symmetric spectra as exist in the case of doped PTTF. This result can be explained in the following way: The PTTF (A.1 and B.1) have a more or less plane structure. Due to this structure an electronic interaction occurs between the TTF units in the polymer, which leads to a reduction in the oxidation (ionization) potential, so that predominantly, an electron transfer from the π-system to the $^3O_2$ occurs, forming a TTF radical cation, which has the same structure as a radical cation formed by doping with iodine or bromine. Possible structural elements are

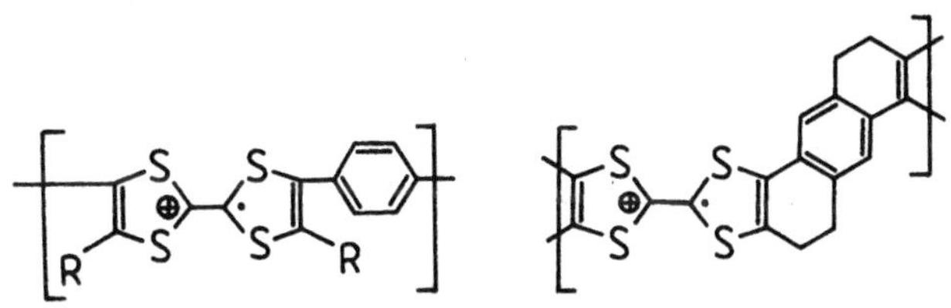

These cation radicals exist also in such polymer samples which predominantly form peroxy radicals with oxygen. This can be seen in the ESR spectra, if we begin to saturate the peroxy radical spectrum by increasing the microwave power. Each asymmetric spectrum in Fig. 2 is a superposition of spectra of peroxy and cation radicals.

3.2. *ESR on doped materials*

In Fig. 3 the number of paramagnetic centres per gram and the electrical conductivity as a function of the iodine content in doped PTTF are shown.

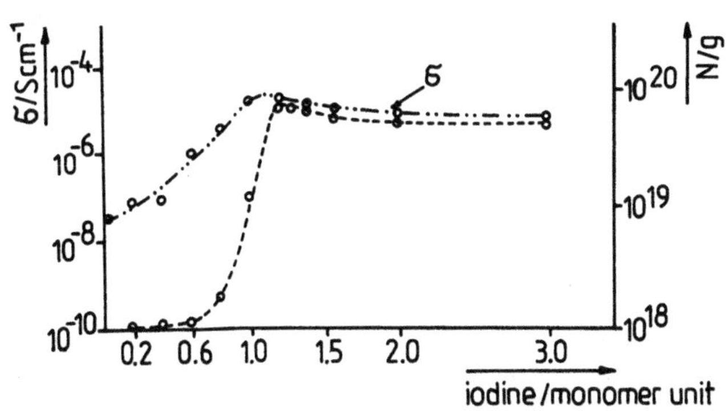

Fig. 3. Number of unpaired spins per gram (right) and electrical conductivity as a function of the iodine content in doped PTTF, compound A.3

The number of paramagnetic centres and the electrical conductivity strongly increase with the increasing iodine content up to a limiting value. This means that in our materials the charge carriers are predominantely species with spin.

The ESR spectra of the iodine doped polymers reveal a single ESR line at $g = 2.0068$ (compound B.1), $g = 2.0072$ (A.2), $g = 2.0074$ (A.3) and $g = 2.0078$ (A.1). These values are in agreement with the g-factor from the TTF radical cation in [23]. The ESR studies on PTTF show that the increase of the number of unpaired spins is associated with an increase of the ESR line width, up to a limiting value.

As a consequence of the nonhomogenous broadening of the ESR lines, it is not possible to determine the real values of the longitudinal relaxation time and of the transversal relaxation time. We have measured these values with the help of the electron spin echo method (ESE). These results will be published later.

In Fig. 3 we can see a difference in the behaviour of the electrical conductivity and the spin concentration at low doping concentration. This could be due to the fact that before doping the PTTF samples already contain a certain number of paramagnetic species, however most of them are localized as peroxy radicals (see 3.1). When starting the doping process the first iodine atoms effect a conversion of the localized unpaired electron spins of peroxy radicals to mobile ones (cation radicals). This produces the increase of conductivity without increasing the spin concentration. Only if all the unpaired electrons of peroxy radicals are converted to mobile charge carriers will the spin concentration increase by further addition of iodine. This result distinguishes itself from those on the compound B.1; in this case we have found an increase in the spin concentration also at low doping concentration (see Fig. 4). This result confirms our notion of the role of peroxy radicals. In the polymer B.1 and other PTTF with planar structure the paramagnetic species are predominantly cation radicals. For that reason doping effects from the beginning an increase in both conductivity and spin concentration.

In Fig. 5 the measured electrical conductivities are shown. As can be seen in Fig. 5 the best conductivities are shown by the compounds B.1 and A.1. These are the polymers with a more or less planar structure.

The structure of the polymers A.1 and B.1 probably affects the conductivity in three ways:

– de facto, no conversion of localised peroxy radicals into mobile cation radicals by doping is necessary

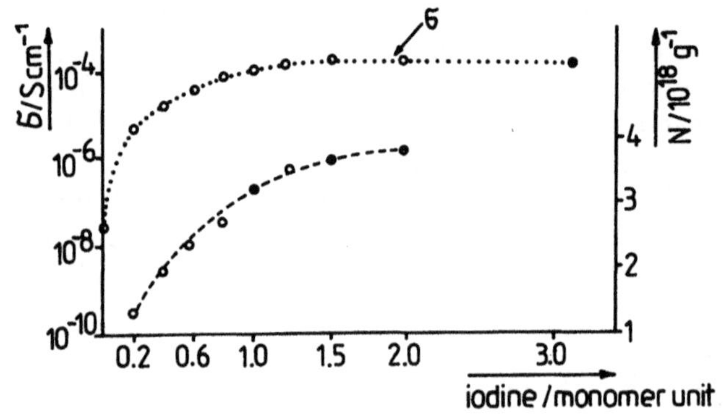

Fig. 4. Number of unpaired spins per gram (right) and electrical conductivity as a function of the iodine content in doped PTHA-TTF, compound B.1

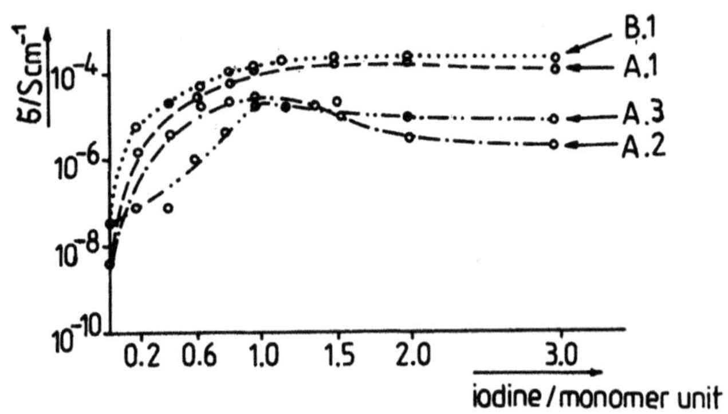

Fig. 5. Electrical conductivity of polymers

– the planar structure is connected with an optimum overlap of π-orbitals promoting the charge transport in single molecules

– the planar structure makes it also possible that in PTTF small stacks or other kinds of ordered areas are formed which promote the charge transport from one polymer chain to the neighbouring chain.

Nevertheless, the samples of the various PTTF are less ordered than the corresponding "monomer" CT-salts of TTF, which show a higher electrical conductivity than the PTTF [19].

All PTTF show a temperature dependence of the conductivity as for semiconductors; the electrical conductivity of the undoped and doped PTTF increases with temperature. The activation energies of conductivity (measured in the temperature range from 10 °C to 70 °C) were determined as 0.37 eV (compound A.3) and 0.15 eV (compound B.1) for iodine doped polymers with 1 iodine atom per monomer unit.

Further investigations of the mechanism of conductivity in PTTF are in progress.

References

1. Hanack M, Pawlowski G (1982) Naturwissenschaften 69:266
2. Hatfield WE (1979) Molecular Metals. Plenum Press, New York
3. (1986) International Conference on Science and Technology of Synthetic Metals, Kyoto, Japan, (1987) publ in Synth Metals 17–19
4. (1987) International Conference on Electrical and Related Properties of Organic Solids, Szklarska Poreba, Poland, (1987) publ in Mat Science 1–2
5. Etemad S (1976) Phys Rev B 13:2254
6. Tomkiewicz Y, Craven RA, Schultz TD, Engler EM, Taranko AR (1977) Phys Rev B 15:3643
7. Kistenmacher TJ, Phillips TE, Cowan DO (1974) Acta Cryst B 30:763
8. Ueno Y, Masuyama Y, Okawara M (1975) Chem Lett p 603
9. Schukat G, Le van Hinh, Fanghänel E, Libera L (1978) J Prakt Chem 320:404
10. Le van Hinh, Schukat G, Fanghänel E (1979) J Prakt Chem 321:299
11. Le van Hinh, Schukat G, Fanghänel E, Libera L (1979) J Prakt Chem 321:475
12. Pittman CU, Narita M, Liang YF (1976) Macromolecules 9:360
13. Watanabe M, Iida T, Sanui K, Ogata T, Kobayashi T (1984) J Polym Sci Polym Chem Ed 22:1299
14. Pittmann CU, Liang YF, Ueda M (1979) Macromolecules 12:355
15. Kossmehl G, Rohde M (1982) Makromol Chem 183:2077
16. Pittman CU, Liang YF, Ueda M (1979) Macromolecules 12:541
17. Thompson WA, Schroeder AH, Kaufman FB (1981) J Vac Sci Technol 18:243
18. Schukat G, Richter AM, Fanghänel E (1987) Sulfur Reports 7:155
19. Trinh vu Quang (1987) Thesis A, Technical University "Carl Schorlemmer", GDR
20. Roth HK, Keller F, Schneider H (1984) HF-Spektroskopie in der Polymerenforschung. Akademie Verlag, Berlin, p 120
21. Carlsen L, Bechgaard K, Jacobsen CS, Johansen I (1979) J Chem Soc Perkin Transact II, p 862
22. Lakshmikantham MV, Garito AF, Cava MP (1978) J Org Chem 43:4394
23. Fitzky HG, Hocker J (1986) Synth Metals 13:335

Received May 15, 1987;
accepted January 30, 1988

Authors' address:

H.-K. Roth
Department of Natural Sciences
Leipzig University of Technology, PSF 66
DDR-7030 Leipzig, G.D.R.

Progress in Colloid & Polymer Science

Progr Colloid Polym Sci 78:79–80 (1988)

Migration of radicals in resist polymers – an example of microscopic transport processes

P. Wünsche

Karl Marx University, Department of Physics, Leipzig, G.D.R.

Abstract: Recent activities of the author in the field of decay reactions of radicals in polymers are described. An advance in the interpretation of the thermostimulated radical processes and reactions in solid polymers has been achieved by the use of a broad distribution of the activation energy which allows correlation with other thermally stimulated processes in the polymers.

Key words: Migration of radicals, resist polymers, thermostimulated processes, activation energy distribution.

The decay reactions of radicals in polymer systems are governed by the migration of reacting species along or perpendicular to the polymer chain. Hence, the investigation of radical reactions in polymers gives us information about microscopic transport processes including the activation energies.

One of the most important advances in the interpretation of thermostimulated processes and reactions in solid polymers is the application of a broad distribution of the activation energy in the discussion of the experimental results. Such broad distributions of the activation energy are correlated with different processes measured with several experimental methods, particularly with electron paramagnetic resonance (EPR), thermally stimulated discharge, and conductivity or luminescence. The comparison of such activation energy distributions gives us the opportunity to discuss the processes in the polymer solid which govern the measured experimental parameter.

As an example in [1] the results of EPR and thermostimulated luminescence (TL) are compared and offer interesting conclusions. The distribution of the activation energy of the decay reaction of radicals in poly-(methylmethacrylate – 20 % methacrylic acid) has the same shape as the TL glow curve of the same polymer. The estimated activation energies in the main parts of the distribution functions are in good agreement if we compare the two methods. The main parts of the distribution functions represent the activation energies of 10.3 kJ/mol (TL), 13 kJ/mol (EPR), and 26.8 kJ/mol (TL); 22 kJ/mol (EPR) in the temperature regions of 120–125 K and 205–210 K, respectively.

This example makes it possible to compare the results of different experimental methods when we discuss the distribution of the activation energy of thermally stimulated processes in polymers. This can be done if the reasons for the two thermally stimulated processes, radical decay and luminescence, are the same. This is only possible if the mobility of polymer segments and side groups stimulates both the radical decay and the detrapping of electrons or ions.

To estimate the distribution function of the radical decay reactions in polymers after γ-irradiation it is necessary to measure the radical concentration in small temperature steps (e.g., 10 degrees). At every temperature the concentration must be measured in dependence on storage time. Starting with such a matrix of experimental results in the preexponential factor of the reaction constant and the value of the distribution function of the activation energy may then be calculated for every temperature [1].

To get a broad distribution of the activation energy the measurement must start at a very low temperature. Hence, the γ-irradiation was performed at liquid nitrogen temperature. The sample was held at this temperature until the EPR measurement began at 100 K. The γ-irradiation followed by the EPR measurement at low temperatures allows better investigation of the primary step of the interaction between the γ-rays and the polymer than at room temperature. During the temperature increase from the liquid nitrogen temperature up to room temperature many radical reactions may occur, e.g., radical decay and radical conversion reactions. Resulting from these may be radical structures in the polymer completely different from the radical structures measured immediately after the γ- irradiation at low temperatures.

As an example of the estimation and discussion of the broad distribution of the activation energy and the effect of thermally stimulated reactions between 77 K and room temperature the EPR measurements on chlorine-substituted alkyl methacrylates are used. The following substances were measured: poly-(-chloroacrylate) (PMCA), poly-(1-chloroethyl methacrylate) (P1CEMA), poly-(2,2,2-trichloroethyl methacrylate) (PtriCEMA), and poly-(1,2,2,2-tetrachloroethyl methacrylate) PteCEMA) [2].

All chlorine-substituted alkyl methacrylates show in the low temperature EPR spectrum as main part radicals were generated by the abstraction of chlorine. Additionally, these substances showed a strongly increased radical concentration relative to the unsubstituted ones. For polymethylmethacrylate the radiation *G* value is about 1 [3], but for all chlorine-substituted alkyl methyacrylates this value is between 3 and 4 [2]. This effect may be used to increase the radiation interaction, or the resist effect.

The primary radicals are not very stable in all substances. Hence, at room temperature another radical structure, the chain break radical or the chain end radical, are measured. In the case of P1CEMA, P2CEMA, PtriCEMA, and PteCEMA this radical has the structure

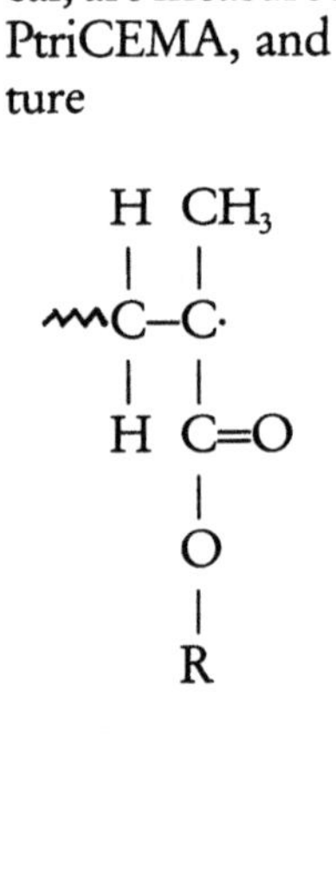

and in PMCA

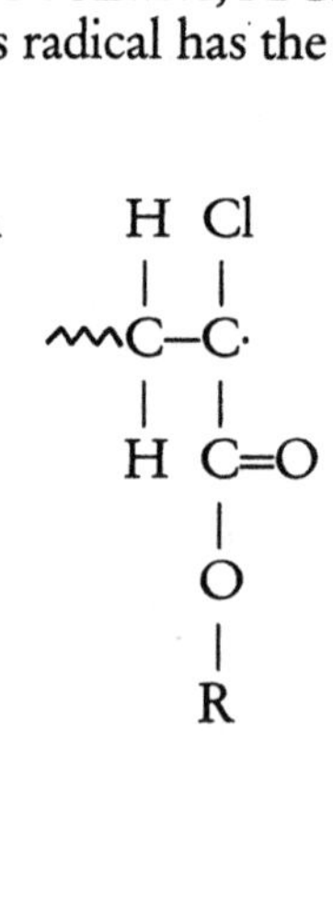

These radical structures are the result of inter- or/and intrachain radical conversions of the primary radicals. This reaction path must be taken into consideration if the resist effect is increased. The different structures of the primary radicals are the reasons for rather large differences in the distribution functions of the decay reactions.

The primary radical of PMCA is situated in the main chain. Therefore, the mobility of this primary radical is connected with the mobility of the main chain. Conversely, the primary radicals of the other substances are generated at the side group. As a consequence, the mobility of these primary radicals is connected to the mobility of the side group. Because the reaction constants of the decay reactions of the radicals are correlated with their mobility, the difference of mobility must have consequence in the distribution of the activation energy. The calculated distribution functions support this. The main part of radicals in PMCA decays has mean activation energies above 50 kJ/mol, but in the case of P1CEMA and P2CEMA, 70 % of the radicals have activation energies of the decay reaction that are lower than 45 kJ/mol. The distribution functions of the activation energy of the radical decay reactions in PtrCEMA and PteCEMA are affected because the side group and main chain mobility are not clearly separated from each other in these substances.

References

1. Wünsche P, Kunze R, Wäsche M (1987) Physica Status Solidi 103:145
2. Wünsche P, Müller S, Lorkowski H-J, received for publication in Journal of Polymer Science A1
3. Wünsche P, Pfeiffer K, Lorkowski H-J, Reinhard J (1985) Acta Polymerica 36:268

Received June 25, 1987;
accepted April 21, 1988

Author's address:

Doz. Dr. P. Wünsche
Sektion Physik
Karl-Marx-Universität
Linnéstr. 5
DDR-7010 Leipzig, G.D.R.

Progress in Colloid & Polymer Science Progr Colloid Polym Sci 78:81–87 (1988)

Charging and charge-detection methods and their use in the characterization of polymer-electret materials

R. Gerhard-Multhaupt and G. M. Sessler

Institut für Elektroakustik, Technische Hochschule Darmstadt, Darmstadt, F.R.G.

Abstract: Recently developed methods for charging and investigating polymer electrets are described and discussed in view of their use for the characterization of the respective sample material. Surface charging with electron beams of relatively low energies may be used to achieve positive charging and to determine the secondary-electron emission yield of the specimen. Laser-induced pressure pulses permit the scanning of charge and polarization profiles as well as the measurement of ultrasonic velocity and attenuation. Poling of piezoelectric polymer electrets may be performed by several techniques including electron-beam charging which leads to polarization zones of predetermined thickness inside the samples; poling-current measurements can be utilized to estimate the ferroelectric polarization in the samples and to model the ferroelectric-switching behavior of the dipoles.

Key words: Electrets, secondary-electron emission, charge and polarization profiles, sound velocity and attenuation, poling techniques.

1. Introduction

In recent years, polymer electrets received increasing attention in science and technology because of their importance for the study of electrical phenomena in polymer materials and their advantages for many new and interesting applications. The present state of the field is documented in a number of book articles [1–5], several reviews [6–12], the proceedings of international conferences [13], and a bibliography [14].

New experimental techniques for the preparation and the investigation of polymer electrets include surface charging by means of electron beams with relatively low energies [15–17], scanning of spatial charge distributions by means of laser-induced pressure pulses [18–24], and poling of piezo- or pyroelectric polymers by means of directly applied fields [25, 26], or of monoenergetic electron beams [27–29]. These methods are briefly described in the following and their application to the study of secondary-electron emission, sound velocity and attenuation, and ferroelectric polarization in polymer materials is discussed.

PCPS HT 035

2. Electron-beam charging of electret surfaces

2.1 Positive and negative surface charging

Electron-beam charging is usually employed in order to inject negative charges into the bulk of electret films [1]; a variation of this technique, which leads to positively charged electrets, is based on the fact that the secondary-electron yield of polymer surfaces often exceeds unity for beam energies around a few hundred eV [15–17]. In order to prevent the emitted secondary electrons from coming back to the sample surface (cf. the right half of Fig. 1), the rear electrode of the one-sided metallized sample is biased with a negative voltage. Then, as schematically shown in the left half of Fig. 1, the secondary electrons are attracted by the grounded parts of sample holder and vacuum vessel and the sample surface charges up positively, provided the effective beam energy leads to a secondary-electron yield above unity [15].

An example of the surface-potential development on a poly(ethylene terephthalate) (PETP) sample during such a charging (and discharging) process is shown

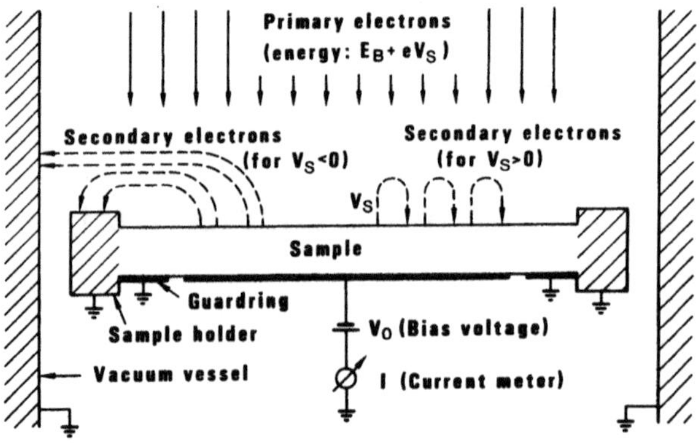

Fig. 1. Diagram of sample, sample holder, and vacuum vessel in a low-energy electron-beam experiment where secondary electron emission is an important effect; back scattering is not shown separately in this figure

in Fig. 2; the potential was calculated from the measured rear-electrode current [17]. After the attainment of a steady-state positive surface charge (in this situation the yield shown in the right part of Fig. 2 is a little below 1 [15]), the bias can be switched to zero which results in a rather high positive surface potential as seen in Fig. 2. If the electron beam would be stopped at this point, one would obtain a positively charged PETP electret; however, since the electron beam continues, it discharges the sample again because its effective energy is now so high that the yield becomes smaller

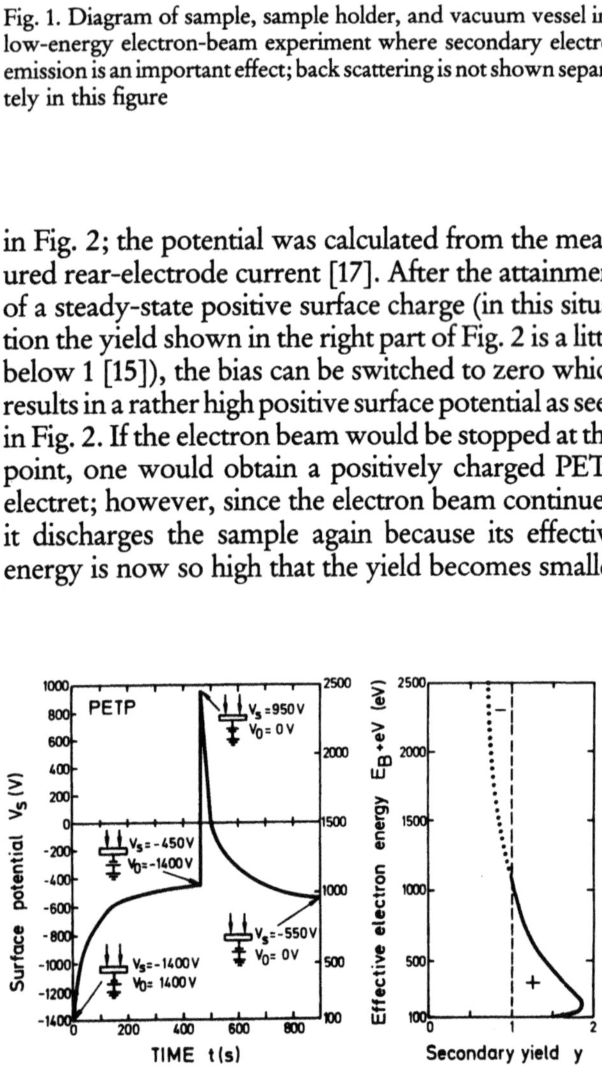

Fig. 2. Development of the surface potential on a PETP sample irradiated with an electron beam of 1.5 keV energy; the sample is first biased to −1400 V and later grounded. The right half of the figure shows the secondary emission yield as calculated from the experiment (solid line) and extrapolated (dotted line), respectively; (+) and (−) signs indicate the resulting polarity of the sample surface if the emitted electrons do not return to the sample (after Ref. [17])

than unity (see the dotted line in the right part of Fig. 2). The discharging process stops again when the effective electron energy corresponds to an emission yield of 1 [15,17]; a negatively charged PETP sample is generated.

2.2 Determination of secondary emission yield

The method just outlined can be used not only to charge or discharge electret samples, but also to determine the secondary-electron yield of polymer surfaces as a function of the primary-electron energy [17]. Actually, the solid line in the right part of Fig. 2 was obtained by this technique [17], which is based on a theoretical analysis of the charging dynamics during irradiation with low-energy electrons [16]. Yield measurements on several electret materials resulted in a single curve (master curve) if the experimental data were normalized to the respective maximum yield and the corresponding electron energy [17].

3. Laser-induced pressure-pulse method

3.1 Charge and polarization profiles

Several non-destructive high-resolution techniques for the determination of spatial charge, polarization, and field profiles in electrets were developed during the last decade [1,8–12]. Today, the most important probing techniques are based on the propagation of an acoustic wave through the sample volume [8–11, 18–24, 30–32]. With laser-light pulses for the generation of short acoustic pulses, coupling and travel-time problems could be avoided; this led to a rapid development of competing methods [12], some of which provide spatial resolutions of about 1 μm without requiring deconvolution [18–24].

The experimental arrangement for the laser-induced pressure-pulse (LIPP) method [18–20,23,24] is shown in Fig. 3: short (70 ps) and energetic (1–10 mJ) laser-light pulses from a mode-locked Nd: YAG laser are absorbed in the front metallization of the sample under investigation or in a graphite layer deposited onto this electrode. Thermoelastic stresses in, and ablation of material from, the absorption layer generate a short pressure pulse which propagates through the sample volume as a compression zone; upon reflection at the sample's rear surface, the propagation direction is reversed and the compression changes into rarefaction. When the pressure pulse passes a charge

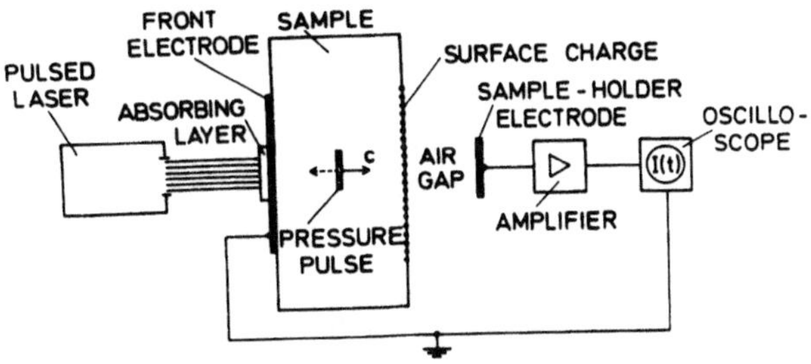

Fig. 3. Diagram of the experimental arrangement used with the laser-induced pressure-pulse method [9,18–24]. The equipment includes a mode-locked Nd:YAG laser (JK Lasers System 2000), a sample holder, a broadband pre-amplifier (B&H 3002DC), and a fast real-time oscilloscope (Tektronix 7104)

and/or polarization layer, the compression/rarefaction-related movement of charges, the strain-induced changes of the dielectric permittivity (electrostriction), and the deformation dependence of the polarization (piezoelectricity) lead to a displacement current which can be detected by means of a fast oscilloscope. For weak signals, an additional pre-amplifier may be inserted.

Several different theoretical treatments of the relation between the detected signal and the desired profile are found in the literature; however, if the different assumptions on the strain dependence of the dielectric permittivity are taken into account, it can be shown that all response equations may be deduced from each other [21]. The current signal generated between the sample electrodes is either proportional to the real and polarization charges or to the electric field inside the sample; thus the temporal development of the signal represents a direct image of the spatial charge, polarization, or field profile [11,21,22]. Typical experimental results obtained on electron-beam-poled poly(vinylidene fluoride) (PVDF) samples will be reported below.

3.2 Longitudinal sound velocity and attenuation

If the charge or polarization profile in a sample is known the signals generated upon repeated scanning of this profile may be used to analyze the propagation of the short laser-induced pressure pulses in the sample material [33–35]. Soft acoustic reflections at the front and rear surfaces of the specimen lead to a train of signals whose relative delays, together with the material thickness, permit calculation of the longitudinal sound velocity. In addition, the relative decreases of the signal amplitudes reveal information on the ultrasonic attenuation in the sample.

For electret samples carrying a surface-charge layer only, the signals are particularly easy to evaluate. In Fig. 4, the above-mentioned signal train obtained on a

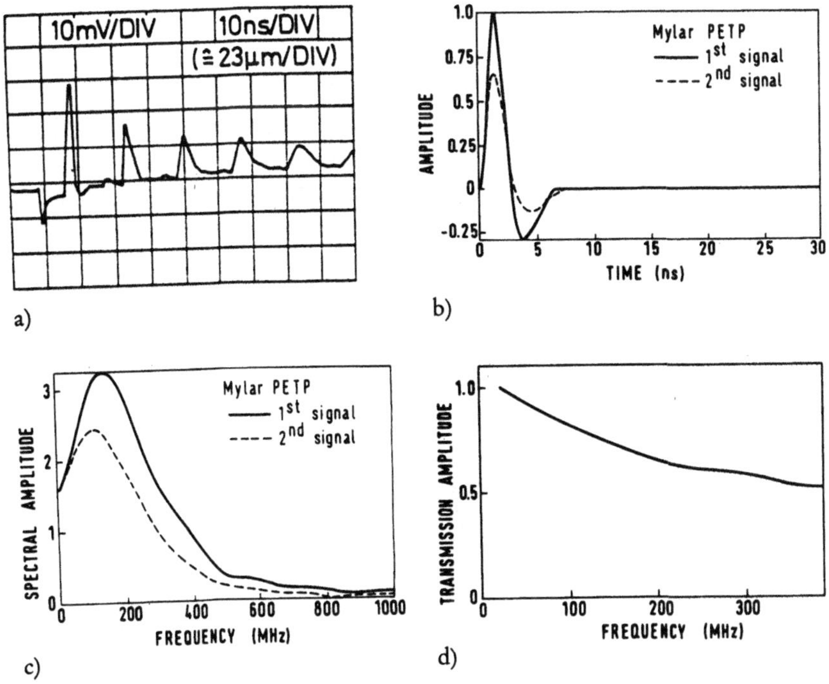

Fig. 4. LIPP experiment on an 18 µm-thick PETP sample positively surface-charged by means of a liquid contact [34]: a) train of signals obtained during propagation and several reflections of the LIPP in the sample; b) the first and second signals normalized to the peak amplitude of the first signal; c) calculated frequency spectra of the two signals; d) relative spectral transmission amplitude calculated from the two spectra

Table 1. Sound velocities and attenuations per wavelength for some commercial polymer films: fluoroethylenepropylene copolymer (FEP), polychlorotrifluoroethylene (PCTFE), polyimide (PI), polycarbonate (PC), poly(vinylidene fluoride) (PVDF), poly(ethylene terephthalate) (PETP), and poly(methyl methacrylate) (PMMA). Values in brackets indicate that only a very small number of experimental results was used for their derivation

Material (Trade name)	Sound velocity (km/s)	Attenuation per wavelength
FEP (Teflon)	1.3	(0.43)
PCTFE (Aclar)	1.7	(0.08)
PI (Kapton)	2.0	0.08
PC (Kimfoil)	2.0	?
PVDF (NSW)	2.2	(0.20)
PETP (Mylar)	2.3	0.09
PMMA (Röhm)	2.9	(0.07)

positively surface-charged PETP sample (a) is shown together with an enlarged image of the first two signals (b), their respective frequency spectra (c), and the resulting frequency-dependent ultrasonic-transmission function (d).

From several such measurements on a number of polymer samples [33–35], it was found that the ultrasonic absorption increases linearly with frequency in the frequency range considered and for all materials studied; therefore, the attenuation per wavelength is a constant. Values of this constant are given in Table 1 together with sound velocities calculated from the respective transit times of the pressure pulses [34]. The frequency-independent attenuation per wavelength in the polymers investigated points towards a hysteresis absorption process whose physical mechanism(s) is (are) far from being understood.

4. Poling techniques for piezo-, pyro- and ferroelectric polymers

Several charging methods may be used to orient the ferroelectric polarization of poly(vinylidene fluoride) (PVDF) [5] and other piezo- and pyroelectric polymer electrets. In order to prevent destructive breakdown during poling it is, however, necessary to limit the current that flows through the sample. This is achieved by means of a current-limiting circuit when the high-field poling technique is employed [25, 26]; the current in a corona discharge is limited by the geometry and the voltages of emitting electrode and grid [36], but it can also be made constant by a feedback technique [37]. A similarly constant current is delivered by an electron beam whose use for the poling of piezo- and pyroelectric electrets was recently demonstrated [27].

4.1 Uniform poling of sample volume

For most applications of piezo- and pyroelectric polymers, a uniform volume polarization is required. This polarization can be achieved by means of several poling techniques provided that the poling field, as well as the poling current, are sufficiently large [11]. An example of the temporal buildup of such a uniform polarization is shown in the LIPP diagrams of Fig. 5, illustrating measurements on PVDF films poled by direct application of an electric field [24]. Similar polarization profiles are found after corona poling with high grid potentials and large sample currents [38], whereas smaller currents often result in non-uniform distributions of the piezoelectricity in a sample [39].

4.2 Non-uniform poling by means of an electron beam

With electron-beam irradiation in a suitable electrical arrangement, the radiation-induced conductivity (RIC) together with the finite electron range leads to the generation of a relatively thin charge layer near the maximum electron range [1]. This charge layer effects the necessary poling field between itself and its image charge in the rear electrode and thus also a polarization zone of pre-determined thickness [27, 29, 38]. Here

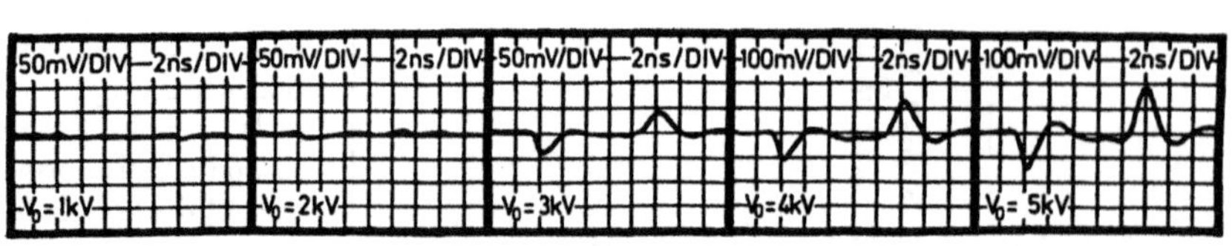

Fig. 5. Piezoelectricity gradients in 25 μm-thick PVDF samples poled at room temperature by electric fields of 40, 80, 120, 160, and 200 MV/m; the voltages indicated in the diagrams were applied via a limiting resistor in order to protect the delicate sensing circuitry [24]

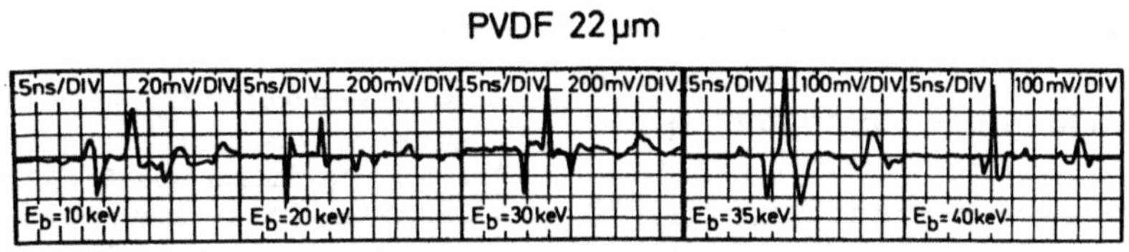

Fig. 6. Piezoelectricity gradients in 22 µm thick PVDF samples poled by electron beams with energies of 10, 20, 30, 35, and 40 keV; the front surface of the samples was unmetallized during poling, while the rear electrode was grounded [29]

again, the current has to be large enough to maintain the poling field during the switching of the ferroelectric polarization.

As demonstrated in Fig. 6, the thickness of the poled volume is determined by the electron-beam energy chosen, since the extension of the unpoled front volume is equal to the respective electron range [27].

In PVDF, the electron range and thus also the thickness of the unpoled volume scale with the 1.5th power of the electron-beam energy [27–29]. Piezoelectric polymer films with a non-uniform volume polarization are useful for flexure-mode or bending devices which can translate a relatively small electrical input into a significant mechanical displacement [28].

4.3 Determination of ferroelectric polarization

With high-field [26], corona [36], and electron-beam [27] poling, the current is influenced not only by the limiting properties of the method itself, but also by the switching properties of the ferroelectric polarization. This is demonstrated in Fig. 7 where the currents observed during a) high-field poling [26], b) positive corona poling [36], and c) electron-beam poling [27] of PVDF are depicted. As seen in Fig. 7, the sample current always exhibits the same characteristic behavior consisting of a peak followed by a relatively flat part which gradually becomes steeper, and finally an almost constant part.

This temporal current behavior can be explained as follows [7, 11]: 1) First the capacitance of the volume to be poled is charged; this also generates the required switching field across all or part of the sample volume. 2) When the ferroelectric polarization orients, a large current is necessary; therefore the current decays very slowly in the beginning and then gradually faster as more and more dipoles become reoriented at successively higher fields. This part is not found in samples without a ferroelectric polarization (dotted and solid lines in Fig. 7b)) or in already poled samples (curve II in Fig. 7c)). 3) The current becomes constant because of leakage through and around the sample.

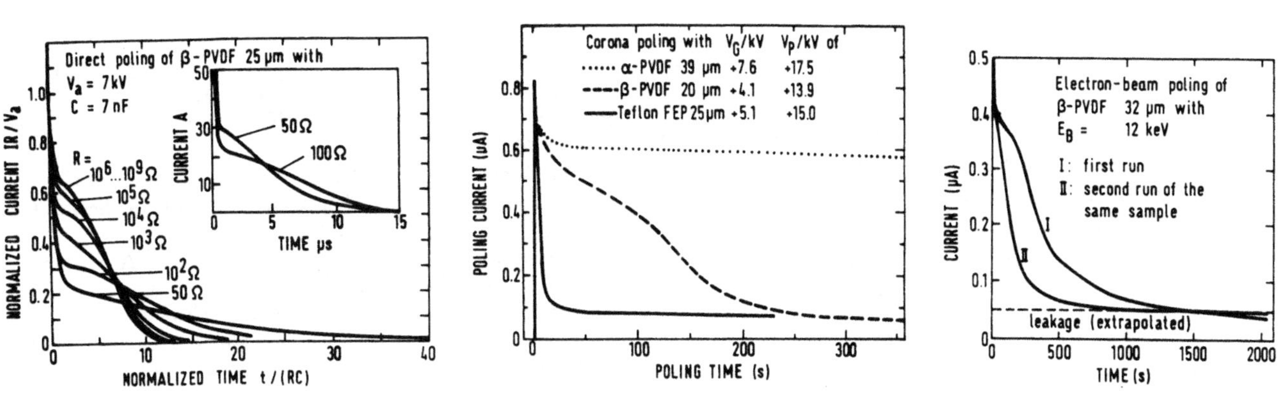

a) b) c)

Fig. 7. Poling currents in β-PVDF during a) direct high-field poling with different resistors in the circuit [26]; b) positive corona poling (Teflon and α-PVDF samples for comparison) [36]; and c) electron-beam poling (two runs on the same sample with compensation of the surface potential by means of a liquid contact in between) [27]

Whereas the qualitative behavior of the current is determined by the poling process itself, its time scale is governed by the current-limiting properties of the method used [7, 11]. This is directly shown in Fig. 7a) where the limiting resistance was chosen between 50 Ω and 1 $G\Omega$, and the time scale changed accordingly [7, 26]. On the basis of the above-described current behavior, the electric-field dependence of the ferroelectric polarization can be determined by means of model calculations [7]. The observed current curves may also be used to estimate the amount of polarization achieved during poling; apart from a simple integration, assumptions on leakage and charging of the sample are required for the estimation [7, 11, 26, 27, 36].

5. Conclusion

Techniques for the preparation and investigation of polymer electrets may also be used to characterize the respective polymer material. Examples briefly discussed here are the use of low-energy electron beams not only for surface charging, but also for the determination of the secondary-electron emission yield, the employment of laser-induced pressure pulses not only for the scanning of spatial charge and polarization distributions, but also for the measurement of sound velocities and ultrasonic attenuation constants, and the application of several poling methods not only for poling, but also for the determination of the ferroelectric polarization.

Other examples are the use of electron-beam charging for the investigation of electron ranges and for the generation of a virtual electrode, which permits mobility or conductivity measurements inside a sample without interface problems [1], and the "reversal" of the well-known heat-pulse technique, which is often employed for the measurement of charge profiles or centroids [1, 40–42], in experiments with known charge distributions from which temperature distributions in electrets and thermal-transport properties of polymeric electret materials may be determined [43].

Acknowledgements

The authors are indebted to Professor Dr. Bernhard Gross, Dr. Heinz von Seggern, Dr. Abdelghafour Berraissoul, and Mr. James E. West for stimulating discussions. Partial support by the Deutsche Forschungsgemeinschaft and by the Stiftung Volkswagenwerk is gratefully acknowledged.

References

1. Sessler GM (ed) (1987) Electrets 2nd Enlarged Edition. Springer-Verlag, Berlin Heidelberg New York
2. Hilczer B, Malecki J (1986) Electrets. Elsevier, Amsterdam
3. Sessler GM (1982) Charge Storage. In: Mort J, Pfister G (eds) Electronic Properties of Polymers. John Wiley & Sons, New York, pp 59–107
4. Wada Y (1982) Piezoelectricity and Pyroelectricity. In: Mort J, Pfister G (eds) Electronic Properties of Polymers. John Wiley & Sons, New York, pp 109–160
5. Lovinger AJ (1982) Poly(vinylidene fluoride). In: Bassett DC (ed) Developments in crystalline polymers – 1. Applied Science Publishers, London, pp 195–273
6. Gross B (1986) IEEE Trans Electr Insul EI-21:249
7. von Seggern H (1986) IEEE Trans Electr Insul EI-21:281
8. Lewiner J (1986) IEEE Trans Electr Insul EI-21:351
9. Sessler GM, Gerhard-Multhaupt R (1984) Radiat Phys Chem 23:363
10. Rozno AG, Gromov VV (1986) IEEE Trans Electr Insul EI-21:417
11. Gerhard-Multhaupt R (1987) Ferroelectrics 75:385
12. Gerhard-Multhaupt R (1987) IEEE Trans Electr Insul EI-22:531
13. Sessler GM, Gerhard-Multhaupt R (eds) (1985) Proc 5th Intern Symp Electrets, IEEE, New York; Das-Gupta DK, Pattullo AW (eds) (1988) Proc 6th Intern Symp Electrets, IEEE, New York
14. Lang SB (1981) Ferroelectrics 32:191; (1981) 34:239; (1982) 45:283; (1982) 46:51; (1983) 47:259; (1984) 61:157; (1985) 62:259; (1986) 67:223; (1987) 74:109
15. Gross B, von Seggern H, West JE (1984) J Appl Phys 56:2333
16. von Seggern H (1985) IEEE Trans Nucl Sci NS-32:1503
17. Gross B, von Seggern H, Berraissoul A (1987) IEEE Trans Electr Insul EI-22:23
18. Sessler GM, West JE, Gerhard R (1981) Polym Bull 6:109
19. Sessler GM, West JE, Gerhard R (1982) Phys Rev Lett 48:563
20. Sessler GM, West JE, Gerhard-Multhaupt R, von Seggern H (1982) IEEE Trans Nucl Sci NS-29:1644
21. Gerhard-Multhaupt R (1983) Phys Rev B 27:2494
22. Gerhard-Multhaupt R, Sessler GM, West JE, Holdik K, Haardt M, Eisenmenger W (1984) J Appl Phys 55:2769
23. Sessler GM, Gerhard-Multhaupt R, West JE, von Seggern H (1985) J Appl Phys 58:119
24. Sessler GM, Gerhard-Multhaupt R, von Seggern H, West JE (1986) IEEE Trans Electr Insul EI-21:411
25. Wang TT, von Seggern H, West JE, Keith HD (1984) Ferroelectrics 61:249
26. von Seggern H, Wang TT (1984) J Appl Phys 56:2448
27. Gross B, Gerhard-Multhaupt R, Berraissoul A, Sessler GM (1987) J Appl Phys 62:1429
28. Fukada E, Sessler GM, West JE, Berraissoul A, Günther P (1987) J Appl Phys 62:3643
29. Berraissoul A, Sessler GM (1987) unpublished results
30. Eisenmenger W, Haardt M (1982) Solid State Commun 41:917
31. Gerhard-Multhaupt R, Haardt M, Eisenmenger W, Sessler GM (1983) J Phys D: Appl Phys 16:2247
32. Holdik K, Eisenmenger W (1985) Charge and polarization dynamics in polymer films. In Proc 5th Intern Symp Electrets, IEEE, New York, pp 553–558
33. Sessler GM, Gerhard-Multhaupt R, West JE (1983) Nondestructive determination of ultrasonic velocity and absorption in

thin dielectrics by laser-induced pressure pulses. In Proc 11th Intern Congr Acoust Vol 2, GALF, Paris, pp 195–198
34. Gerhard-Multhaupt R, Sessler GM, West JE (1985) Ultrasonic velocity and absorption in thin polymer films. In Conf Proc Ultrason Intern 85, Butterworth, Guildford, Surrey, pp 317–322
35. Sessler GM, Gerhard-Multhaupt R, West JE (1986) Ultrasonic attenuation in the thickness direction of thin polymer films at frequencies up to 1 GHz. In Proc 12th Intern Congr Acoust Vol 2, Toronto, pp G7–5
36. Gerhard-Multhaupt R (1984) Polarisation of poly(vinylidene fluoride). In Proc 4th Intern Conf Diel Mater Meas Appl, IEE Conf Publ 239:53
37. Gross B, Giacometti JA, Leal Ferreira GF, Oliveira Jr. ON (1984) J Appl Phys 56:1487
38. Sessler GM, Berraissoul A (1987) Ferro Electrics 76:489
39. Gerhard-Multhaupt R, West JE, Sessler GM (1986) unpublished results
40. Collins RE (1975) Appl Phys Lett 26:675
41. Collins RE (1977) Rev Sci Instrum 48:83
42. Collins RE (1980) J Appl Phys 51:2973
43. Gerhard-Multhaupt R, Xia Z-F, Berraissoul A (1988) Determination of temperature distributions in electret foils. In: Proc 6th Intern Symp Electrets, IEEE, New York, pp 18–22

Received June 25, 1987;
accepted April 21, 1988

Authors' addresses:

Dr. R. Gerhard-Multhaupt
Heinrich-Hertz-Institut für Nachrichtentechnik
Einsteinufer 37
D-1000 Berlin 10, F.R.G.

Prof. Dr. G. M. Sessler
Institut für Elektroakustik
Technische Hochschule Darmstadt
Merckstr. 25
D-6100 Darmstadt, F.R.G.

Progress in Colloid & Polymer Science Progr Colloid Polym Sci 78:88–92 (1988)

Charge carrier photogeneration in polymers

S. Nešpůrek and V. Cimrová

Institute of Macromolecular Chemistry, Czechoslovak Academy of Sciences, Prague, Czechoslovakia

Abstract: The intrinsic charge carrier photogeneration in polymers is depicted as a multistage process which can be described within the frame of the ballistic model. Thermalization of the autoionized excited state proceeds in an isotropic quasi-continuum and therefore an initial distribution of radii of charge transfer states can be assumed. The dissociation of electron-hole pairs is in a good agreement with predictions based on the Onsager theory of geminate recombination. The pair separation distances, typically 2 to 3 nm, were observed in several polymers.

Key words: Charge carrier photogeneration, ballistic model, thermalization, dissociation of electron-hole pair, Onsager geminate recombination, poly(N-vinyl-carbazole), poly-(phenylacetylene).

Introduction

The potential use of polymers as photogenerators of free charge carriers in light-sensitive electronic devices, low-cost photovoltaic cells, as well as in electrophotographic recording systems has stimulated interest in their electronic properties. Photogeneration has been intensively studied for a long time, for example, in poly(N-vinylcarbazole) [1–5], polydiacetylenes [6, 7], polyacetylene [8, 9], polyphenylacetylene [10, 11], and in other polymers (see reviews [12–15]), as well as in molecularly doped polymers [16–19]. Several models, e.g., [20–22], were used to explain the photogeneration in organic materials. A major drawback in the application of any of the theoretical models to experimental results is the complex character of photoconduction processes in organics. Apart from the differences between the intrinsic and extrinsic process which are not always easily distinguished experimentally, complications also arise due to exciton-exciton, exciton-photon, or exciton-charge carrier interaction and trapping and space charge effects. Moreover, excited molecular states can dissipate their energy by fluorescence, phosphorescence, and radiationless deactivation and these processes therefore compete with photocarrier generation.

As found in poly(N-vinylcarbazole) (PVCA), the photogeneration efficiency strongly depends on the electric field [1, 2]. The field dependence was in agreement with predictions based on the Onsager theory [20]. Separation distances r_o of the bound electron-hole pairs have been reported by Pfister and Williams [1] as 2.2 and 3.0 nm for excitation into the first and third excited singlet states, respectively. The primary quantum yield η_o was found to be 0.14 charges per photon and independent of wavelength. Later, Borsenberger and Ateya [2] obtained 2.6 nm and 3.0 nm for the values of r_o with $\eta_o = 0.11$ charges per photon at 296 K on amorphous PVCA films, supplied the value of r_o for the second excited singlet state $r_o = 2.8$ nm, and determined the dependence of the primary quantum yield on temperature. The measurements were carried out with electric fields not exceeding $F = 10^8$ V m^{-1}.

Two questions should be answered concerning the above-mentioned problem: (i) is the Onsager formula suitable to describe the quantum efficiency vs applied voltage relationship for other types of polymers as well?; (ii) can the Onsager formula be used over a wider range of applied voltages (in particular for higher voltages)? To answer these questions we performed

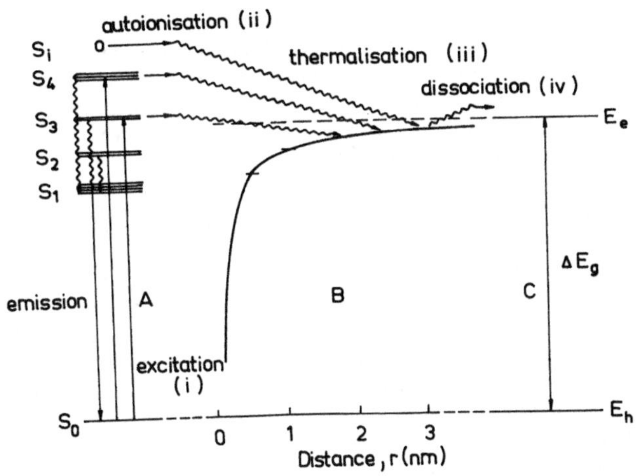

Fig. 1. Photogeneration steps according to the ballistic model: A: neutral electronic singlet states; B: charge transfer states of bound electron-hole pairs; C: ionized states; E_e: conduction electron level; and E_h = conduction level

measurements, not only on amorphous PVCA, but also on crystalline and plasma-polymerized samples of PVCA and samples of poly(phenylacetylene). The voltage dependence of the photogeneration efficiency was measured up to the electric fields $F = 3 \times 10^8$ V m^{-1}. As will be shown below, the conventional Onsager formula fits our experimental results reasonably well, although it fails at higher fields ($F > 10^8$ V m^{-1}).

Theoretical

The intrinsic charge carrier photogeneration in polymers can be depicted as a multistage process which involves (Fig. 1) (i) photoexcitation to a neutral molecular electronic state; (ii) autoionization of the excited state resulting in the creation of a localized hole and a hot quasi-free electron; (iii) thermalization of the electron due to fast inelastic scattering leading to the formation of a Coulomb field-bound geminate electron-hole pair (so-called charge-transfer (CT or pair state); and (iv) thermal dissociation of the electron-hole pair into free carriers by Brownian motion subject to a combination of Coulomb and applied fields. The fraction of absorbed photons that results in bound thermalized pairs is the primary quantum yield η_o which is often assumed to be independent of the applied. Under this assumption, the overall field dependence of photogeneration is governed by the efficiency of the dissociation step (iv) which may be described in terms of the Onsager [20] solution to the Smoluchowski equation. The overall photogeneration efficiency η may be calculated as

$$\eta\,(\vec{F}, T) = \eta_o \int F(\vec{r}, \vec{F}, T)\, g(\vec{r})\, d\tau \qquad (1)$$

where $f(\vec{r}, \vec{F}, T)$ is the dissociation probability of bound pairs separated by a distance $r = |\vec{r}_e - \vec{r}_h| = |\vec{r}|$; $\vec{r}_e$ and $\vec{r}_h$ are the position vectors of the electron and the hole, respectively; $g(\vec{r})$ represents the initial spatial distribution of bound pairs; and $d\tau$ is the volume element. Assuming that the distribution is spherically symmetrical and that all pairs are separated by the same distance r_o, we have $g = (4\pi r_o^2)^{-1}\,\delta(r - r_o)$.

The dependence of the potogeneration efficiency on the applied electric field F and on temperature T can be described by a modified Onsager formula (20) which can be rewritten in terms of Bessel functions [23] in the form

$$\eta(r_o, F) = \eta_o \left[1 - \frac{kT}{eFr_o} \sum_{j=o}^{\infty} I_j\left(\frac{e^2}{4\pi\varepsilon kTr_o}\right) I_j\left(\frac{eFr_o}{kT}\right)\right] \qquad (2)$$

where $I_j(x)$ is a modified Bessel function of the first kind, e is the unit charge, and other symbols keep their usual meaning. The Coulomb energy $E_b(r_o)$ of the two charges separated by the distance r_o can be approximated with a remarkably good accuracy as $E_b(r_o) = e^2/4\pi\varepsilon r_o$ (the expression is valid in the zero-field approximation).

The efficiency of dissociation depends on the initial distance between the oppositely charged carriers. The distance, according to the thermalization step, depends on the excess of excitation energy, thus forming the basis of the wavelength dependence of η. The process of separation involves the loss of the excess kinetic energy of the carrier pair by photon emission. In materials with mean free paths comparable with the interatomic spacing, the motion during this process is expected to be diffusive and the separation distance r_o at the end of a thermalization time t_t will be given by the relation [24] $r_o = (Dt_t)^{1/2}$, where D is the diffusion coefficient. From the theory of relaxation in inelastic scattering, the rate of loss of energy in an amorphous material is expected to reach its maximum at a phonon frequency ν_p multiplied by the typical phonon energy. The excess kinetic energy E_k to be dissipated is then taken as the excess energy over the local Coulomb potential [24], $E_k = h\nu - E_g + e^2/4\pi\varepsilon r$, where E_g is the energy gap. The time required to dissipate this energy is then $t_t = E_k/h\nu_p^2$. The electron-hole separation distance r_o at the end of the thermalization process is then $r_o = R_T(E_k/kT)^{1/2}$, with $R_T = (kTD/h\nu_p^2)^{1/2}$. When the thermalization of the autoionized excited state proceeds in an isotropic quasi-continuum, one should assume a distribution of radii of charge transfer states to exist. Thus, the g-function in Eq. (1) should be written in the form [25, 26]

$$g(r) = \frac{1}{\pi^{3/2}\alpha^3} \exp(-r^2/\alpha^2) \quad (3)$$

where α is the dispersion parameter. It may be demonstrated that the parameter r_o, formally describing the initial separation distance in the Onsager model (cf. Eq. (2)), is then field-dependent: at high fields it tends to α, while at low fields it becomes approximately 2α [27].

Experimental

1. Samples

PVCA was obtained by radiation polymerization and purified by multiple precipitation from benzene solutions into methanol. Amorphous films were prepared from cyclohexene-benzene solutions on a nickel substrate. To obtain a crystalline material, the amorphous films were heated for 3 h to 573 K under vacuum (10^{-3} Pa). The degree of crystallinity was 76%.

Plasma-polymerized films were deposited in a system consisting of two vertical parallel aluminium electrodes placed in a glass cylindrical vessel. Prior to deposition the reaction volume was evacuated by a diffusion pump to a base pressure of 10^{-3} Pa. Argon was then admitted to a pressure of 27 Pa and a discharge was excited by an ac signal supplied from an RC generator (frequency 40 kHz, power 9W, deposition time 7 min). N-vinyl-carbalzole monomer vapors were introduced into the plasma region from a resistively heated stainless steel evaporation source located 30 mm below the electrodes.

Poly(phenylacetylene) (PPA) was prepared by coordination polymerization of phenylacetylene, catalyzed by the metathetic $WOCl_4$ + Ph_4Sn catalyst dissolved in a benzene/dioxane mixture using a sealed glass vacuum apparatus (initial pressure of around 10^{-4} Pa). The resulting PPA was precipitated from the reaction mixture by methanol under vacuum, re-precipitated twice under nitrogen using the solvent system toluene-methanol, washed with methanol, and dried in vacuo at 320 K. This cis-transoidal form was predominantly obtained [28]. PPA films were prepared by spinning from a solution in xylene.

2. Measurements

The measurements of the photogeneration efficiencies were carried out using the technique of emission-limited photoinduced discharge [29, 30]. With this technique, a sample is charged to the initial voltage $U_o = Q_o/C$ (C being the capacitance per unit area) by a corona and then discharged upon illumination. The surface potentials were measured with a rotary electrodynamic condenser electrometer [31]. To minimize the influence of trapped-space charge in sample bulk, the samples were adapted in dark for 2 h between measurements. The photoinduced discharge quantum efficiency η' was calculated from formula [29]

$$\eta' = -(1/e\phi)(dQ/dt)_{Q_o} = -(1/e\phi)C(dU/dt)_{U_o} \quad (3)$$

where ϕ is the absorbed photon flux density and $(dQ/dt)_{Q_o}$ represents the rate of change of the surface charge density. Under emission-limited conditions and under the assumption that "Schubweg" is at least equal to the sample thickness, the photoinduced discharge quantum efficiency η' is independent both of sample thickness and of the intensity of light, and is equal to the photogeneration efficiency η.

For comparison, the PPA films were also tested by the technique of pulse photoconductivity. Samples SnO_2 | SiO | PPA | SiO | Au were illuminated by an Xe 457 micropulser (Xenon Corp.) in vacuo (10^{-4} Pa) and the photogenerated charge was measured as the current integral.

Results

The shape of the quantum efficiency-electric field relationship at room temperature for the positively charged surface of the crystalline PVCA films is shown in Fig. 2. Over the accessible range of electric fields, the results are consistent with predictions of the Onsager theory. The full line 1 represents the best theoretical fit to experimental points measured on crystalline PVCA with the following parameters: the effective charge carrier separation length r_o = 2.5 nm, the primary quantum efficiency of autoionization $\eta_o = 0.13$ charges per photon. The photogeneration quantum efficiency at zero electric field was 10^{-4} charges per photon. It is also apparent from Fig. 2 that at higher voltages the experimental values of quantum efficiences lie above the curve calculated using the common Onsager formula with a discrete value of r_o. The shape of the experimental dependence suggests that in the photogeneration process at higher fields, separation distances shorter than r_o predominate. The full line 2 represents the best theoretical fit of the data using a distribution of states that participate in the photogeneration mechanism. This in contrast with the usual ballistic description of thermalization where a single, definite initial charge separation distance r_o is introduced. The fit was obtained with the following parameters: the dispersion parameter $\alpha = 1.3$ nm, the primary quantum efficiency of autoionization $\eta_o = 0.6$ charges per photon.

Similar dependences were observed also for other types of polymers. The characteristic parameters of the fitting are summarized in Table 1.

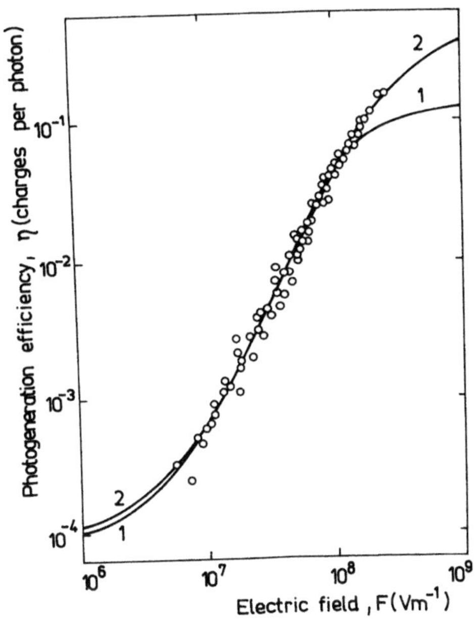

Fig. 2. Field dependence of the photogeneration efficiency in crystalline poly(N-vinylcarbazole) for 254 nm illumination. The data are representative of three samples with different thicknesses 1, 2, and 4μm. The solid line 1 is calculated from the conventional Onsager theory with $r_o = 2.5$ nm, $\eta_o^r = 0.13$ pairs per photon; line 2 is calculated using the distribution function (3) with $\alpha = 1.3$ nm ($\eta_o^\alpha = 0.6$ pairs per photon)

Conclusions

Although some details of the photogeneration mechanism in polymers are still not well understood, the intrinsic charge carrier photogeneration proceess and, particularly, the thermal dissociation of the electron-hole pairs can be described by the Onsager theory of geminate recombination (with the exception of polyacetylenes where the exciton [32] or soliton [33] mechanisms in quasi-one-dimensional medium [34, 35] seem to be more appropriate, and also for polydiacetylenes in which this mechanism has also been put in doubt [36]). In the region of higher fields the experimental values of quantum efficiencies cannot be usually fitted by curves based on the fundamental requirement of the diffusion theory, (in particular a definite initial charge separation distance r_o), because in the photogeneration process separation distances shorter than r_o predominate. Thus, a whole series of distributed states then participate in the photogeneration process. Under this assumption, we can get a very good agreement of the experimental results with the theoretical dependence of the photogeneration quantum efficiency on the electric field calculated on the basis of the Onsager theory.

Although the ballistic model represents a good enough description of the charge carrier photogeneration, some experimental data obtained need further and more detailed study. While it is true that in molecular crystals or organic liquids the geminate recombination of a coulombically bound geminate electron-hole pair is a diffusive process, (completed within typically 10 – 100 ps [37, 38]), a delayed field collection experiment on poly(N-vinylcarbazole) indicated a pair lifetime as long as 100 s [39]. This long-time recombination behavior is controlled by traps that extend to an energy $\gtrsim 0.6$ eV below the center of the distribution of intrinsic hopping states [40]. Possible candidates for this are incipientsandwich dimers active in excimer formation. The dynamic character of this trapping process is deduced from the thermally activated be-

Table 1. Characteristic parameters of photogeneration in poly(N-vinylcarbazole) (PVCA) and poly(phenylacetylene) (PPA)

	r_o (nm)	η_o^r	α (nm)	η_o^α	λ (nm)	
PVCA – amorphous	2.9	0.11	1.6	0.4	254	A
	2.7	0.14	1.4	0.80	355	A
PVCA – crystalline	2.5	0.13	1.3	0.6	254	A
	2.4	0.15	1.2	0.83	355	A
PVCA – glow-discharge polymerization	2.5	0.013	1.3	0.06	254	A
PPA – amorphous	3.5	0.02	1.7	0.16	254	A
	3.0	0.002	1.5	0.01	355	A
	2.0	0.02	–	–	280	B

A = data from the measurements of electrophotographic discharge; B = data from the pulse photoconductivity measurements [28]; λ = wavelength of light havior of η_o (ca. 0.05 eV). As follows from the thermally stimulated luminescence measurements [41], the self-trapping of charge carriers could be important in the photogeneration process.

Acknowledgement

The authors are indebted to Dr. L. Martinů for the preparation of the plasma-polymerized films.

References

1. Pfister G, Williams DJ (1974) J Chem Phys 60:2416
2. Borsenberger PM, Ateya AI (1978) J Appl Phys 49:4035
3. Okamoto K, Haya A (1984) Bull Chem Soc Jpn 57:1626
4. Regensburger PJ (1968) Photochem and Photobiol 8:429
5. Pearson J (1977) Pure and Appl Chem 49:463
6. Seiferheld V, Ries B, Bässler H (1983) J Phys C Solid State Phys 16:5189
7. Donovan KJ, Wilson EG (1986) J Phys C Solid State Phys 19:L357
8. Orenstein J, Vardeny Z, Baker GL, Eagle G, Etemad S (1984) Phys Rev B30:786
9. Kiess H, Keller R, Baeriswyl D, Harbeke G (1982) Solid State Commun 44:1443
10. Kang ET, Ehrlich P, Anderson WA (1984) Mol Cryst Liq Cryst 106:305
11. Kang ET, Ehrlich P, Bhatt AP, Anderson WA (1984) Macromolecules 17:1020
12. Stolka M, Pai DM (1978) Adv Polym Sci 29:1
13. Pennwell RC, Ganguly BN, Smith TW (1978) Macromol Rev 13:63
14. Biswas M, Das SK (1982) Polymer 23:1713
15. Biswas M, Uryu T (1986) J Macromol Sci Rev Macromol Chem Phys C26:249
16. Borsenberger PM, Contois LE, Hoestery DC (1978) Chem Phys Letters 56:574
17. Borsenberger PM, Contois LE, Ateya AI (1979) J Appl Phys 50:914
18. Goliber TE, Perlstein JH (1984) J Chem Phys 80:4162
19. Mort J, Pfister G (1979) Polym – Plast Technol Eng 12:89
20. Onsager L (1938) Phys Rev 54:554
21. Bounds PJ, Siebrand W (1980) Chem Phys Letters 75:414
22. Noolandi J, Hong KM (1979) J Chem Phys 70:3230
23. Mozumder A (1974) J Chem Phys 60:4300
24. Knights JC, Davis EA (1974) J Phys Chem Solids 35:543
25. Chance RR, Braun CL (1976) J Chem Phys 64:3573
26. Mozumder A (1974) J Chem Phys 60:4305
27. Nešpůrek S, Cimrová V (1985) 30th IUPAC International Symposium on Macromolecules, The Hague, Abstracts p 451
28. Pfleger J, Nešpůrek S, Vohlídal J (1988) Mol Cryst Liq Cryst, in press
29. Chen I, Mort J, Tabak MD (1972) IEEE Trans ED-19:413
30. Mort J, Chen I (1975) Applied Solid State Science. Academic Press, New York, 77
31. Nešpůrek S, Ulbert K (1975) Čs čas fyz 25A:144
32. Orenstein J, Vardeny Z, Baker GL, Eagle G, Etemad S (1984) Phys Rev B30:786
33. Tani T, Grant PM, Gill WD, Street GB, Clark TC (1980) Solid State Commun 33:499
34. Vardeny Z, Strait J, Moses D, Chung TC, Heeger AJ (1983) J Phys Colloq 44 C3:403
35. Shank CV, Yen R, Fork RL, Orenstein J, Baker GL (1982) Phys Rev Lett 49:1660
36. Moses D, Sinclair M, Heeger AJ (1987) Phys Rev Lett 58:2710
37. Shein LB, Anderson RW, Enck RC, McGhie AR (1979) J Chem Phys 71:3189
38. Braun CL, Scott T (1983) J Phys Chem 87:4776
39. Mort J, Morgan M, Grammatica S, Noolandi J, Hong KM (1982) Phys Rev Lett 48:1411
40. Stolzenburg F, Ries B, Bässler H (1987) Material Science 13:259
41. Tale IA, Butlers PI, Pospíšil J, Nešpůrek S (1987) Fifth International Seminar on Polymer Physics, High Tatras

Received Juneä 26, 1987;
accepted April 21, 1988

Authors' address:

S. Nešpůrek
Institute of Macromolecular Chemistry
Czechoslovak Academy of Sciences
162 06 Prague 616, Czechoslovakia

Self-trapping of charge carriers in polymers: A comparative study of poly(p-phenylene) and poly(N-vinylcarbazole)

P. Butlers[1]), I. Tale[1]), J. Pospíšil[2]), and S. Nešpůrek[3])

[1]) Institute of Solid State Physics, Latvian State University, 063 Riga, USSR
[2]) Department of Polymer Physics, Faculty of Mathematics and Physics, Charles University, Prague, Czechoslovakia
[3]) Institute of Macromolecular Chemistry, Czechoslovak Academy of Sciences, Prague, Czechoslovakia

Abstract: The electronic relaxation processes in poly(p-phenylene) (PPP) and poly(N-vinylcarbazole) (PVCA) were studied. The fractional thermostimulated luminescence gave evidence of two different thermoactivated processes in PPP and PVCA, taking place in different temperature regions. Fractional thermostimulated luminescence results have been discussed in terms of self-trapping of charge carriers. In PVCA the self-trapping of charge carriers proceeds via formation of dimer cation radicals.

Key words: Self-trapping of electronic excitation, fractional thermostimulated luminescence, hole drift mobility, poly(p-phenylene), poly(N-vinylcarbazole).

Introduction

Organic polymers in which self-trapping of electronic excitation is observed can be divided into two groups. Firstly, there are the so-called conducting polymers (polyacetylene) where self-trapping results from Paierls instability of the quasi-onedimensional conjugated polymer chain. It is well known that in such polymers self-trapping occurs by the formation of different kinds of solitons, namely kinks and antikinks in trans-polyacetylene [1] and polarons, bipolarons and polaron excitons in other systems [2] (Fig. 1a). Generally, such self-trapped electronic excitations represent confined king-antiking pairs in different charge states and forming deep local energy levels inside the gap. A polymer with saturated bonds in the backbone and with aromatic pendant groups (polystyrene, poly(N-vinylcarbazole), and some others) should be referred to the second group. We have shown that in the latter case the self-trapping of electronic excitations occurs by the formation of excimer or dimer radical ion, respectively, which is limited by the thermoactivated torsional vibrations of the polymer pendant groups.

Materials and Methods

The aim of this paper is to achieve a detailed picture of the electronic relaxation processes in two different systems: in a quasi-onedimensional conjugated polymer, i. e. poly(p-phenylene) (PPP), and in poly(N-vinylcarbazole) (PVCA), which represents a polymer with saturated bonds in a chain and with aromatic side groups. The method of fractional thermostimulated luminescence (FTSL) was used and made possible the determination of the mean activations energy $\langle E\rangle$ and the mean frequency factor $\langle s\rangle$ of the trapped charge carriers. The FTSL is based on the sample temperature superposed on a certain heating function [3]. The basic relations are

$$\langle E\rangle = -\frac{d(\ln I)}{d(l/kT)}, \qquad (1)$$

where I is the intensity of thermostimulated luminescence, and T is the temperature of one measuring cycle, and

$$\langle s\rangle = \frac{dT}{dt}\left(\frac{E}{kT^2} - \frac{1}{I}\frac{dI}{dT}\right) \exp\left(\frac{\langle E\rangle}{kT}\right). \qquad (2)$$

Additional information of the relaxation process is obtained from measurements of the spectral distribution of thermostimulated luminescence (TSL)) and the photoluminescence, as well as from the luminescence decay kinetics and its temperature dependence.

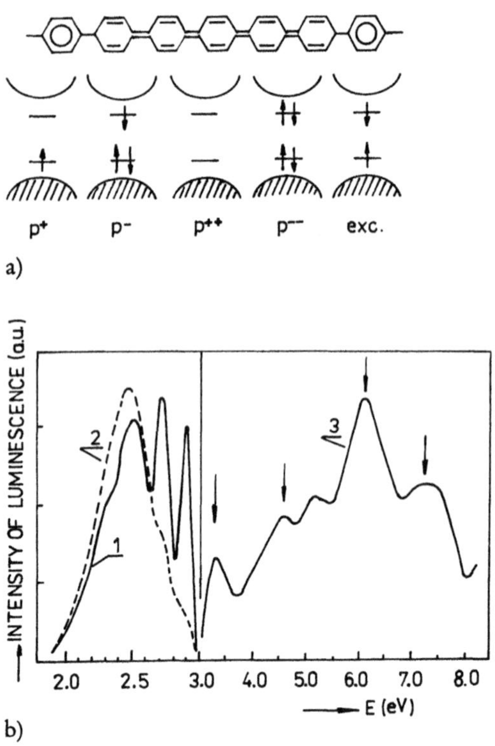

Fig. 1. a) The schematic diagram of different kinds of polaron excitations. b) The photoluminescence spectrum of poly(p-phenylene) at nitrogen temperature (curve 1) and room temperature (curve 2). Curve 3 shows the luminescence excitation spectrum

Results and discussion

The spectral distribution of photoluminescence of the PPP thin films was recorded at a liquid nitrogen and room temperatures. We found at low temperature photoluminescence with a well pronounced vibrational structure, in which the local mode with the energy 0.19 eV is distinguished, coinciding with one of the polaron local modes [4]. The luminescence excitation spectrum has maxima at photon energies 3.2, 4.6, 6.1 and 7.2 eV, correlating with the π-electron transition energy in the conjugated chain of the polymer, as well as with the energies in the π-electron system of an individual phenyl ring. The spectral distribution of photoluminescence in PPP at room temperature is similar to the exciton luminescence reported in pentaphenyl molecular crystals [5]. This suggests that photoluminescence in PPP originates from the annihilation of the self-trapped exciton, the hole component of which is a molecular radical cation or hole polaron, localized at the five phenyl rings along the polymer chain (Fig. 1b).

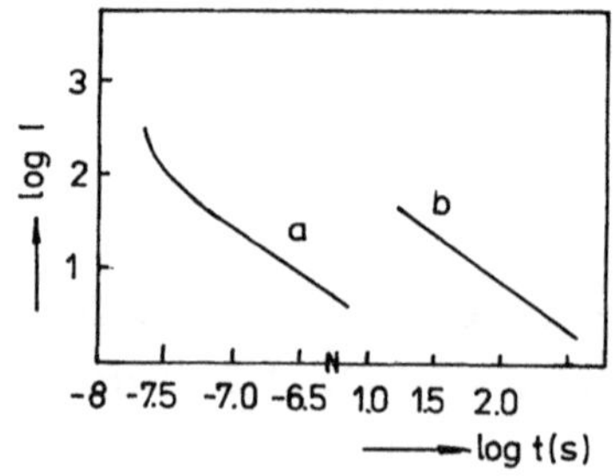

Fig. 2. The luminescence decay kinetics after excitation by pulse of N_2 laser at 80 K (curve a) and excitation by X-rays at 120 K (curve b)

The luminescence decay kinetics (see Fig. 2) after excitation by a short pulse (10 ns) of the nitrogen laser at 80 K has a prolonged component which can be approximated by Becquerel's law

$$I \sim t^{-\alpha} \tag{3}$$

where the exponent α is approximately one. The same plot is observed after excitation of the polymer by X-rays at 120 K and a subsequent jumpwise cooling to the liquid nitrogen temperature. This indicates the tunnelling interchain recombination mechanism of electron and hole polarons with the formation of self-trapped excitons in the polymer chain.

Two stages of the electronic relaxation process have been distinguished in X-ray irradiated PPP samples by FTSL (see Fig. 3). The low temperature stage is characterized by a continuous distribution of the activation energy; the frequency factors could be a result of recombination in the closely situated electron and hole, limited by their thermoactivated migration,

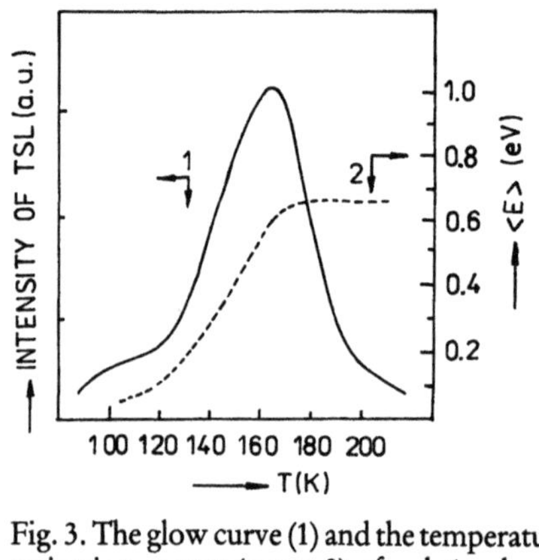

Fig. 3. The glow curve (1) and the temperature dependence of mean activation energy (curve 2) of poly(p-phenylene)

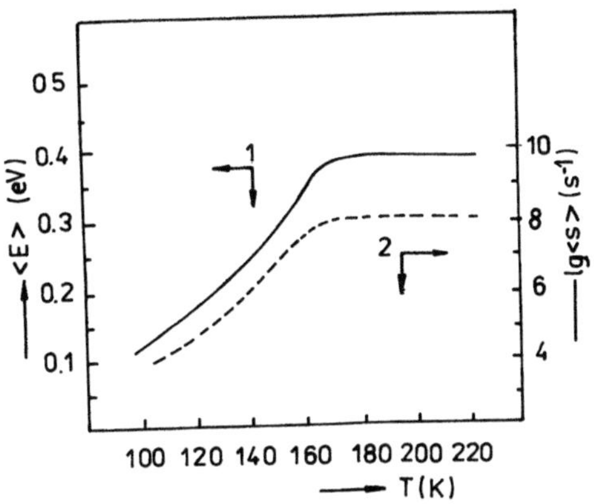

Fig. 4. The temperature dependence of mean activation energy (curve 1) and mean frequency factor (curve 2) of poly(N-vinylcarbazole)

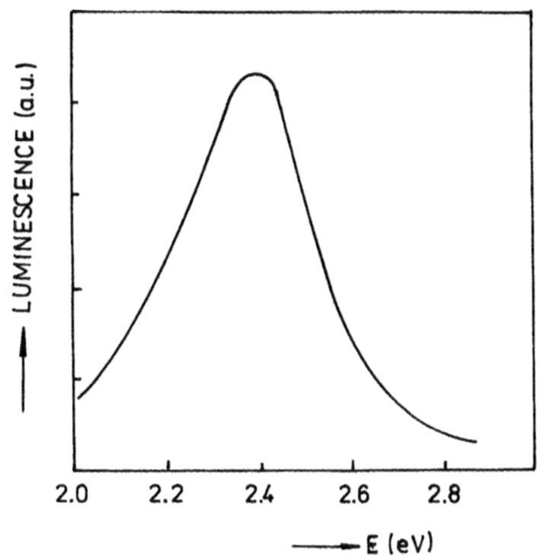

Fig. 5. The spectral distribution of thermostimulated luminescence of poly(N-vinylcarbazole)

incorporating the electron tunnelling between pairs of localized states of both polaron and bipolaron (the so-called interpolaron hopping). The high temperature stage of FTSL is characterized by a constant activation energy 0.65 eV, determined by recombination limited by the interchain hopping of the self-localized charge carriers.

Two temperature regions of stimulation of the electronic relaxation process in X-ray irradiated PVCA samples can also be observed (see Fig. 4). Starting at a temperature of c. 160 K, the activation energy and the frequency factor of the relaxation process are practically stable, being 0.4 eV and 10^{-8} s^{-1}, respectively. The activation energy in the high temperature region is close to the activation energy of the hole drift mobility. Lower values of the frequency factor suggest that the recombination luminescence is controlled by the diffusive migration of the trapped holes. The spectral distribution of TSL (see Fig. 5) consists of a luminescence band with a maximum near 2.4 eV which coincides with the spectral characteristics of the phosphorescence of excimers. This fact suggests that the relaxation process ends by the formation of a triplet state of an excimer and that is a dimer radical cation. The length of the thermoactivated hopping of such a localized hole can be estimated by means of an equation for the pre-exponential factor of mobility

$$\mu_o = \frac{ea^2}{6kT} \nu_{ph} \exp\left(-\frac{2a}{R_B}\right), \tag{4}$$

where $\nu_{ph} \approx 10^{13}$ s^{-1} is the phonon frequency, $R_B \simeq \hbar/\sqrt{2mE}$ is Bohr's radius of the wave function of a localized hole with the bonding energy 0.4 eV, the mass m is approximately equal to the mass of a free electron, and $\mu_o \simeq 10^{-6}$ m^2 V^{-1} s^{-1} [6]. Such a simple estimate of the length of the hole thermoactivated jump yields the value of about 0.4 nm, which is close to the distance between the neighbouring carbazole groups. Consequently, the self-trapped hole which forms a dimer radical cation in PVCA at temperature above 170 K migrats diffusively and ends by the radiative recombination with the trapped electron. It can be shown that such a self-trapped hole migration is limited by the thermoactivated torsional vibrations of the polymer pendant groups. According to the reference data, dielectric losses corresponding to the relaxation of PVCA occur at 210 K (at 10 kHz); an activation energy of c. 0.35 eV was observed. This indicates that the time of γ-relaxation at room temperature is about 10^{-6} s, which is equal to the time needed for the self-localization of an excess charge (hole) in the polymer. We can assume that the process of γ-relaxation in PVCA includes both the simple torsional vibration of carbazole pendant group about the C-N bond and the torsional wagging motions of the carbazole groups [7].

Conclusion

The reported investigations show that the self-trapping of charge carriers can occur both in polymers with conjugated backbone groups and those with saturated aromatic groups in the main chain. In the first case, the self-trapping of charge carriers proceeds via formation of dimer cation radicals.

References

1. Su WP, Schrieffer JR, Heeger AJ (1980) Phys Rev B 22:2099
2. Brazovskii SA, Kirova NN (1981) Soviet Phys JETP Lett 33:6
3. Tale I (1981) Phys Status Solidi A66:65
4. Yaniger SI, Rose DJ, Mc Kenna MP, Eyring EM (1984) Macromolecules 17:2579
5. Matvejeva EN, Medvedediev MN, Rubina OG, Shapranov MD (1963) Izv AN USSR Ser Phys 27:763
6. Pfister G, Griffiths GH (1978) Phys Rev Lett 40:659
7. Pochan JM, Hinman DF, Nash R (1975) J Appl Phys 10:4115

Received February 1, 1988;
accepted April 21, 1988

Authors' address:

Dr. J. Pospíšil
Faculty of Mathematics and Physics
Charles University
V Holešovičkách 2
180 00 Praha 8, CSSR

Progress in Colloid & Polymer Science Progr Colloid Polym Sci 78:97–108 (1988)

Charge transport in polymers characterized by geometrical restrictions

I. Chudáček

Charles University, Department of Polymer Physics, Prague, Czechoslovakia

Abstract: The main results of our investigations are described covering the charge carrier transport in polyvinylcarbazole thin layers doped with monomer-like molecules N-vinylcarbazole. With the enhancement of the monomer concentration two steps are observed in the carrier mobility, the conductivity, and trapping properties. This behavior can be analyzed by the introduction of the transport network determined from the molecular superstructure of the polymer system by the introduction of the local hopping space. The fractal approached was used for the characterization of the transport network. We propose the introduction of transport tubes to describe the carrier transport in disordered sytems.

Key words: Charge transport layer, disordered system, fractal dimension, hole mobility, transport network, polyvinylcarbazole.

Introduction

The charge carrier transport in polymers has been measured by many authors. Fundamental works on polymer model system polyvinylcarbazole (PVK) were performed by Pai [1] and Gill [2]. The authors found that the charge transport was carried out by holes with very low drift mobility $\mu_D = 10^{-12}$ m^2/Vs, (for reviews see Gill [3], Brehmer [4], Mort and Pfister [5]). In order to explain the diffusive motion of charge carriers, the main difficulty is to characterize the effects of the structural disorder. The system disorder produces local variations of the wavefunction overlaps between neighboring hopping sites (HS), which control the exchange rate of a charge carrier ("off diagonal" or "geometrical" disorder). The other consequence of a geometrical disorder is the spatial fluctuaction of the electrostatic interaction valuable for a charge carrier, or for an excitation localized at a HS with its environment [6] ("diagonal" or "energetical" disorder). On the other hand each disorder produces electronic trapping-centers and the observed dispersion in the time-of-flight t_F from one electrode to the opposite can be given by the different trapping time for carriers. When the distribution of trapping times is constant in time we have a Gaussian transport controlled by traps [7, 8] and μ_D is not time dependent during the flight of the carier. When $\mu_D = \mu_D(t)$, then we have a continuous time random walk model [9], which is influenced by the geometrical and energetical disorder. The resulting distribution $g(t_F)$ of t_F is in this case stochastic.

It is now accepted that between HS these occurs a transfer of charge by carrier hopping among the localized states associated with the HS. The nature of these HS is a function of the nature of the chromophoric group of the polymeric monomer [2] (the carbazole group in PVK), or with doping molecules [2, 10]. In these last systems it is possible to chemically control the polarity of the moving charge and the value of the sample conductivity.

The charge transport has been extensively studied by the time of-flight (TOF) technique [11], where the current induced by an injective sheet of carriers drifting in a uniform field of a capacitively charged film is resolved in time. It has been found that the molecular mobility influences the drift mobility μ_D of holes, namely its field dependence [12, 13]. A study by Bässler [14] shows that the polymeric system with the charge transport can be represented by model systems with disorder modulated carrier motion not affected by deep trapping.

The aim of this paper is to use the above mentioned results and concepts and also our results with the system where PVK is doped by the monomeric low molecular mass N-vinylcarbazole (NVK [15]). We are going to interpret the observed effects by the assumption that the transport of charge carriers is more influenced by geometrical restrictions than by energetic interactions. It is now possible to characterize a given "amorphous" state for a disordered system using the fractal dimension [16,17]. Using this modern fractal approach we discuss the transport phenomenon in amorphous PVK.

2. The charge carrier transport controled by one type of traps

It is known [18] that each HS is only responsible for one type of localized energy level W_h in the corresponding energy diagram of the polymer we discuss. In the following we dont take into account the small blurring of W_h and other local energy levels caused by density fluctuactions. The possibility of the carriers to hop to a neighboring HS is given by the reciprocal value of the time of life τ_h in one HS. This time of life value is also controlled by temperature. In the interval of temperatures between the start of gamma movements (or beta) and the glass transition temperature T_g, the drift mobility of the carriers is given by thermal activated hopping [18]. This means that a potential barrier W_h appears between neighboring HS.

Let us to assume that the traps in question are only of one type. Corresponding energy levels to traps are W_T. The relative concentration of traps versus HS is $[T]$, the life time of holes in traps is τ_T and the activation energy necessary for the trapped hole to leave is W_T. Now the kinetics equations for non-equilibrium photogenerated holes may be written. The electron transport is not considered because the electrons are captured by acceptors in the system [2] and we assume a low concentration of acceptors.

If only a small charge is in motion then the local electric field does not change and the drift mobility and velocity of holes can be obtained from simple formulas

$$v_D = \frac{d}{t_F}, \mu_D = \frac{d^2}{U_o \cdot t_F}. \tag{1}$$

where d is the space separation between electrodes and U_o the applied voltage across the layer.

The best electric field (E) and temperature (T) dependence of μ_D was proposed by Gill [2], and was often measured by different authors, i.e.,

$$\mu_D = \mu_{Do} \cdot \exp\left(-\frac{W_D - \beta\sqrt{E}}{k\,T_{ef}}\right). \tag{2}$$

where

$$\frac{1}{T_{ef}} = \frac{1}{T} - \frac{1}{T_o}$$

and μ_{Do}, T_o are material constants, k is the Boltzmann constant, and W_D is the activation energy of the drift transport.

Now we consider the concentration $p(t)$ of holes actually sitting in HS and the concentration of holes captured in traps $[T^+]$. We can write for certain time t, smaller than t_F, equations

$$\frac{dp}{dt} = \frac{[T^+]}{\tau_T} - \frac{p}{\tau_V}, \frac{d[T^+]}{dt} = \frac{p}{\tau_V} - \frac{[T^+]}{\tau_T} \tag{3}$$

where τ_V and τ_T are the life times of holes on HSs and traps. The rate of the hole trapping and detrapping are $(\tau_V)^{-1}$ and $(\tau_T)^{-1}$. The number of holes P photogenerated during the excitation time t_g given by a light pulse of intensity I is

$$P = \alpha\, I_g\, t_g, \tag{4}$$

where α is the quantum efficiency of the electron-hole pairs' photogeneration. For time $t < t_F$ the charge conservation law gives

$$p(t) + [T^+] = P. \tag{5}$$

For simplicity we take these assumptions

$$[T] \gg [T^+], p(0) = P, [T^+(0)] = 0, \tau_T > \tau_V > \tau_{VF}$$

where τ_{VF} is the life time of holes in HS for time $t > t_F$ which take into account the escape of holes throught the opposite electrode.

An analogue set of kinetic equations (3) may be written for time $t > t_F$, but Eq. (5) is not more valid. In this case the charge conservation law gives relation

$$p(t) + [T^+(t)] = P\left(1 - \int_0^t g(t_F)\, dt_F\right) \tag{6}$$

where

$$\int_0^\infty g(t_F)\, dt_F = 1;$$

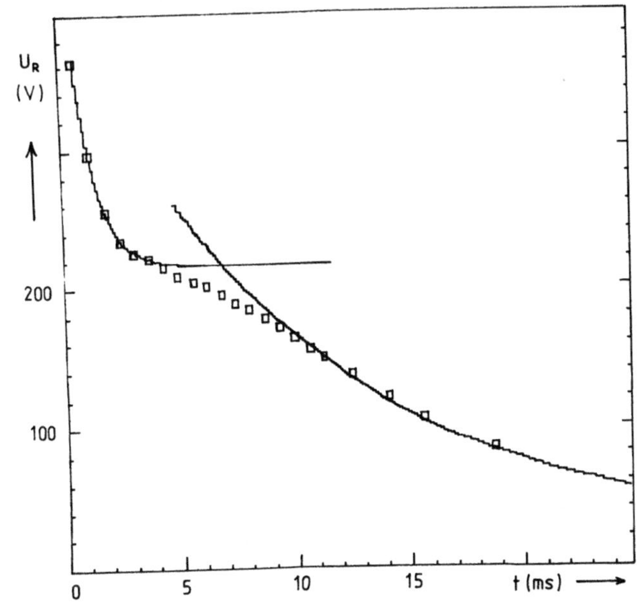

Fig. 1. The photoresponse $U_R(t)$ potential drop on loading resistance R after a delta excitation pulse for pure PVK at room temperature. The electric field across the layer is $E = 3.10^7$ V/m

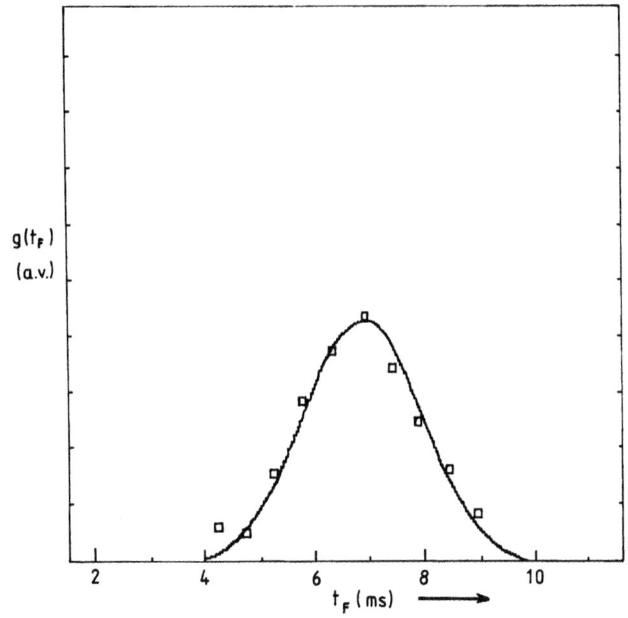

Fig. 2. The distribution of time-of-flight $g(t_F)$ in the case of Fig. 1.

$g(t_F)$ is the distribution of the time-of-flight. The solution of the kinetic equations (3) has the form

$$p(t) = \frac{P}{(\tau_T + \tau_V)}\left[\tau_V + \tau_T \cdot \exp\left[-\left(\frac{1}{\tau_V} + \frac{1}{\tau_T}\right)t\right]\right]$$

$$= \frac{P}{\tau_T}\left[\tau_V + \tau_T \cdot \exp\left(-\frac{t}{\tau_V}\right)\right] \tag{7}$$

for time $t < t_F$. The solution of the kinetic equations for time $t > t_F$ is not easy because $g(t_F)$ is unknown. For larger time the right side of Eq. (6) goes to zero and the time dependence of $p(t)$ is given approximately by

$$p(t) \doteq p(t_F) \cdot \exp\left(-\frac{t - t_F}{\tau_{VF}}\right). \tag{7}$$

In Fig. 1 the time photoresponse is shown as a voltage $U_R(t)$ across some load resistance R, which is proportional to $p(t)$ (see Eq. (32)). It seems that our model is well-suited for the photoresponse of a thin layer of PVK cast from solution. In Fig. 1 we can see a difference of experimental p/exp/ (□) and calculated p/cal/ (line) from Eq. (7) hole concentration. Because p/cal/ for $t < t_F$ does not take into account the escape of holes from the layer, the difference p/cal/ − p/exp/ must be proportional to the number of carriers for the time at which the flight is accomplished. That means this difference is proportional to the distribution $g(t_F)$. For the same reasons p/cal/ − p/exp/for $t > t_F$ is also proportional to $g(t_F)$. We have relations

$$\int_0^t g(t_F)\, dt_F = C\,(p_{\text{cal}} - p_{\text{exp}}),\ \int_0^\infty C \cdot g(t_F)\, dt_F = P \tag{8}$$

where C is a normalization constant. In Fig. 2 the integral of $g(t_F)$ is pictured. It follows from this figure that the $g(t_F)$ is not very dispersive for PVK cast from solution.

We assume that the reciprocals of the life times $(\tau_V)^{-1}$, $(\tau_T)^{-1}$, which give us the rates of a hole's trapping an detrapping, we have for $T \ll T_g$ an Arrhenius-like temperature dependence

$$\frac{1}{\tau_V} = \exp\left(-\frac{W_h}{kT}\right), \frac{1}{\tau_T} = \exp\left(-\frac{W_T}{kT}\right). \tag{9}$$

From measurements obtained at different sample temperatures the Arrhenius plot gives us

$$W_h = W_T = 0.41 \text{ eV}; \quad W_D = 0.56 \text{ eV}.$$

These values has been determined from the extrapolation to $E = 0$ for pure PVK. This result may be

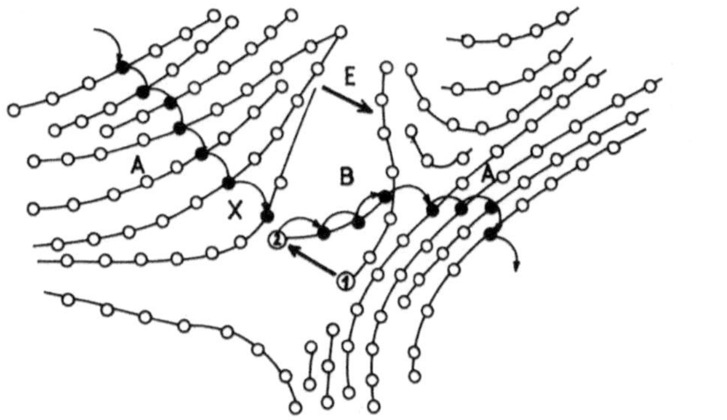

Fig. 3. The hopping of one carrier. The carrier hops in A region and stops its movement in X when the chain of B region is in position 1. After a conformation change: $1 \rightarrow 2$, the carrier passes through B and continues in the next A region. ●, ○ visited and not visited HS

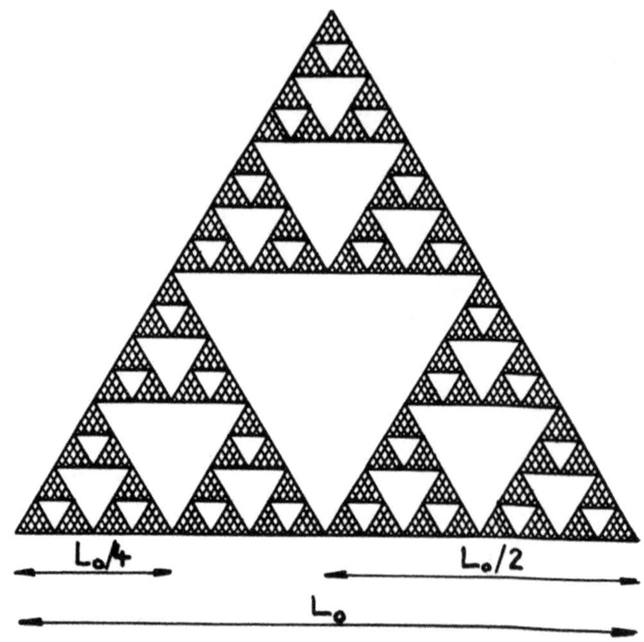

Fig. 4. One type of Sierpinski gasket

interpreted so that the traps in our case are of the conformation type [5]. In Fig. 3 we present a carrier hopping under the influence of an applied field E in a polymer sample. At the beginning the carriers are in the area A, where the hopping is unencumbered. At the interface boundary with area B (which is not as dense as area A), the carriers do not find a convenient HS. However, after some time and under the influence of molecular movements, a new conformation of polymer chains occurs. This raises the possibility of the forward prolongation of the carrier drift in the field direction.

As the conclusion of our experimental results we can affirm that in polymer systems like PVK the charge carrier transport is determined by the existence of two types of regions: the high density (A) and low density (B). At the interface of A and B areas conformational type traps exist where the carriers must wait. This waiting time is necessary for an appropriate change of the local conformation of polymer chains. The existence of high- and low-density areas in solid polymers is an old theoretical and experimental result e.g., theories of aggregates, and a meander model [19]. It is not surprinsing that this fact plays an important role as well in the carrier transport of polymers.

3. The geometrical approach to the transport problem

In any transport problem we have some initial distribution of moving particles and after the application of external forces during some interval of time another distribution of these particles is established. The particles move in a space that is more or less structuralized. To resolve the transport problem two ways can be followed;

i) the particle movement is governed by the applied forces and existing geometrical obstacles are taken considered as passive forces;

ii) the particle movement is mainly given by the geometrical structure of the space where the transport occurs. The applied forces "curve" the transport space.

The second approach (ii) will be used in this paper. This case is, for example, valid when you drive your car rush hour. In this case, the time that is needed to reach your destination is not a function of the power of your car, but it is mainly limited by the "geometrical restrictions" of the traffic.

Because the movement has no symmetry and is, in general, disordered, it will be convenient to employ the fractal approach [20]. In the beginning we introduce geometrical fractals and their dimension. Considere an unusual two-dimensional object: the Sierpinski gasket (Fig. 4) [20]; we show a photograph of the presented triangle; on a length scale L_0 it appears to be solid, but as we enlarge the negative we detect that it has holes of triangular shape. At scale L_0 we observe density 3/4, at scale $L_0/2$ density $(3/4)^2$, at scale $L_0/4$ density $(3/4)^3$, and so on. If we consider that the mass

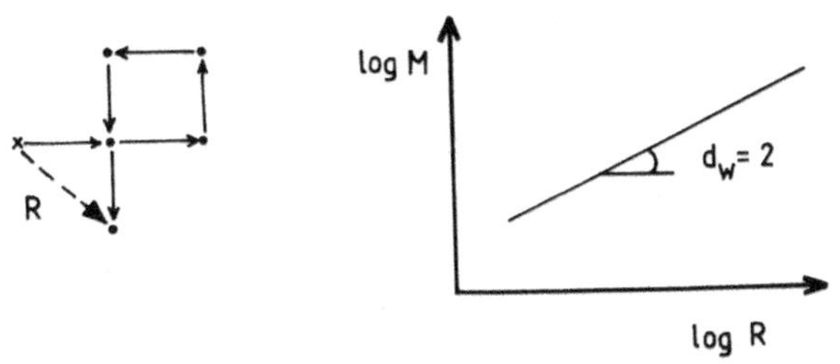

Fig. 5. A random walk for $M = t = 6$ steps [20]

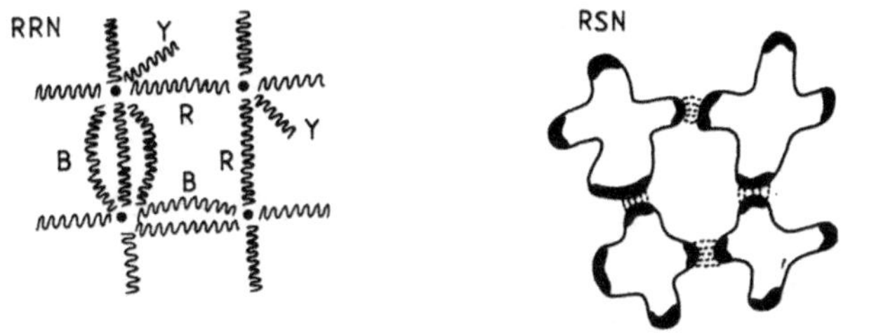

Fig. 6 The conductivity of a general two-component random mixture: (a) RRN-R, B, and Y are "red", "blue", and "yellow" bonds; (b) RSN: . . . "anti-red bonds" [20]

M is proportional to the occupied surface S we have between two neighboring scalings the relation

$$M(L) = 3\, M(L/2) \tag{10}$$

and we define the dimension of this fractal by the equation

$$M(L) = 2^{d_f}\, M(L/2) \tag{11}$$

or for a universal case

$$M(\lambda L) = \lambda^{d_f}\, M(L) \tag{12}$$

For our example $2^{d_f} = 3$, we have $d_f = \log^3/\log^2 =$ 1585.

A second class of fractals is the statistical ones. The simplest example of this type is an N-step random walk on a square lattice (Fig. 5). The walker takes one step per time unit so that $N = t$. At each step there are four choices for the walker for the direction of the next step. This corresponds so that at time t there are a total of $4\,t$, distinct configurations. Most of these configurations are not fractals, however, certain averaged quantities obey relations identical to those valid for the geometrical fractals. Now we use the distance Rafter t steps from the origin as the scale. The average values are $L(t = 1) = 1$, $L(2) = \sqrt{2}$, and $L(3) = \sqrt{3}$ .... In general we have

$$L(t) = \sqrt{t}\,. \tag{13}$$

To better imagine the outlined concept, a monomer unit is placed after each step and the steps correspond to chemical bonds between monomers. In this case the walker traces out a "free flight" polymer which ignores the impenetrability of given monomers. The total mass M after t steps is proportional to t; it is obtained from Eq. (13)

$$t \sim M(L) \sim L^2. \tag{14}$$

Hence, for the random walk, we can write the functional equation

$$M(L) \sim L^{d_w} \tag{15}$$

where the fractal dimension of the random walk is $d_w = 2$.

The transport of charge carriers in a polymeric system can be simplified by the model of an "ant in a labyrinth", or by the "termite problem" [21]. In both cases the transport occurs in a space which is a mixture of good (A) and poor (B) conductive regions as depicted in Fig. 3. The respective electrical conductivity is valid in that $\sigma_A \gg \sigma_B$. For $\sigma_B \to 0$, $\sigma_A \neq 0$ we have a random resistor network (RRN) represented by an ant (= the carrier) in a labyrinth. For $\sigma_A \to \infty$, $\sigma_B \neq 0$ we have a random superconductive network (RSN) represented by the termite (= the carrier) in a wood labyrinth. By the change of the time scale it is possible to demonstrate that the "transport physics" in A and B regions is the same and the observed processes occur only at different time scales [22].

For the transport behavior the mean point is the existence of a multitude of transport centers through which the transport occurs. In our case the transport centers are hopping sites (HS) that must be organized in a network placed between the electrodes. For the RRN case Fig. 6 we have a transport network where the nodes are, of course, the HS. These nodes are convected among them selves by only one bond (designated the red bond), or more than one bound (blue bonds). Another possibility is dangling (yellow) bonds

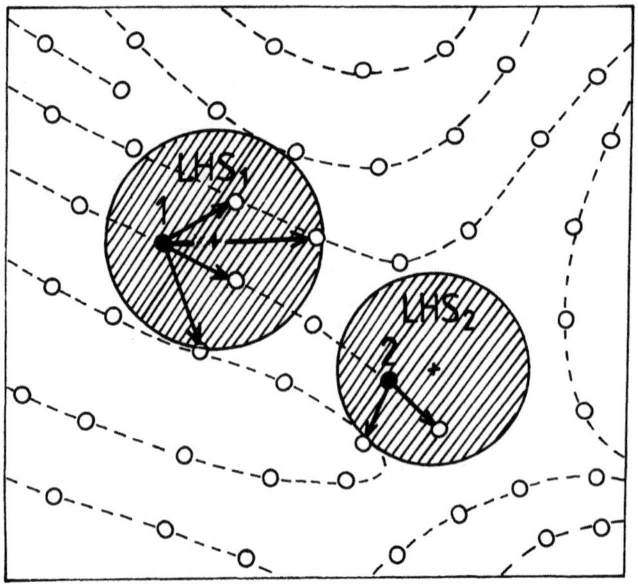

Fig. 7. The representation of two local hopping spaces (LHS) for two HS of a polymeric structure. The arrows show the possible hops from HS 1 and 2

[20]. These bonds represent the possibilities between HS. The red bonds are very important for transport because through them pass many carriers (these bonds are heated, therefore they are red). In the RSN case the transformation of point nodes to clusters of HS occurs. At the places of transport active points of the cluster surface (unscreened perimeter [20]) are localized the so-called anti-red bonds which are the complement of red bonds.

It is convenient to understand the resistance between two points as the "mass" of 1 ohm resistors that we would place in series between the two points in order to obtain the same resistance. This lot of resistors N_R is proportionate to the distance L between the two points raised to the power of d_R, which is the fractal dimension of our resistor network. We have

$$\text{Resistance } \overline{XY} \sim N_R\,(\overline{XY}) \sim L^{d_R} \tag{16}$$

where $d_R = -d_V$ for RSN and $d_R = d_W - d_f$ for RRN; d_U is the fractal dimension of the unscreened (stick out) perimeter of the HS cluster present between X and Y; d_f is the fractal dimension of the transport network, and d_W the fractal dimension of the random walk between X and Y.

4. The geometrical interpretation of the carrier transport

In order to use the above introduced fractals for the explanation of our charge transport results, one must transform the superstructure (the physical structure) of the polymer system in the transport network. For this purpose we must at first determine the HS. Secondly we must build the transport network as a number of connections between HS which represent the possible carrier local hops. To realize this we introduce the notion of the local hopping space (LHS): this is the space around each HS that determines the possible hops from the given HS to the neighboring HS. All HS present in the LHS can be targets for carriers which hop from the given HS (Fig. 7).

The frequency of hops w is given by the formula characterized the hopping process [5]

$$w \sim \varrho^2 \cdot \exp\{-\varrho/\varrho_o\} \tag{17}$$

where ϱ is the distance between HS and ϱ_o is the charge carrier delocalization in the HS. The biggest w occurs for $\varrho_m = 2\varrho_o$. The value of ϱ_o is a partial function of the intensity of the applied electric field E. We assume that this dependence is approximately linear and we can write for the largest hopping distance equation

$$\varrho_H = \varrho_{Ho} + \gamma E \cos\alpha \tag{18}$$

where γ is a constant and α is the angle between the hop direction and E. We put the coordinates origin on the given HS and take into account that $tg\alpha = y/x$. We obtain the analytical expression for the border of the LHs

$$\sqrt{x^2+y^2} = \varrho_{Ho} + \frac{\gamma E x}{\sqrt{x^2+y^2}} \tag{19}$$

for small E we have

$$\varrho_{Ho} > \gamma E \cos\alpha = \frac{\gamma E x}{\sqrt{x^2+y^2}}$$

and from Eq. (19) we obtain

$$x^2 + y^2 - 2\gamma E x = \varrho_{Ho}. \tag{20}$$

This is the equation for a circle which has its center in the point $x_c = \gamma E$, $y_c = 0$ and its radius is equal to $R = \sqrt{(\gamma E)^2 + \varrho_{Ho}^2}$. In Fig. 8 we show a polymer bundle

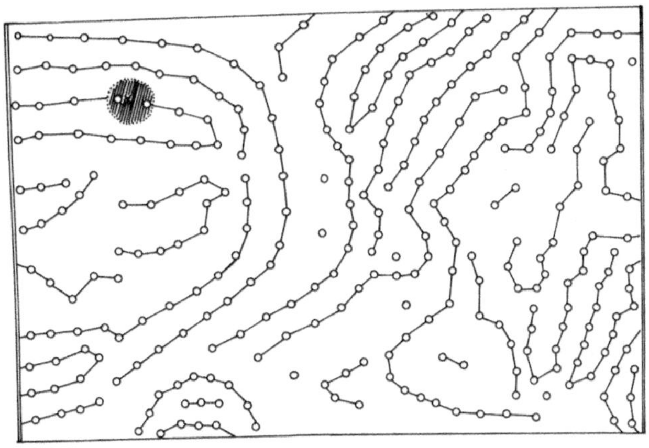

Fig. 8. A polymeric bundle-like molecular superstructure with the acting LHS

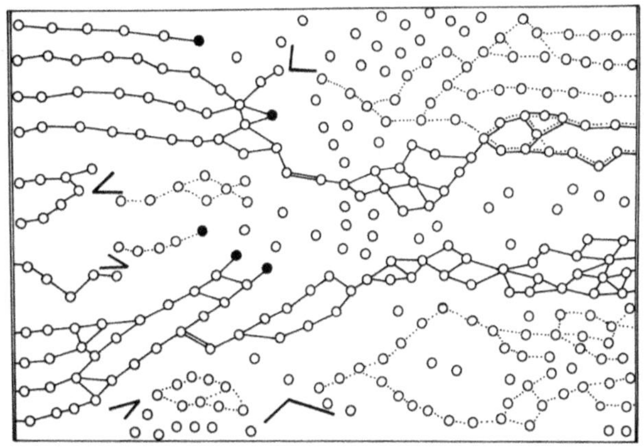

Fig. 9. The polymeric structure of Fig. 8 transformed by the action of LHS on the transport network: =, - red and blue bonds, ● the ends of yellow bonds (the space charge), < the conformation trap ... bonds after <

structure with superfolding. In some places we can see the conformation traps and observe areas of both *A*- and *B*-types. Let us transform this polymer structure in the transport network by the above described process (Fig. 9). The transport network contains the same *A* and *B* regions as the polymer structure. We can see two red bonds, many blue ones, and some yellow. Several paths between electrodes are independent of the conformation which has conformational traps. These paths are practically independent of the intensity of the molecular motion and its affect on temperature. At high temperatures the conformational traps act as bridges and the conductivity in region *B* becomes relatively high. At the border, between *A* and *B*, the ends of dangling (yellow) bonds are frequently located. Obviously, these places have localized charge carriers which jointly create a space charge.

It is now possible in our model to determine the time-of-flight t_{Fj} for a given transport path j

$$t_{Fj} = n_{Aj} \cdot \tau_A + n_{Bj} \cdot \tau_B \,; \quad n_{Aj} = \frac{L_{Aj}}{a}\,, \; n_{Bj} = \frac{L_{Bj}}{b} \tag{21}$$

where n_{Aj}, n_{Bj} are the numbers of visited HS during the j path, L_{Aj}, L_{Bj} are the parts of the path j in the A and B region, τ_A, τ_B the respective lifetimes, and a, b the respective distances between neighboring HS. And so we introduce

$$\frac{\tau_A}{\tau_B} = f_C\,, \quad \frac{L_{Bj}}{L_{Aj}} = \lambda_j$$

where f_C is the mean frequency of the creation of "conductive" states in configuration traps of the B regions. From Eq. (21) we obtain for $a = b$ and $v = a\tau_A$ the relation

$$t_{Fj} = \frac{L_{Aj}}{v_A}\left(1 + \frac{\lambda_j}{f_C}\right) = t_{Hj}\left(1 + \frac{\lambda_j}{f_C}\right) \tag{22}$$

where t_{Hj} is the hopping time of the carrier during t_{Fj}. And we have

$$t_{Hj} = t_{Fj} - n_{Tj}\,\tau_{Tj} = t_{Fj} - n_{Tj}\,\tau_T \tag{23}$$

where n_{Tj} is the number of visited traps during t_{Fj} and τ_{Tj} the time of life in them. Because we assume only one type of trap, we come to the mean values n_T and τ_T. The introduced parameter λ_j is now expressed using Eq. (23)

$$\lambda_j = \frac{L_{Bj}}{L_{Aj}} = \frac{n_{Tj}\,\tau_T\,f_C}{t_{Hj}} \tag{24}$$

and the ratio

$$\frac{\lambda_j}{f_C} = \frac{n_{Tj}\,\tau_T}{t_{Hj}} = \theta_j \tag{25}$$

gives us the relation of time spent by the carrier in the traps to its hopping time. We can transform (22) using (24), (25)

$$t_{Fj} = t_{Hj}\,(1 + \theta_j),\; t_F = (1 + \theta)\,t_H = \frac{1}{J}\sum_{j=1}^{J} t_{Fj} \tag{26}$$

Table 1. Transport diagram

Column / Line	1	2	3	4	5	6	7	8	9	10	11	12	13
1	0	0	0	0	0	0	0	0	0	0	0	0	0,1,2
2	1	1	1,2	1	0	1	1,2	1	1	1	1,2	1,3	–
3	2,3	2,4	–	2,3	1,2	2,3	–	2,3	2,3	2	–	2,3,4	–
4	–	3,5	3,4	–	–	3,4	–	3,4,5	3,4,5	–	3,4,5	–	–
5	–	4,5,6,7	4,5,6,8	–	–	–	4,5,6,8	4,5,6,7	4,5,6,7	6,7,8,9	4,5,6,8,10	–	–
6	–	–	5,6,7,9	–	–	–	5,6,7,9	–	–	–	5,6,7,9,11	–	–
7	–	7,8	6,7	8,10	–	–	–	6,7,8,10	–	–	7,8,10	6,8,9,11,12	–
8	–	8,9	–	9,11	–	–	7	8,9,11	–	–	–	7,9,10,12,13	–
9	–	9,10	–	10,12	–	–	8	9,10,12	–	–	–	8,10,11,13,14	–

$$t_H = t_F - n_T \tau_T \tag{27}$$

and J is the number of paths.

Now we chose a discrete time scale and rewrite Eq. (6) as

$$\frac{p(t_i)}{p} = 1 + \sum_{(j)}^{i} g(t_j) - \frac{1}{P} [T^+ (t_i)]. \tag{28}$$

In order to study the carrier transport in the transport network shown in Fig. 10 we use Eq. (28). This network has the same fractal structure as the Sierpinski gasket which was shown in Fig. 4. As a computer experiment we used few carriers that are in the first line at time $t = 0$ (Table 1). In the transport diagram of our transport network all HS are in the matrix where columns and lines are the representation of the transport network. We assume that the transport occurs in the terms of time scale in the form of steps of equal time intervals. The numbers placed on each HS designate each time step that the given HS is occupied by a carrier. There are three rules for the modeling of the transport process of hopping carriers:

i) after each time-step the carriers must hop to the next line forward in the same column or to the left or right neighboring column;

ii) this hop is possible only when at least one of the hopping places is vacant;

iii) when the carrier comes to a place where HS doesn't exist in the forward direction, then it must wait for one time-step and after that it moves to the neighboring column in the same line;

In our transport diagram we have nine lines. The gasket thickness is surmounted by the fastet carriers in eight time intervals. This occurs for two carriers. The "time table" of hopping carriers is shown in Table 2. In Fig. 11 the time dependence of $p(t_i)$ and the distribution $g(t_{Fi})$ are pictured using values from Table 2. In our example a stochastic distribution of t_{Fi} and $p(t_i)$ is practically time independent. This is shown by a small amount of red bonds in regards to the carrier concentration. However, this behavior is not observed in our experiments. We will show that this finding is caused by the fact that, in reality, such a high concentration of moving carriers does not take place.

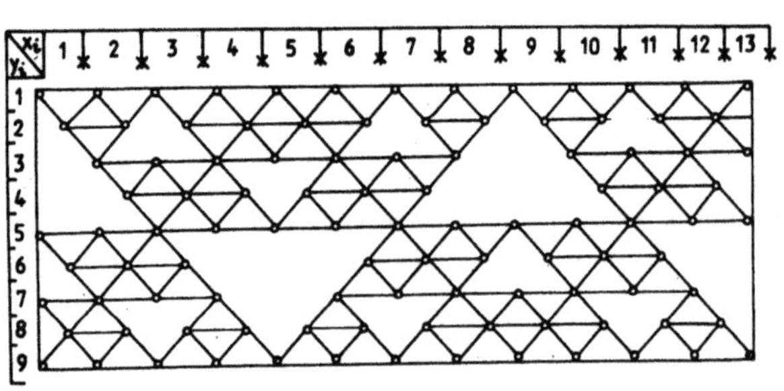

Fig. 10. A transport network with Sierpinski gasket from Fig. 4. Each HS has the coordinates x_i, y_i

5. The charge transport in PVK:NVK system

For a better understanding of the charge carrier transport we studied a polymeric system by the TOF method and changed the molecular superstructure without introducing new types of HS. This is possible

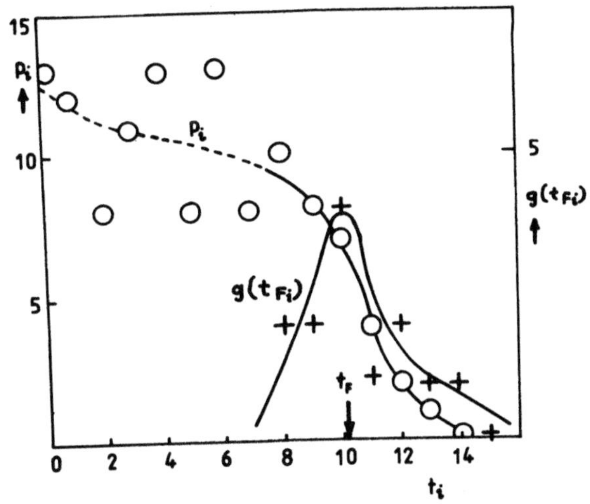

Fig. 11. The number of hopping carriers p_i and their time-of-flight distribution $g(t_{Fi})$ for transport network of Fig. 10

Table 2. The characteristics of the transport

Time step	$p(t_i)$	$T(t_i)$	g_i	$\sum_{(j)}^{i} g_j$
0	13	0	0	0
1	12	1	0	0
2	8	5	0	0
3	11	2	0	0
4	13	0	0	0
5	8	5	0	0
6	13	0	0	0
7	8	5	0	0
8	10	1	2	2
9	8	1	2	4
10		0	4	8
11	4	0	1	9
12	2	0	2	11
13	1	0	1	12
14	0	0	1	13

by the doping of the polymer by its monomer. In our case we used NVK as monomer. (The experimental conditions may be found in [15].)

Also we measured the two exponentials of Eq. (7) shown in Fig. 1. In terms of measured voltages we can rewrite Eq. (7) as

$$U_{R1}(t) = U_{RS} + U_{Ro} \cdot \exp\left[-\left(\frac{1}{\tau_V} + \frac{1}{\tau_T}\right)t\right], \; t < t_F$$

$$U_{R2}(t) = U_{RF} \cdot \exp\left[-\frac{G(t-t_F)}{t_F}\right], \; t > t_F \qquad (29)$$

where U_{RS} is the voltage for time $t \to \infty$, U_{RF} is the voltage at $t = t_F$, and $U_{RS} + U_{Ro}$ is the voltage at $t = 0$. In the case of the drift of a small charge we have relations

$$U_{Ro} = \frac{e \cdot S \cdot R \cdot d \cdot P}{t_F} \cdot \frac{\tau_T}{\tau_V + \tau_T}$$

$$U_{RS} = \frac{e \cdot S \cdot R \cdot d \cdot P}{t_F} \cdot \frac{\tau_V}{\tau_V + \tau_T} \qquad (30)$$

where S is the layer surface, R the loading resistence, and e the elementary charge. From the fitting of $U_R(t)$ we determine t_F, and the lifetimes τ_y and τ_T as functions of the concentration c of NVK monomers.

In Table 3 we summarized the values of t_F, τ_V, τ_T and P obtained from the above mentioned fitting. From the values of t_F, τ_V, and τ_T we can calculate the value of n_T from the relation

$$(n_T + 1)\,\tau_T + n_T \tau_V > t_F > n_T \tau_T + (n_T - 1)\,\tau_V \qquad (31)$$

Table 3. Experimental characteristics of the system PVK:NVK

NVK Concentration (weight %)	$\tau_V \cdot 10^4$ (s)	$\tau_T \cdot 10^4$ (s)	$t_F \cdot 10^4$ (s)	$P \cdot 10^{19}$ (m^{-3})	$t_H \cdot 10^4$ (s)	n_T	θ
0	19.6	50.1	80.0	3.5	29.9	1	1.7
11	1.62	3.05	11.5	3.7	5.4	2	1.1
16	2.41	3.51	9.5	5.3	2.48	2	2.8
20	15.6	13.9	27.0	6.8	13.1	1	1.1
25	16.5	5.57	13.6	7.4	8.03	1	0.7
33	5.3	3.14	9.3	5.7	6.16	1	0.5
50	3.19	2.49	4.0	2.4	1.51	1	1.6

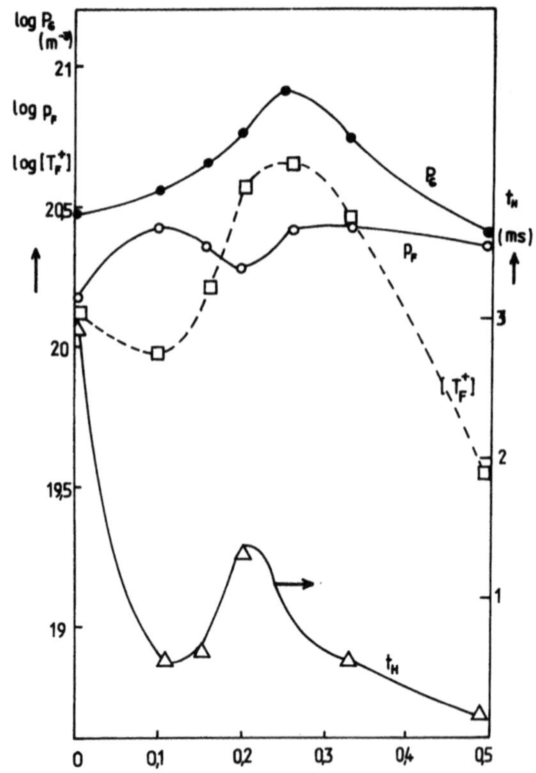

Fig. 12. Concentration dependence of P_G, $[T_F^+]$, t_H, and p_F from the fitting of experimental results

where n_T is an integer. Now we can calcualte t_H and θ from Eq. (27). We can calculate the layer conductivity σ_F at time t_F from equation

$$\sigma_F = \frac{U_{RF} \cdot d}{U_o \cdot R \cdot S} . \tag{32}$$

From the concentration dependence of the drift mobility μ_D (calculated from t_F via Eq. (1) and σ_F, we determine the charge concentration of holes p_F at time t_F and compare it with the initial concentration in the generation layer P_g (Fig. 12). The difference $P_g - p_F$ gives us the concentration of trapped holes $[T^+(t_F)[$ at time t_F.

The first thing which we determine is the mutual relation between the concentration of carriers and HS. For an undoped PVK layer we have the initial carrier concentration in the photogeneration layer equal to $0.5 \cdot 10^{20}$ m^{-3}, and because the thickness of this layer is in our sample around 0.1 micron, we obtain for the specific surface concentration C_s of carrier at least

$$C_s = P_g \cdot 10^{-7} = 0.5 \cdot 10^{20} \cdot 10^{-7} = 5 \cdot 10^{12} \text{ m}^{-2} .$$

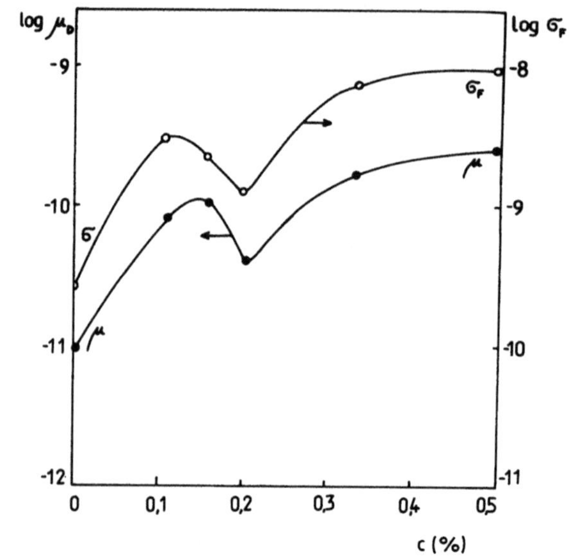

Fig. 13. The concentration dependence of μ_D and σ_F from the fitting of experimental results (pure PVK, $T = 300$ K, $E = 3 \cdot 10^7$ V/m

The distance between PVK chains is about 2 nm [23] and for the specific surface concentration of chains C_c we obtain the value

$$C_c = (2 \cdot 10^{-9})^2 = 2.5 \cdot 10^{17} \text{ m}^{-2}.$$

In this manner we have found that the number of chains available for one carrier is

$$\frac{C_c}{C_s} = 200{,}000 .$$

This means that one carrier has two hundred thousand possible polymeric chains to move in the direction of the applied electric field. We see that it is very unlikely that thee agglomeration of charge carriers occurs as in our earlier example.

A second interesting point is that each carrier crosses only one trap region (B), independent of the concentration of NVK. Thirdly, time t_H devoted to the intrinsic hopping process is concentration dependent. In Fig. 13 we show the simultaneous concentration dependence of μ_D and σ_F for the same sample at room temperature for electric field $E = 3.10$ V/m. Both entities have an analogue behavior that gives us the practical concentration independence of p_F (see Fig. 12). The mean feature of these curves is the depletion at NVK concentration $C = 20$ wgh%. This phenomenon occurs move often than at this temperature and field.

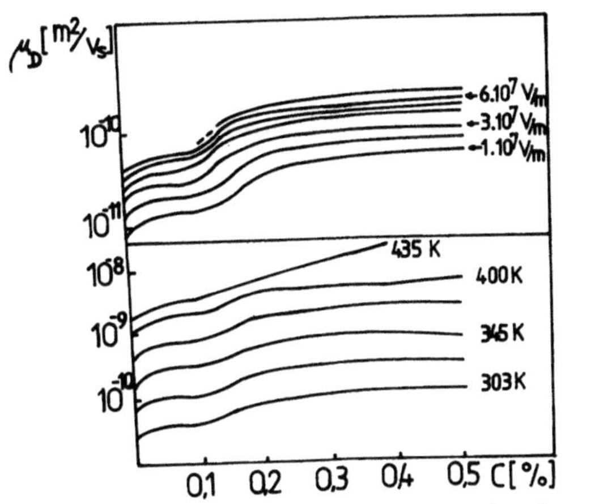

Fig. 14. From TOF method the concentration dependence of μ_D at different temperature ($E = 3 \cdot 10^7$ V/m) and fields ($T = 300$ K) [15]

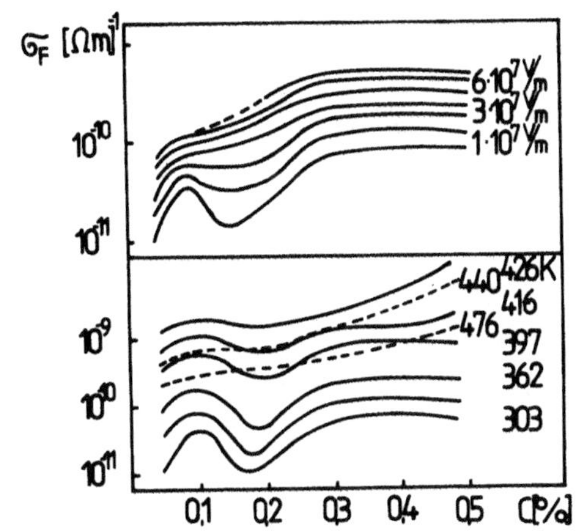

Fig. 15. From photoresponse shape the concentration dependence of σ_F at different temperatures ($E = 3 \cdot 10^7$ V/m) and fields ($T = 300$ K) [15]

We see in Figs. 14 and 15 that at higher temperature and field we observe the two steps in μ_D and σ_F concentration dependence, but no as strongly expressed. From the concentration dependence of μ_D, σ_F, p_F, and t_H it seems reasonable to assume that with the enhancement of monomeric HS we obtain from the point of view of charge carriers transport two types of polymeric materials or two types of transport network:

i) at low concentration below 10 wgh% we have an amorphous polymer with a given concentration of traps, which diminish with the growth of NVK concentration C. This corresponds to the enhancement of $\mu_D \cdot \sigma_F$, and p_F with C;

ii) at about 20 wgt% starts the growth of the other polymeric structure. This point probably corresponds to the formation of a large number of small crystallization germs which destroy the previous relatively high homogenity of the amorphous state. At the interface between amorphous and crystalline phases, new types of traps [24] are produced which cause the lowering of μ_D, σ_F, and p_F and the enhancement of $[T^+(t_F)]$. At the concentration $C = 50$ wgt% the layer was practically crystallized, as can be seen in Fig. 16. When we have crystalline PVK:NVK layer, then μ_D, σ_F, and p_F increase, but t_H decreases.

To summarize our experiment, enhancement of the concentration of free HS (monomeric NVK moleculels) leads to higher homogenity in the density of the system. The density is grows with the filling of empty volume by monomeric HS. But at some concentration of monomeric HS clustering occurs that leads to the start of the crystallization of the layer. At this moment the disorder of the system is enhanced. When nearly the whole layer has been crystallized we have a high order in the geometry of HS.

It seems that the large change in value of t_H does not correspond with this concept. At a high order we have $t_H = 0.1$ ms and for the beginning disorder we have $t_H = 3$ ms. This result does not explain a probable short cut of the original meander path in the amorphous phase. This discrepancy is possible to explain by the probability of reverse hops in those places in the system where the intensity of the field E is in the direction of possible hops very low, and the drift is in this case very weak.

6. Conclusions

We have seen that in PVK, and of course in similar systems, the charge carrier runs through a number of hopping sites organized in high and low density

Fig. 16. The photographs of PVK:NVK thin layer ($d = 1\,\mu$m) cast from solution on a steel support; (a) $C_{NVK} = 0$ (amorphous state), (b) $c = 0.5$ (crystalline state)

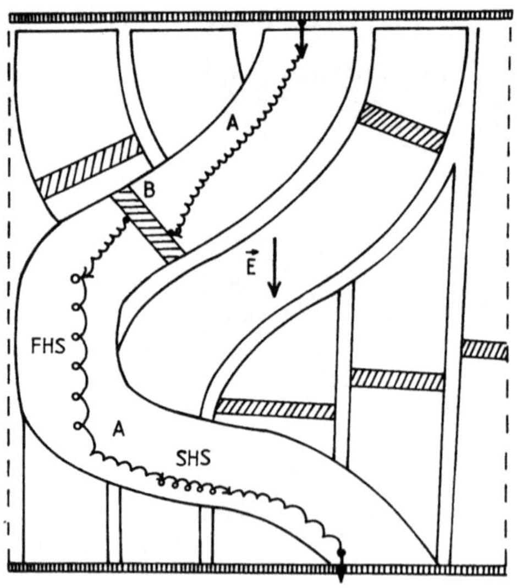

Fig. 17. Schema of proposed transport tubes in the thin polymeric layer between electrodes; A, B: good and poor conductivity regions FHS, SHS: fast and slow HS

regions which are more or less disordered. Each carrier hops alone and therefore, we propose the introduction of transport tubes for the modeling of charge carrier transport in disordered systems. Each carrier moves in one tube which has a bundle meander form with respect to $\vec{E}$ (Fig. 17). The HS of the A region are of two types: the fast, with small carrier lifetime, and the slow, with large lifetime. The fast HS are localized in places where the drift is large. This occurs when the active HS are aligned in the direction of the applied field. The slow HS are localized in other places in the A region. The apparent enhancement of the waiting time of the carrier in slow HS is given by reverse hops with regard to the direction of the applied field which occurs for these HS. Each transport tube has one or more trapping areas (B region) where the hopping frequency is very small compared to the same entity for HS in the A region.

The introduced transport tube is a structure element of the transport network and it is not directly connected with the molecular superstructure of the given polymeric system. If we put free monomeric HS into the polymer, we observe the formation of bridges between the transport tubes in a mathematical model. These bridges permit short cuts that bypass paths which are not advantageous and therefore, the growth of the carrier mobility. Because the monomeric HS fall also into B regions, these become not so empty and the trap becomes not so deep.

The small crystalline clusters are obstacles for the carrier hopping. When these clusters fill the empty space between transport tubes then the transport becomes ideal, practically only via the crystalline state, which is interrupted by some small amorphous regions.

Acknowledgement

The author is indebted to Prof. H. Biederman and Dr. M. Marvan for reading the manuscript and to Mr. Z. Chmel for technical assistance.

References

1. Pai DM (1970) J Chem Phys 52:2285
2. Gill WD (1972) J Appl Phys 43:5033
3. Gill WD (1976) In: Mort J, Pai DM (eds) Photoconductivity and related Phenomena. Elsevier, Amsterdam, p 303
4. Brehmer L (1981) Acta polym 32:415
5. Mort J, Pfister G (eds) (1982) Electronic Properties of Polymers. Wiley, New York, p 215:
6. Bässler H (1981) Phys Stat Sol (b) 107:9
7. Noolandi J (1977) Phys Rev B 16:4466
8. Schmidlin FW (1977) Phys Rev B 16:2362
9. Sher H, Montroll EW (1975) Phys Rev B 12:2455
10. Stolka M, Yanus JF, Pai DM (1984) J Phys Chem 88:4707
11. Mort J, Pfister G (1979) Polym Plast Technol Eng 12:89
12. Abkowitz M, Pai DM (1986) Philsoph magaz B 53:193
13. Chudáček I, Naumenko G (1985) unpublished results
14. Bässler H (1984) Philosoph magaz B 50:347
15. Chudáček I, Naumenko G, Urban J (1986) In: Sedlacek B (ed) Charge Transport in Thin Films of Polyvinylcarbazole: N-vinylcarbazole, Morphology of Polymers. W de Gruyter, Berlin, p 689:16
16. Stanley HE, Ostrowsky N (eds) (1985) On Growth and Form. Nijhoff-Kleuver, Dordrecht
17. Pietronero L, Tosatti E (eds) (1985) Fractal in Physics Proceed of 1985, Trieste Confer, North Holland, Amsterdam
18. Meier H (1975) Dark and Photoconductivity of Organic Solids. Weinheim, Verlag Chemie
19. Pechhhold W (1984) Makromol Chem, Suppl 6:163
20. Stanley HE, Form, An Introduction to Self-Similarity and Fractal Behavior, in cit 16
21. de Gennes PG (1980) J Phys, Paris, Colloq 41:C3–C17
22. Hong DC, Stanley HE, Coniglio A, Bunde A (1985) Phys Rev B 32:615
23. Griffiths CH, van Laeken A (1976) J polym Sci Polym Phys Ed 14:1433
24. Nešpůrek S, Slavínská D, Klein-Szymanská B (1979) Europ Polymer J 15:965

Received November 13, 1987;
accepted April 21, 1988

Author's address:

Prof. Dr. I. Chudáček
Faculty of Mathematics and Physics
V Holešovičkách 2,
180 00 Praha 8, CSSR

Progress in Colloid & Polymer Science Progr Colloid Polym Sci 78:109–112 (1988)

Charge carrier transport in polyethylene studied by surface potential kinetics

H. von Berlepsch

Akademie der Wissenschaften der DDR, Institut für Polymerenchemie "Erich Correns", Teltow-Seehof, G.D.R.

Abstract: Based on measurements of the isothermal $U(t)$ and thermally stimulated surface potential decay TSCD, the charge carrier transport in HDPE is investigated. A simple mathematical model for describing the isothermal surface potential kinetics is presented, involving hopping transport of injected charge carriers being modulated by shallow and deep traps at the surface and in the bulk. The model parameters charge carrier mobility, hopping distance, trap concentration, trap depth and attempt-to-escape frequency were determined by curve fitting. Additional details of the energetic structure of traps are obtained from the TSCD investigations. The dependence of mobility and lifetime on the field strength is discussed.

Key words: Polyethylene, surface potential kinetics, charge carrier mobility, trap parameters.

1. Introduction

In spite of a multitude of experimental and theoretical investigations of charge transport in polymers, particularly in PE, various questions have been left unanswerd (for review see [1]).

As yet the charge carrier mobility in PE has not been completely elucidated. The mobility values determined by different nonpulse techniques are in the range from 10^{-14} m^2 V^{-1} s^{-1}. However, the values obtained by pulse methods are of the order of 10^{-8} m^2 V^{-1} s^{-1}. The presence of traps considerably affects the electric properties of polymers. So far, however, the nature of the trapping centres and their energetic and kinetic parameters remain unknown in most cases.

The study of the surface potential kinetics has proved to be a powerful tool in investigating the charge carrier transport processes [2–5]. This method enables a convenient determination of the charge carrier mobility and of trap parameters. In the investigations described in this paper the charge carrier transport process in high density polyethylene (HDPE), chosen as a model material, was determined by this method.

2. Experiment

2.1. Samples and experimental details

For the measurements carefully characterised samples of HDPE films with a thickness of 15–30 μm (standard thickness $L = 30$ μm), containing no additives, having a molecular weight of $M_w = 1.4 \times 10^5$, a degree of crystallinity of 60%, and two branching points per 10^3 C atoms [6], were used. The measurement of the surface potential decay was carried out in the open-circuit configuration by the field mill method [7]. The samples were coated on the back surface by evaporation of gold or aluminium. The excess charge was introduced into the sample by exposure to a positive or negative point corona discharge (air gap 10^{-2} m, corona voltages up to 8 kV). After corona charging for a time $t_c = 10$ s the isothermal potential decay (temperature range 22–77 °C) was measured for 30 minutes. Then the sample was heated at a uniform rate of $\beta = 2$ °C min^{-1} to the melting point obtaining the TSCD curve. The measurements were carried out in air at relative humidity 60–80%.

2.2. Experimental results

A family of $U(t)$ curves for various initial potentials U_o is shown in Fig. 1.

Under the experimental conditions of these investigations, particularly at both polarities, a cross-over of the $U(t)$ curves with different initial potentials U_o was observed. The $U(t)$ kinetics can be expressed by the empirical relationship containing five parameters [8]. Generally, the

$$U(t) = \begin{cases} U_o - \dfrac{a}{\alpha_i}\left(\dfrac{t}{t_T}\right)^{+\alpha_i}, & t < t_T \\ U_o - a\dfrac{(\alpha_i + \alpha_f)}{\alpha_i \alpha_f} + \dfrac{a}{\alpha_f}\left(\dfrac{t}{t_T}\right)^{-\alpha_f}, & t > t_T \end{cases} \quad (1)$$

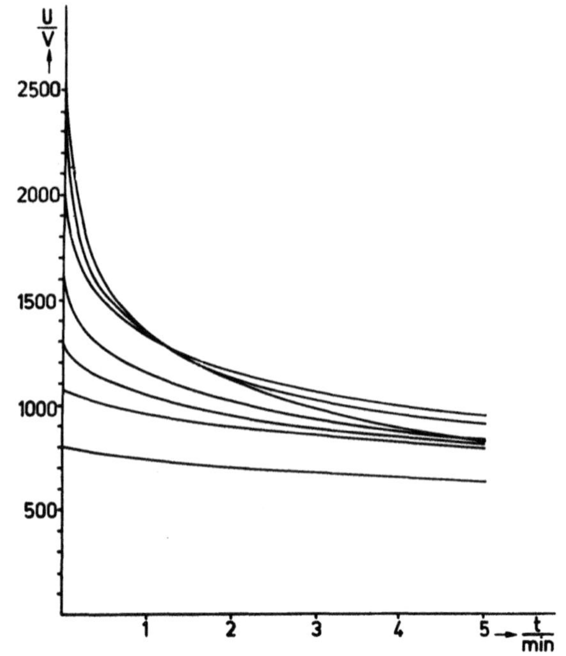

Fig. 1. Family of $U(t)$ curves for various initial potentials U_o. Sample HDPE/Au, negative corona, $L = 20$ µm, $T = 50$ °C

four parameters t_T, α_i, α_f and a are functions of the electric field strength and temperature. The characteristic time t_T obeys approximately $t_T \sim (1/E_o)^{1/\alpha_E}$ where $\alpha_E = 0.46$, in the region of field strength investigated, i. e. $E_o = U_o/L = (3 \times 10^7 \text{ to } 1.5 \times 10^8)$ Vm^{-1}. The temperature dependence obeys the exponential relationship $t_T \sim \exp(\Delta/kT)$ within the region 22–77 °C at an activation energy of $\Delta \approx 0.4$ eV. The decay parameters α_i and α_f were found to depend approximately linearly on the field strength and temperature, respectively, in the corresponding regions investigated. For the parameter a, which represents a measure of the field dependence of the decay rate, the relationship $a \sim E_o^3$ was found to hold. Within the limits of error no differences in polarity were observed. This also holds for the TSCD curves.

3. Theory

The decay of the surface potential $U(t)$ can be explained as a result of trap-controlled hopping of charge carriers injected by corona charging under the action of their own electric field. This physical model leads to a system of partial differential equations which can by solved analytically by some simplifying suppositions and yield a four parameter expression of the surface potential $U(t)$ [8]. The basic concept of this model consists in the following classification of localised electronical states being involved in the transport.

1. Two types of surface states are discerned, namely states of low energy depth and states of high energy depth being characterised by corresponding detrapping times which, as compared with the transit time, are very much smaller or higher, respectively. In the transport equations they are represented by only one model parameter, the injection parameter ϱ_o. Thus $\varrho_o \sigma_o$ denotes the fraction of the total surface charge $\sigma_o = \varepsilon \varepsilon_o U_o/L$ injected on charging into the shallow states or immediately into the bulk, while $(1-\varrho_o)\,\sigma_o$ designates the fraction trapped all over the measuring time in the deep surface states.

2. Localized transport states in the bulk. In these states the transport of the injected charge carriers proceeds by hopping. They are characterized by the drift mobility $\mu_d = 2 \times 10^{-8}$ m^2 V^{-1} s^{-1} at $T = 50$ °C and $E_o = 3 \times 10^7$ Vm^{-1} as determined by the TOF method [9].

3. In the bulk we suppose the existence of energetically shallow traps, being in thermal equilibrium with the transport states corresponding to the θ_S-factor $\theta_S \ll 1$, modifying μ_d to the effective mobility $\mu = \theta_S \mu_d$.

4. Further, we assume the existence of deep traps in the bulk not in thermal equilibrium with the transport states. In the transport equations they are described by the approximation of a single discrete energy level and characterised by their trapping rate ω and their release rate into the transport states, r.

As the solution of the transport equations for the times $t < t_{tr}$, with t_{tr} being the shallow-trap-modulated transit time, the expression for the surface potential

$$U(t) = U_o - \frac{\mu}{2}\left(\frac{U_o}{L}\right)^2 (1-(1-\varrho_o)^2)\,\frac{1}{R} \times \left(rt + \frac{\omega}{R}(1-e^{-Rt})\right) \qquad (2)$$

results, where $R = r + \omega$. Based on this formula the experimental decay curves can be described theoretically. For determining the model parameters ϱ_o, μ, ω and r a computerised fitting of the curves in required.

4. Theoretical results and discussion

As a consequence of the particular interrelation of the model parameters in Eq. (2) the fitting procedure yields independent values of the parameters r and ω

only, while the remaining parameters ϱ_o and μ are interrelated. In order to avoid this difficulty the injection parameter was determined in an additional experiment by the change in the decay rate at repeated charging of the sample [10]. As shown in Fig. 2 the value of ϱ_o proved to depend on the field strength. The $\varrho_o(E_o)$ dependence estimated, which can be interpreted by assuming a trap saturation of the deep surface states, explains simultaneously the occurence of the cross-over effect. For the effective concentration (per unit area) of the deep surface states, the value $N_{DS} = 2.6 \times 10^{15}$ m^{-2} was obtained. Based on the values of ϱ_o the fitting procedure yields for the effective mobility independent of the polarity the average value of $\mu = 3 \times 10^{-15}$ m^2 V^{-1} s^{-1} at $E_o = 5 \times 10^7$ Vm^{-1} and $T = 50$ °C. In the regions of E_o and T investigated μ was found to depend on field strength and temperature according to the expression

$$\mu = \mu_d \theta_S = \frac{1}{6} \frac{e\lambda^2 r_o}{kT} e^{-(2\lambda/r_o)} e^{-(W/kT)} \times \frac{\sinh(e\lambda E_o/2kT)}{eE_o/2kT} \frac{N_h}{N_S} e^{-(E_S/kT)} \tag{3}$$

being typical for field dependent hopping between localized states [11]. Fig. 3 illustrates the E_o dependence of the estimated values, the full line representing the best fit with the hopping distance of $\lambda = 1.9$ nm. As to the order of magnitude this value agrees with that of the average distance of crystallites in semicrystalline HDPE. Thus, the transport process can be conceived as hopping between crystallites [12]. The temperature dependence of μ can be described by an Arrhenius

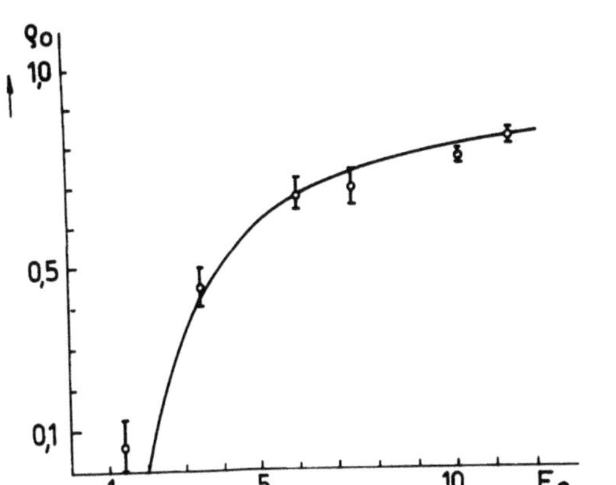

Fig. 2. Field strength dependence of the injection parameter $\varrho_o(E_o)$. Sample HDPE/Au, negative corona, $L = 30$ μm, $T = 45$ °C

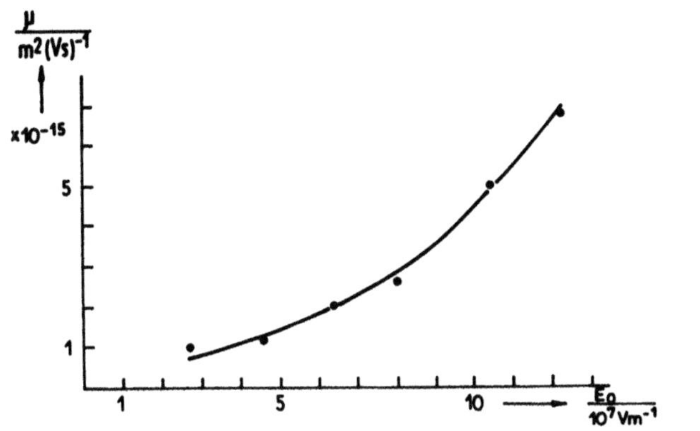

Fig. 3. Dependence of the effective mobility μ on electric field strength. Sample HDPE/Au, negative corona, $L = 30$ μm, $T = 50$ °C

law, the energy of activation being $\Delta_\mu \approx 0.6$ eV and the pre-exponential factor $\mu_o = 9 \times 10^{-6}$ m^2 V^{-1} s^{-1} for both polarities. Assuming the characteristic activation energy for hopping of electronic charge carriers in the transport states W to be negligible as compared with the energetic depth of traps, interacting with the transport states, Δ_μ is interpreted as the depth of a discrete shallow trap $E_S \approx 0.6$ eV. Based on the condition of the thermal equilibrium, which is approximately satisfied at typical measuring times, the relationship $\mu_o = \mu_d N_h/N_S$ will hold, where $N_h \approx 10^{26}$ m^{-3} is the concentration of the transport states and $N_S \approx 3 \times 10^{23}$ m^{-3} the concentration of the shallow traps. The factor $\theta_S (= \mu/\mu_d)$ is $\approx 10^{-7}$. For the phonon frequency and the localization radius we obtain the values $\nu_o = 2 \times 10^{13}$ s^{-1} and $r_o = 0.4 \times 10^{-9}$ m, respectively. The r_o value is one order of magnitude higher than Bohr's radius for carbon atoms. Based on the picture of hopping between the crystallites it seems reasonable to suggest a relatively weak localization of the wave function within the ordered regions.

The reciprocal trapping rate of the deep trap (lifetime) $\omega^{-1} \equiv \tau$ depends on the field strength, as illustrated in Fig. 4. The temperature dependence of the lifetime obeys approximately an Arrhenius law, the activation energy being $\Delta \approx 0.45$ eV. The quantity τ is identical with the characteristic time t_T as determined in the experiment. The dependence of τ on the field strength and temperature can be interpreted, based on

$$\tau^{-1} = N_D S_D v \sim \sinh(e\lambda E_o/2kT) \tag{4}$$

as the result of a corresponding dependence of the effective transport velocity $v = \mu E_o$ [12]. With a capture cross-section of $S_D = 10^{-19}$ m^2 of the deep traps,

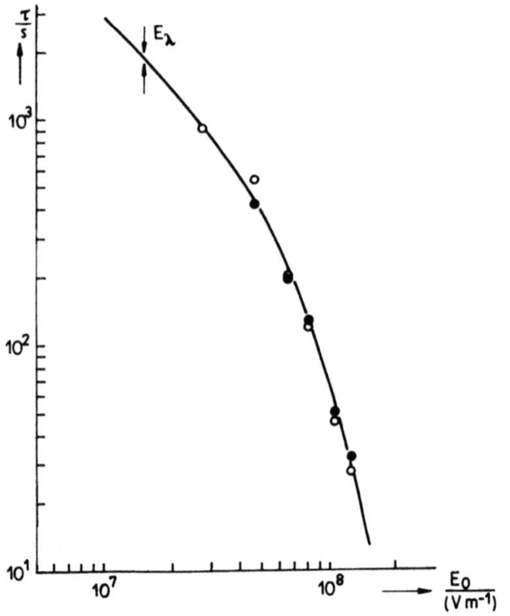

Fig. 4. Dependence of the lifetime on electric field strength. E_λ is defined by $E_\lambda = kT/e\lambda$. Values of τ obtained immediately by fitting (●) and those calculated from the μ-values (○) by Eq. (4). Sample HDPE/Au, negative corona, $L = 30$ µm, $T = 50$ °C

being supposed to be neutral in the unoccupied state, a trap concentration of $N_D = 1.8 \times 10^{23}$ m^{-3} follows. The temperature dependence of the release rate leads to the trap depth of $E_D \approx 1$ eV, and to the attempt-to-escape frequency of $\nu \approx 10^{10}$ s^{-1}.

The theoretical results were supplemented by the TSCD measurements. Figure 5 shows open circuit TSCs for films charged to different U_o values. The curve are the derivatives of their corresponding TSCDs. The reproducibility of the peak positions among the different samples (U_B values) is poor. Three broad peaks are found at roughly 130 °C, 100 °C and 65 °C. The 130 °C peaks is only resolved at the lowest initial potential $U_B = 67$ V. According to Fig. 2 injection into the bulk is negligible at this field strength and only deep surface traps are occupied. Therefore we conclude that the 130 °C peak represents the deep surface traps. The 65 °C peak is due to the shallow bulk traps. Its maximum position corresponds with that of the α-relaxation. Thus these traps are obviously located in the crystalline regions [1]. The peak is relatively weak pronounced because it is partially heated out during the preceding isothermal potential decay at $T = 47$ °C. The third peak at $T \approx 100$ °C is attributed to the deep bulk traps, which are assumed to be caused by defects in the amorphous-crystalline interlayers or

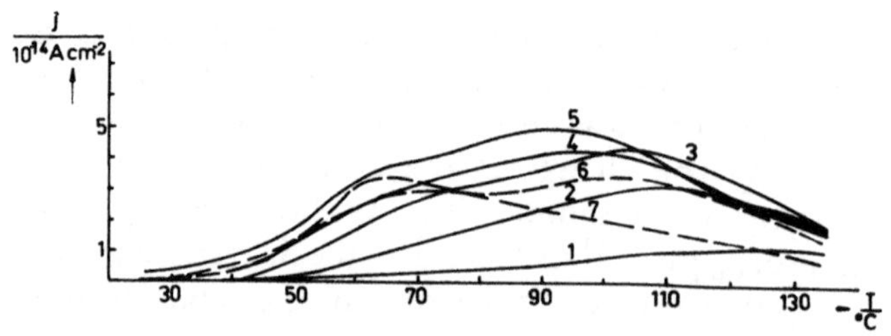

Fig. 5. Open circuit TSCs of differently charged samples. Samples HDPE/Al, positive corona, $L = 20$ µm, equivalent voltage bias: $U_B(1 \ldots 7) = (67, 175, 250, 285, 350, 245, 200)$ V

amorphous regions (cavities) in HDPE. Owing to the cross-over effect (Fig. 1) an increase of U_o is accompanied by a decrease of $U_{res} \equiv U_B$. In the TSC spectrum this manifests in a decreasing intensity of the 100 °C peak [13].

Acknowledgements

The author is indebted to Drs. L. Brehmer, S. Nešpůrek and M. Pinnow for valuable comments.

References

1. Wintle HJ (1983) Engineering Dielectrics II A, Electrical Properties of Solid Insulating Materials. ASTM, Washington, pp 239–354
2. Ieda M, Sawa G, Shinohara U (1967) Jpn J Appl Phys 6:793
3. Wintle HJ (1972) J Appl Phys 43:2927
4. Perlman MM, Sonnonstine TJ (1975) Proc Int Symp on Electrets and Dielectrics. São Carlos, pp 337–355
5. Toomer R, Lewis TJ (1980) J Phys D 13:1343
6. Brehmer L, Dahme A, Weigel P, Pinnow M (1979) Potsdamer Forsch B 19:40
7. Gubkin AN, Mitronina VS, Sergienko VF, Subbotin MJ (1959) Prib i Tekhn Eksper 4:113
8. von Berlepsch H (1985) J Phys D 18:1155
9. Brehmer L, Pinnow M, Kornelson M, von Berlepsch H, Hanspach J (1983) J Electrostat 14:19
10. von Berlepsch H (1985) Phys Status Solidi a 92:285
11. Bagley BG (1970) Solid State Commun 8:345
12. von Berlepsch H (1985) Phys Status Solidi a 90:K97
13. von Berlepsch, Pinnow M (1988) Phys Status Solidi a 105:485

Received June 25, 1987;
accepted April 24, 1988

Author's address:

Dr. H. von Berlepsch
Institut für Polymerenchemie "Erich Correns"
Akademie der Wissenschaften der DDR
Kantstr. 55
DDR-1530 Teltow-Seehof, G.D.R.

Progress in Colloid & Polymer Science Progr Colloid Polym Sci 78:113–115 (1988)

Thermoluminescence and ESCA investigations in polymer resist layers

R. Kunze, M. Wäsche, and E. Linke

Akademie der Wissenschaften der DDR, Zentralinstitut für physikalische Chemie, Berlin, G.D.R.

Abstract: Radiation chemical processes in poly(methacrylate) and chlorinated polyphenylmethacrylate after ion- and electron-beam irradiation are analyzed by means of TL- and ESCA-measurements. The results are in accordance with proposed models for degradation and crosslinking mechanisms of the polymers. A correlation is found between the TL intensity and the sensitivity of positive resists based on poly(methacrylates).

Key words: Polymer resist layers, polymethacrylate, chlorinated phenylmethacrylate, thermoluminescence.

1. Introduction

In recent years there has been increased interest in the radiation chemistry of polymers and the use of polymers as resist materials in submicrometer lithography. New requirements for the resists has arisen, especially concerning the transition to short-wave radiation such as deep-UV radiation, electron- and ion-beam radiation, and the transition to ion-beam-assisted etching.

In this work we report on thermoluminescence (TL) and ESCA investigations of selected polymers and give an analysis of radiation-induced degradation and crosslinking after irradiation with electron and ion-beams.

The equipment used for the TL-measurements was described in detail in [1]. The polymer samples were excited with 10 keV-electrons at $T = 80$ K. The luminescence was measured with a photomultiplier directly after the excitation at a constant temperature $T = 80$ K and during the thermal scan from $T = 80$ K to $T = 300$ K (glow curve).

The ESCA spectra of the polymers were measured with a spectrometer AEI-Kratos ES 200B. The photoelectrons were excited with Al-Kα radiation.

The ion-irradiation in the ESCA-experiments was performed with argon-ions of 1 keV, as ESCA is a surface sensitive method.

2. Poly(methacrylate) polymers

All poly(methacrylates) and their copolymers undergo mainchain scission when irradiated; they are positive resists. Figure 1 shows the plot of the sensitivity (D_S) of various poly(methacrylates) resist vs TL-intensity (I_{TL}). The correlation is based on the dependence on the two quantities on the rate of radical production [2]. Theoretical considerations provide the relation

$$I_{TL} \sim \sqrt{D_s}$$

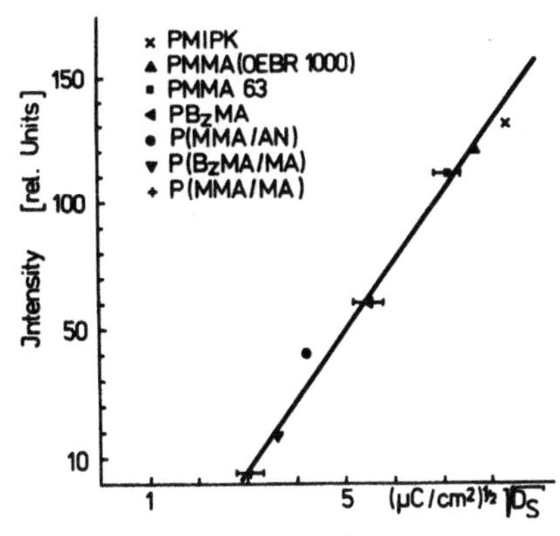

Fig. 1. TL intensity as a function of D_S

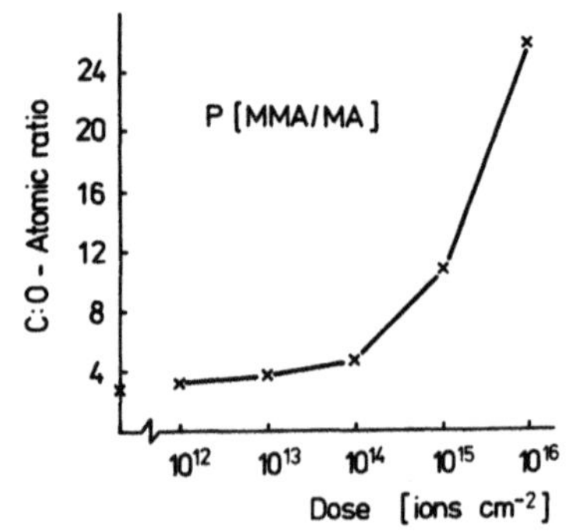

Fig. 2. C : O atomic ratio as a function of ion dose

which is well confirmed by our investigations. The sensitivity of p-resists can be characterized by TL-measurements.

Changes in the C : O atomic ratio as a function of the ion dose are studied by ESCA. During ion bombardment the oxygen content of the P(MMA/MA)[1]) surface decreases. The Cl_s core level intensities show a loss of mostly oxygen species from the surface of the polymer.

For ion doses of 10^{14} ions cm^{-2} a strong degradation of the P(MMA/MA) structure is observed (Fig. 2). The radiation induced degradation of P(MMA/MA) results from main chain scission induced by side group elimination [3].

At argon ion doses above 10^{15} ions cm^{-2} the ESCA-spectra show a carbon enrichment of the polymer surface layer. The results obtained are in good qualitative agreement with AES results of PMMA under electron beam irradiation [4].

3. Chlorinated polyphenylmethacrylates

Poly-p-chlorphenylmethacrylate crosslinks after electron or ion-beam irradiation, i.e., it belongs to the *n*-resists. TL-investigations on polyphenylmethacrylates and halogenated polyphenylmethacrylates showed that the dissociative electron capture by the Cl-atoms is indicated in a pronounced peak in the temperature range around 150 K [5]. The intensity of this peak depends on the chlorine content and also on the possibility of chlorine abstraction (Fig. 3). Therefore, this peak can be used as a quantity for the change of chlorine content after irradiation.

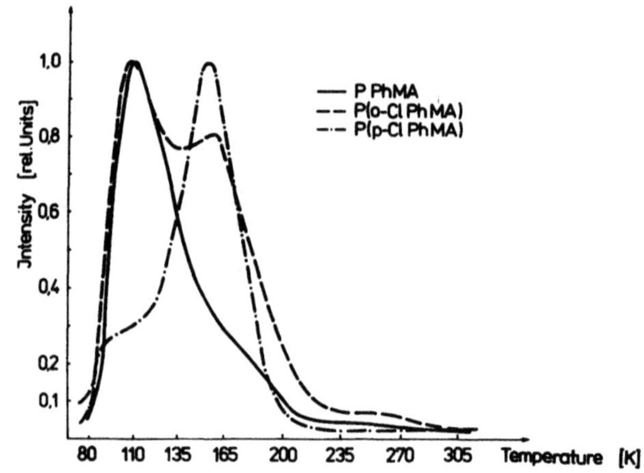

Fig. 3. Glow-curves of various PPhMA (heating rate 6 K/min)

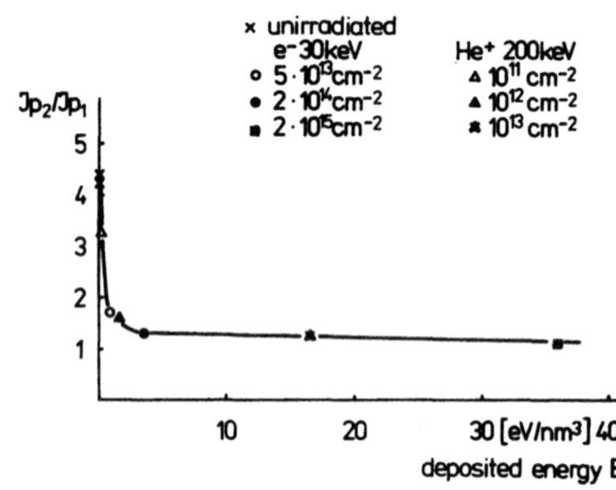

Fig. 4. Ratio of the intensity of the second peak (I_{P_2}) to the first peak (I_{P_1}) as a function of deposited energy E_A

Figure 4 represents the ratio of the intensity of the peak at about 150 K (I_{P_2}) to the intensity of the first peak at about 110 K (I_{P_1}) as a function of the deposited energy E_a after pre-irradiation of the polymer samples with He^+-ions (200 keV) and electrons (30 keV). In the range of 10^{11}–10^{12} ions/cm^2, resp., 10^{13} e^-/cm^2, the ratio of I_{P_2}/I_{P_1} decreases significantly indicating a strong chlorine removal in the polymer. With an increasing pre-irradiation dose the influence of the Cl-atoms in the TL-process is lowered and the chlorine removal rate shows saturation.

The influence of ion-beam exposure on the polymer samples was also investigated by the ESCA-method [6]. The results obtained by ESCA, especially the atomic ratio Cl : C (Fig. 5), show a chlorine abstraction. As can be seen from Fig. 5, an intense chlorine removal is already found for small argon ion doses. For high ion doses this effect is saturated and at doses above 10^{16} ion/cm^2 the polymer surface layer shows a relative carbon enrichment.

---

[1]) P(MMA/MA): Copolymer of PMMA with PMA in the ratio of 70/30.

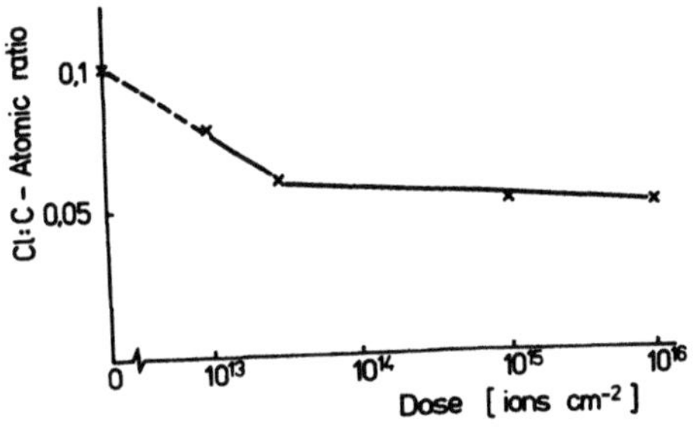

Fig. 5. Cl:C atomic ratio as a function of ion dose for P(p-ClPhMA)

TL- and ESCA-results confirmed the model mechanism [7] for the crosslinking of halogenated aromatic polymers, which assumes the abstraction of chlorine atoms as the dominating process during irradiation.

References

1. Kunze R, Linke E (1983) Isotopenpraxis 19:383
2. Kunze R, Pfeiffer K, Wäsche M, Linke E (1986) In: IV. Symp Phys Grundlagen zu Bauelementetechnologie der Mikroelektronik, Frankfurt (O)
3. Hiraoka H (1977) IBM J Res Develop 21:121
4. Samoto K, Shimizu R, Hashimoto H (1985) Jap J of Appl Phys 24:482
5. Kunze R, Wäsche M, Linke E (1986) Polym Bulletin 15:285
6. Wäsche M, Gross T, von Lampe I (1985) Polym Bulletin 14:147
7. Imamura S, Tamamura T, Sukegawa K, Kogure O, Sugawara S (1984) J Electrochem Soc Solid State Sci and Technol 131:1122

Received August 25, 1987;
accepted April 21, 1988

Authors' address

R. Kunze
Akademie der Wissenschaften der DDR
Zentralinstitut für physikalische Chemie
Rudower Chaussee 5
DDR-1199 Berlin, G.D.R.

Progress in Colloid & Polymer Science Progr Colloid Polym Sci 78:116–118 (1988)

Thermally stimulated depolarization currents in hydrated casein solid samples

P. Pissis and A. Anagnostopoulou-Konsta

Physics Department, National Technical University Athens, Greece

Abstract: Thermally stimulated depolarization current measurements on hydrated solid samples of the protein casein in the temperature range 77–300 K and the water content range 1–24 % (w/w) show two dispersions, a low-temperature dispersion in the range 100–170 K, and a high-temperature dispersion in the range 190–290 K. The high-temperature dispersion was studied in detail using several experimental techniques offered by the TSDC method. It consists, for high water contents, of two complex bands numbered I and II in the order of increasing temperature. Band I is attributed to water-assisted relaxation of polar side chains and band II to space-charge relaxation.

Key words: Biopolymers, hydration properties, dielectric relaxation, depolarization current.

1. Introduction

Dielectric measurements on hydrated solid samples of the protein casein by means of the thermally stimulated depolarization current (TSDC) method in the temperature range 77–300 K show two dispersions [1]. The first dispersion in the range 100–170 K (low-temperature dispersion) has been attributed to the reorientation of loosely bound water molecules. The second dispersion in the range 190–290 K (high-temperature dispersion) has been found to shift to lower temperature and to increase in magnitude with increasing water content, these changes being especially pronounced in the water content range 8–10 %, where the primary hydration layer is completed [2]. In this work we study in detail the complex high-temperature dispersion. We use several experimental techniques offered by the TSDC method [3] to distinguish between dipolar and space charge contributions to the dispersion and to resolve it into individual peaks.

2. Experimental

The TSDC method, its equations, the methods of evaluating the activation energy for the dipole reorientation, W, the pre-exponential factor in the Arrhenius equation, τ_0, and the contribution of a mechanism to the static permittivity, $\Delta\varepsilon$, as well as the experimental apparatus used have already been described [3, 4].

The casein powder (Sigma, St. Louis) was compressed to cylindrical pellets of 13 mm diameter and about 1 mm height. The water content h, defined as the ratio of weight of the sorbed water to the weight of dry casein, was varied between 1 and 24 %

Typical values of the experimental conditions were 3 kV/cm for the polarizing field, 300 K for the polarization temperature, 5 min for the polarization time and 3 K/min for the heating rate.

3. Results and discussion

Figures 1 and 2 show the effect of different electrode arrangements on the high-temperature dispersion for two different water contents, 9.8 and 16.4 %. It follows from the figures that the use of totally blocking (teflon) or partially blocking (brass) electrodes reveals the complex character of the dispersion (band I in Figure 2) and that in the case of highly hydrated samples a contribution to the dispersion (band II in Figure 2) is eliminated when using totally blocking electrodes. It is reasonable to attribute this contribution at the high temperature side, which is eliminated with totally blocking electrodes, to space-charge polarization connected with dc conductivity. The strong increase of dc conductivity with increasing water content [5] explains why this contribution appears only in plots measured on highly hydrated samples and increases in magnitude with increasing water content [1].

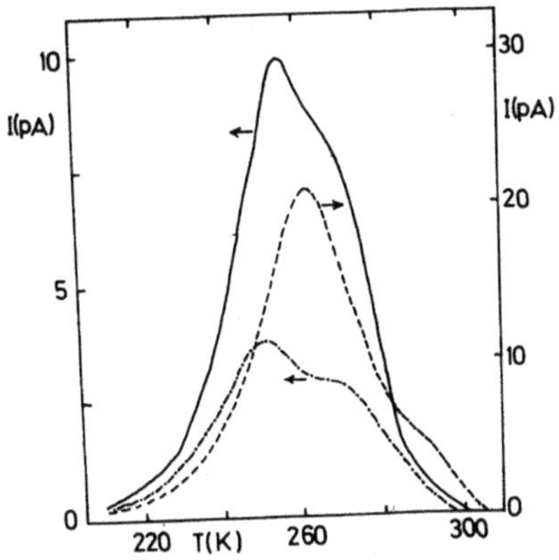

Fig. 1. The temperature dispersion of casein samples with $h = 9.8\,\%$ for three different electrode arrangements: direct contact of the sample with the brass electrodes of the measuring capacitor (——), sample provided with silver paste contacts (– – –), and with thin insulating teflon foils inserted between the sample and the capacitor plates (–·–·–)

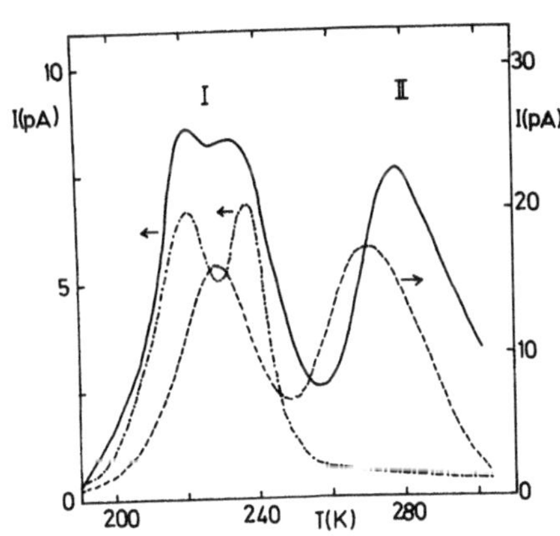

Fig. 2. The same as Fig. 1 for samples with $h = 16.4\,\%$

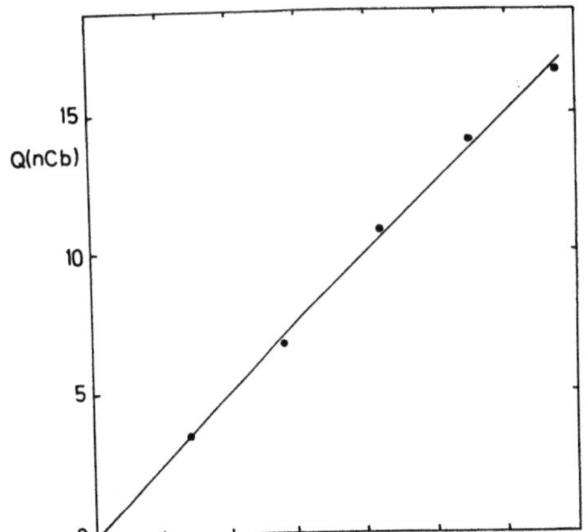

Fig. 3. Linear dependence of depolarization charge on polarizing voltage for a sample with $h = 19.8\,\%$ in the totally blocking (teflon) electrode arrangement

The results of the measurements with different electrode arrangements are consistent with the attribution of the contribution, which is present independently from electrode arrangement (band II in Fig. 2, for example), to dipolar polarization. This hypothesis is supported by the results of measurements with different polarizing fields. Figure 3 shows the depolarization charge Q of this contribution as a function of polarizing voltage for a sample with $h = 19.8\,\%$ in the case of totally blocking (teflon) electrodes. The linear dependence suggests dipolar polarization [3].

The complex high-temperature dispersion was analyzed into simpler responses by means of the fractional polarization technique, the partial heating technique and the thermal sampling technique [3]. Figure 4 shows the complex high-temperature band for a sample with $h = 14.3\,\%$ provided with silver paste contacts and two responses isolated in this band by the fractional polarization technique: during cooling down to 77 K the sample was polarized only in the temperature ranges 273–263 K and 243–233 K.

Figure 5 shows the responses isolated in band I of Fig. 4 by means of the thermal sampling technique. This technique consists of "sampling" the relaxation process within a narrow temperature range by means of polarization at T_p, followed by depolarization at T_d, a few K lower than T_p. The polarization time was 2 min, the depolarization time 2 min and the temperature window $T_p - T_d = 5$ K. The polarization temperature was scanned in steps of 5 K. Attempts to experimentally also analyze band II by the thermal sampling technique were less successful: the thermal sampling responses were broad and their peak temperatures outside the intervals (T_p, T_d). These results provide

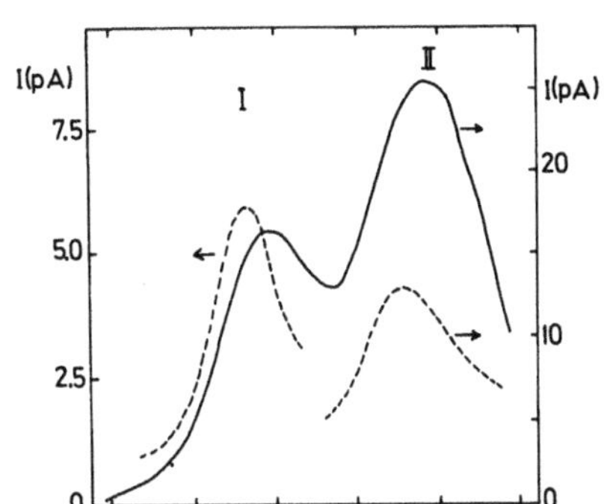

Fig. 4. The high-temperature dispersion measured on a sample with $h = 14.3\,\%$ provided with silver paste contacts (——) and two responses isolated by the fractional polarization technique (– – –)

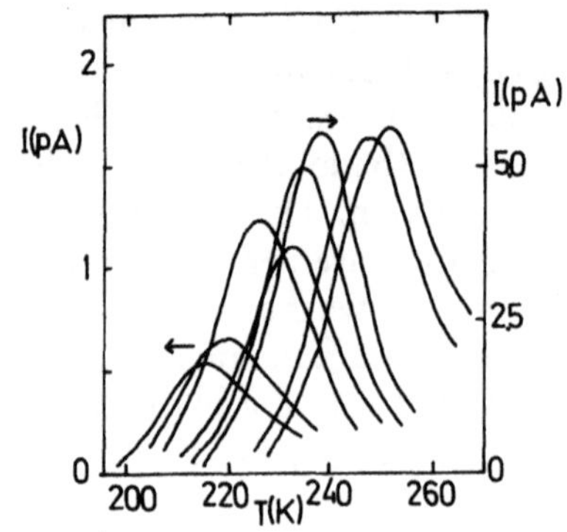

Fig. 5. Thermal sampling responses isolated in band I of Fig. 4

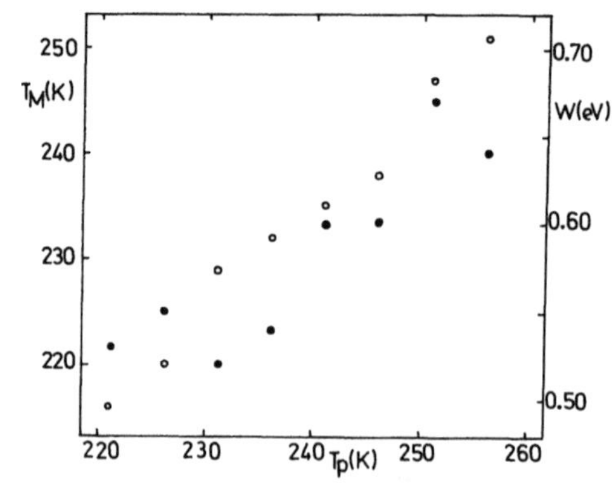

Fig. 6. Peak temperature T_M (○) and activation energy W (●) of the thermal sampling responses of Fig. 5 as functions of the polarization temperature T_p

further evidence that band I is due to dipolar polarization and band II to space charge polarization [6].

Figure 6 shows the peak temperature T_M and the activation energy W, calculated by the graphic integration method [3, 4], of the thermal sampling responses of Fig. 5. The unsystematic variation suggests that several relaxation mechanisms contribute to band I. The mean activation energy of about 0.60 V is lower than the activation energy of thermal sampling responses isolated in the central region of band II, $W = 0.85$ eV. Partial heating experiments support this finding.

The results reported above make reasonable to attribute band I to water-assisted relaxation of polar side chains of the protein macromolecules and band II to relaxation of space charges, probably ions.

References

1. Anagnostopoulou-Konsta A, Pissis P (1987) J Phys D 20:1168–1174
2. Kent M, Meyer W (1984) J Phys D 17:1687
3. Vanderschueren J, Gasiot J (1979) Field-induced thermally stimulated currents, In: Bräunlich P (ed) Thermally stimulated relaxation in solids. Springer, Berlin pp 135–223
4. Pissis P, Diamanti D, Boudouris G (1983) J Phys D 16:1311
5. Pethig R (1979) Dielectric and electronic properties of biological materials. J Wiley and Sons, Chichester
6. Pissis P, Anagnostopoulou-Konsta A (1985) In: Sessler GM, Gerhard-Multhaupt R (eds) Proceedings of the 5th International Symposium on Electrets, Heidelberg, IEEE Service Center, Piscataway, pp 842–847

Received April 24, 1987;
accepted April 21, 1988

Authors' address:

P. Pissis
National Technical University Athens
Physics Department
Zografou Campus
GR-157 73 Athens, Greece

Thermally stimulated current studies on polyimide film

L. Qingquan

Department of Electrical Materials Engineering, Harbin Institute of Electrical Technology, Harbin, China

Abstract: The thermally stimulated current in polyimide film was measured under different conditions. These results, with the addition of steady state *dc* conduction, have clearly shown that high temperature TSC peak in the range of 160–180 °C is possibly due to depolarization of space-charge of ions, bound into Frölichs two potential well, and generated through ionization of unimidized polyamic acid.

Key words: Thermally stimulated current, polyimide film, depolarization.

1. Introduction

Because polyimide possesses predominant electrical insulating properties, some researchers [1–3] have focused their attention on and discuss electrical behaviors in extensive temperature ranges (although TSC technique has been useful tool in the study of dipole and space charge relaxation in polymers). Tanaka [4] has carefully investigated the TSC characteristics in polyimide film and suggested the low temperature peak (β) at 50 °C is due to dipole depolarization and that the high temperature (α) at 173 °C comes from trapped electron space charges.

The main purpose of the present paper, based on some results obtained from TSC and *dc* conduction measurement, is to examine the nature of the α peak in polyimide film.

2. Experimental

Du Pont Kapton polyimide film of 50 μm thickness was sheared 40 × 40 mm^2, and used as a specimen. The specimens were first cleaned with alcohol, then two-side-metallized with aluminum. The diameter of Al-electrodes contacting with the specimen is 20 mm. The metallized films were dried at 70 °C for 24 h and at 370 °C for 2 h and referred to as specimen PI-1 and PI-2. Since polyimide film is sensitive to humidity all specimens were kept in a drier at room temperature before measurement. Furthermore, the films were previously purified with N,N-dimethylformamide (DMF), then Al-metallized and called specimen PI-3.

The poling parameters used in TSC measurement are as follows: poling field E_p = 2–70 kV/mm, poling time t_p = 15 min, 1 or 2 h, poling temperature T_p = 170 °C, heating rate $b = dt/dt$ = 60 K/h. The currents were measured with a model FJ-365 electrometer.

3. Results and discussion

Figure 1 shows short-circuit/bias field E_b = 0/TSC spectra of specimens PI-1 and PI-2 poled with E_p = 50 kV/mm at T_p = 170 °C, then rapidly cooled to room temperature with liquid nitrogen, and meas-

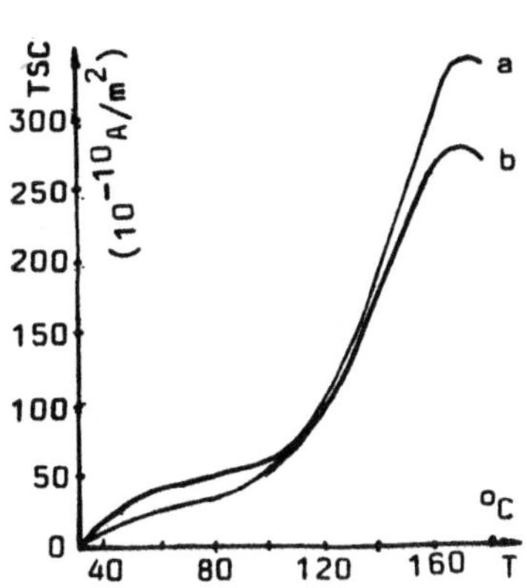

Fig. 1. Short-circuit TSC spectra of specimen PI-1 (a) and PI-2 (b) with E_p = 50 kV/mm at T_p = 170 °C

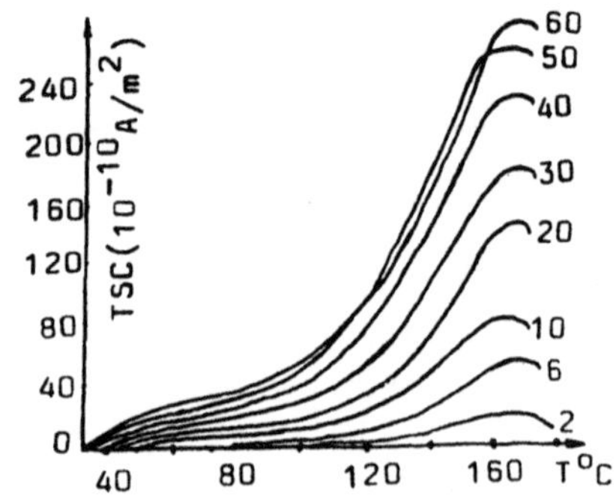

Fig. 2. TSC spectra of specimen PI-2 at different poling fields (kV/mm)

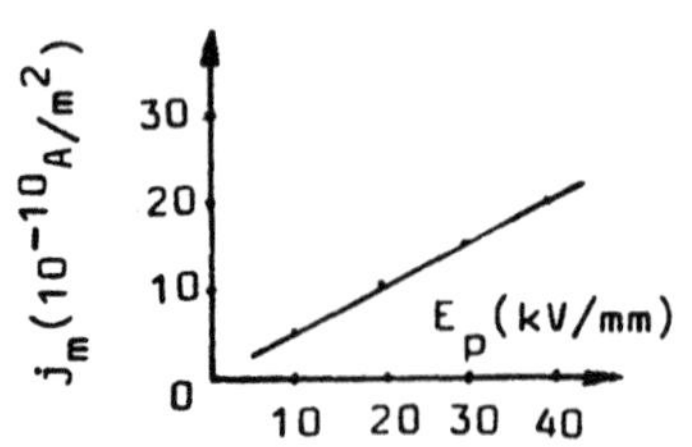

Fig. 3. β peak current as a function of poling field

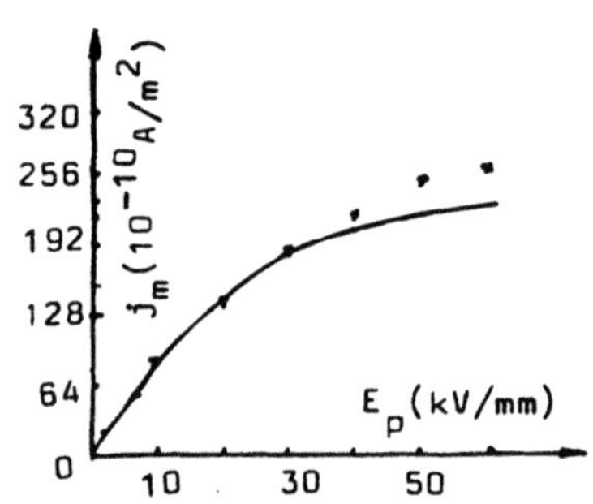

Fig. 4. α peak current as a function of poling field. (× – experimental value; solid curve – theoretical value

ured. It can be seen from Fig. 1 that there is a shoulder near 70 °C (β) and a pronounced peak near 170 °C (α) in both TSC curves. The TSC spectra at different poling fields for specimen PI-2 are shown in Fig. 2.

In order to determine the mechanism of peak generation we have plotted the peak current j_m as a function of poling field E_p in Figs. 3 and 4. The linear poling field dependence of the β peak shows that the β peak should be related to unimidized polyamic acid in polyimide film. Such a result agrees with Tanaka's conclusion. It is also noteworthy that at poling field E_p (= 2 kV/mm) this is much lower than that of the threshold (10 kV/mm) [5], i.e., in the absence of electron injection process the α peak can also be observed. That, in addition to our later discussion, proves the probable ionic property of the α peak. Both α and β peaks of specimen PI-2 in Fig. 1 are lower than those of specimen PI-2 in Fig. 1 which in part demonstrates that the unimidized polyamic acid has further produced imidization reaction generally proceeding in the range 150–300 °C. Consequently, the area of TSC curve can be used to assess the imidization extent.

It is well known that there are a lot of chemical and physical defects in polymers and that some of them can form electron or hole traps [6] localized in the "forbidden band" used in the band model. Some others rise to ionic traps [7] to the two-potential well applied to analyze ion conduction and polarization by thermal activation. The saturation of polarization approximately equal to the peak value, as shown in Fig. 4, implies that the polarization increases less than linearly with increasing field strength. According to Fröhlich's two-potential well [8], the polarization p is given as follows:

$$p = A \tanh (qdE_p/kT) \qquad (1)$$

where A is a constant independent of E_p, q is the electrical charge of carrier, d is half of carrier hopping distance, E_p is poling field, k is Boltzmann constant. The function $\tanh (qdE_p/kT)$ indicates the saturation behavior of polarization with increasing E_p. In our case $q = 1.6 \times 10^{-19}$ C, $d = 15 \times 10^{-10}$ m, $E_p = 20$ kV/mm, and $T = 443$ K (peak temperature). We assume that the value of the α peak is proportional to p and select a proper value of A to fit an experimental value; of course its magnitude never influences the tendency of curve change. After that we calculate the dependence of j_m on E_p (solid curve in Fig. 4). It can be seen that both experimental and theoretical curves are nearly consistent.

By measurement we observed that the α peak can not be ascribed to a single relaxation. Then, with peak-cleaning technique, an isolated α peak is acquired as shown in Fig. 5. In terms of initial rise method the migration activation energy of carrier is evaluated to be 1.21 eV. Fig. 6 shows the temperature dependence of steady-state dc conduction current and from the slope of the Arrhenius plot, we calculated the activation energy of dc conduction to be 1.31 eV. Because the specimen is short-circuited during TSC measurement, the α peak should arise from the self-migration of

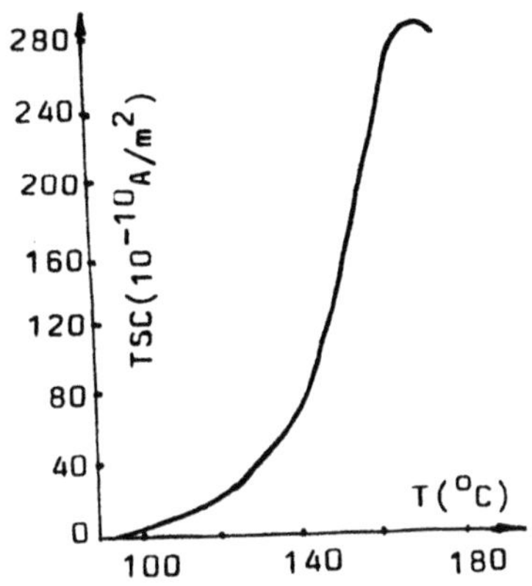

Fig. 5. Isolated α peak of specimen PI-2

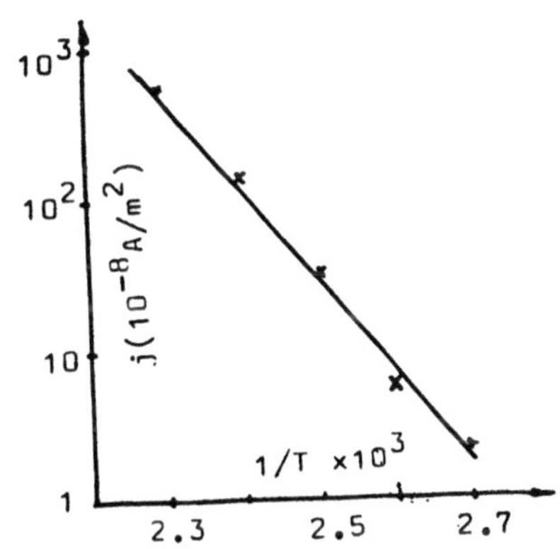

Fig. 6. Temperature dependence of dc conduction current

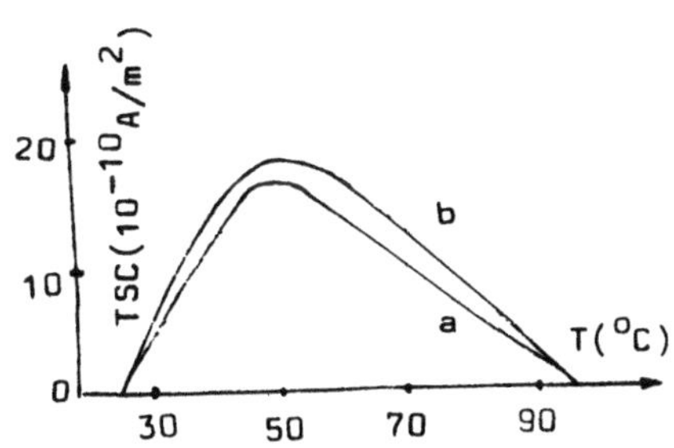

Fig. 7. TSC spectra for specimen PI-2 at $E_p = 70\,\mathrm{kV/mm}$ at $T_p =$ 25 °C. (a) $T_p = h$, (b) $T_p = 3\,\mathrm{h}$

space charges. Naturally, such a self-motion also forms dc conduction. This explains why both the activation energies responsible for the peak and dc conduction are nearly equal.

Figure 7 presents the TSC spectra for specimen PI-2 at $E_p = 70\,\mathrm{kV/mm}$ and $T_p = 25\,°\mathrm{C}$ for various poling

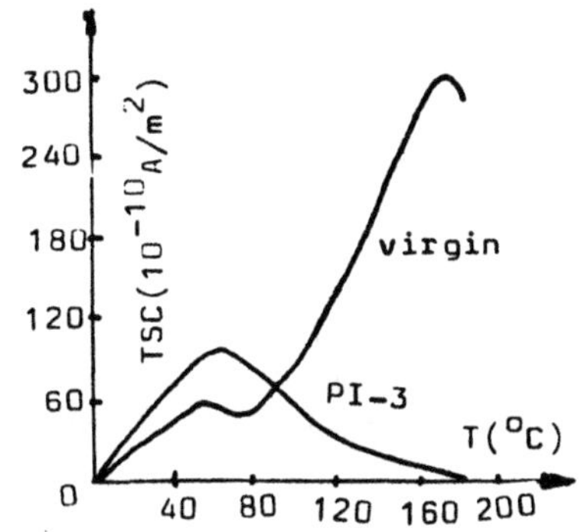

Fig. 8. TSC spectra for specimen PI-3 and virgin at $E_p = 70\,\mathrm{kV/mm}$

times. Although such a poling field is high enough to promote Schottky emission, the α peak in all the curves disappears, whereas, when the poling field (= 2 kV/mm) is too low to form electron injection, the α peak still exists as shown in Fig. 2. It seems that the α peak is independent of the release of trapped electrons.

By analyzing IR spectra of polyimide film, it has been seen that they contain a large amount of ions generated by dissociation of impurities and carboxylic acid groups and such ionic carriers that can contribute to TSC. The TSC spectra of specimens PI-3 and Virgin at $E_p = 70\,\mathrm{kV/mm}$ are shown in Fig. 8. The peak for specimen PI-3 almost disappears, which manifests that in film treated with the solvent mentioned above, the ionic concentration is lowered significantly. In addition, it should be noted that N,N-dimethylformamide cannot solve the unimidized polyamic acid in film, but only lowers it. When we obtain purified film which contains more polar molecules, then the peak becomes higher.

4. Conclusion

By analyzing the results of TSC and dc conduction of Kapton film, the α peak which likely comes from depolarization of bounded ions is explained by the following: (1) the saturation behavior is possibly predicated by Frölich's two-potential well, and is more responsible for an ionic-bounded state than for an electronic one, since the latter commonly belongs to Coulombic traps; (2) even at poling fields lower that those of the threshold poling field of electron emission, the α peak still exists; (3) activation energy of the α peak is approximately equal to that of dc conduction; and (4) the α peak for the purified polyimide film basically disappears.

References

1. Wrasidio W (1973) J Polym Sci Polym Phys Ed 11:2143
2. Sacher E (1978) IEEE Trans Electr Insul EI-13, 94
3. Sacher E (1979) IEEE Trans Electr Insul EI-14, 85
4. Tanaka T (1978) J Appl Phys 49:784
5. Hashimoto T, Shiraki M, Sakai T (1975) J Polym Sci Polym Phys Ed 13:2401
6. Partridge RH (1965) J Polym Sci A2 3:2817
7. Backer RE Jr (1976) Pure and Appl Chem 46:157
8. Frölich H (1958) Theory of Dielectrics. Clarendon, Oxford

Received April 24, 1987;
accepted April 21, 1988

Author's address:

L. Qingquan
Department of Polymer Physics
Charles University
V Holešovičkách 2
180 00 Praha 2, CSSR

Progress in Colloid & Polymer Science Progr Colloid Polym Sci 78:123-125 (1988)

Thermally stimulated currents in thin 1,4-cis polyisoprene films

J. Świątek, S. Tkaczyk, and A. Kwiatkowska

Institute of Physics, Pedagogical University, Częstochowa, Poland

Abstract: The TSC (thermally stimulated currents) method has been applied to study the trap structure of thin 1,4-cis polyisoprene films. The obtained spectra in the temperature range 100 K–350 K have been analyzed numerically by fitting Gauss distributions to TSC peaks. Five groups of traps of the energy depths 0.43, 0.46, 0.50, 0.66 and 0.77 eV have been distinguished.

Key words: Thermally stimulated currents, trap structure, 1,4-cis polyisoprene, crystallization, viscoelastic transition.

1. Introduction

The thermally stimulated current (TSC) method is a convenient tool for the study of electronic properties of crystalline as well as polycrystalline and amorphous high-resistivity materials. Peaks occuring in the TSC spectra, particularly in amorphous solids, can be interpreted in terms of corresponding maxima in the energy dependence of the gap states [1], or as current peaks associated with the thermal release of non-equilibrium carriers from localized states in the energy gap. For chalcogenide glasses, for example, it was demonstrated by Agarwal et al. [2, 3] and Müller [4, 5] that the TSC peaks observed in these materials do not contain any information about a distribution of states in the energy gap, but that they are connected with a discharge of various electrode and sample capacitances through the temperature-dependent resistance of the material. The aim of this work was to determine the trap structure of thin 1,4-cis polyisoprene films.

In literature [6–11] one may find a large number of models which relate experimental TSC spectra to trapping state parameters such as the energy depth, capture cross-section, attempt-to-escape frequency factor, and to their concentration. Each of these methods gives only approximate values of trap parameters. An important factor in the analysis of TSC spectra is a precise determination of the energy depth of trapping states. Its value is crucial for evaluation of other parameters of localized states and estimation of the total number of states participating in TSC. For this reason methods which allow us to isolate peaks from a TSC spectrum are valuable. Most widely used here is the "thermal cleaning" procedure proposed by Nicholas and Woods [12]. We applied a "computer cleaning" of a TSC spectrum to obtain undisturbed shapes of TSC peaks.

2. Experimental

The TSC measurements were performed for 1,4-cis polyisoprene of average viscosic molecular weight $1.5 \cdot 10^6$. Samples were prepared by centrifugal spinning from a solution in toluene on a glass wafer covered with a metal electrode. Electrodes of gold and aluminium were deposited by vacuum evaporation to produce a sandwich configuration. One of these contacts (Au) was on ohmic contact to inject charge carriers into a sample and fill the trapping centers at low temperature. Aluminium played the role of noninjecting contact to prevent injection of carriers during TSC measurements. Current-temperature characteristics were measured using a picoammeter and registered with an X-Y recorder. To control the temperature a Copper-Constantan thermocouple was employed. All measurements were performed in a vacuum of 10^{-6} Torr at a temperature range of 100 K–350 K. The experimental set up was described elsewhere [13, 14]. For TSC measurements the sample was polarized by an electric field of about: $10^7 \frac{V}{m}$ for 30 min and then cooled in liquid nitrogen. At this temperature the polarizing

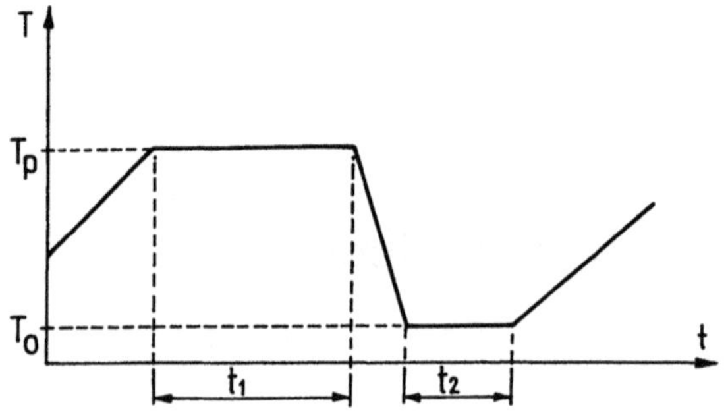

Fig. 1. The scheme of a heating-cooling process: T_p — temperature fo polarization; t_1 — time of polarization; T_0 — the lowest temperature of the experiment; t_2 — time of freezing: The heating rate for the whole temperature range was 4 K/min

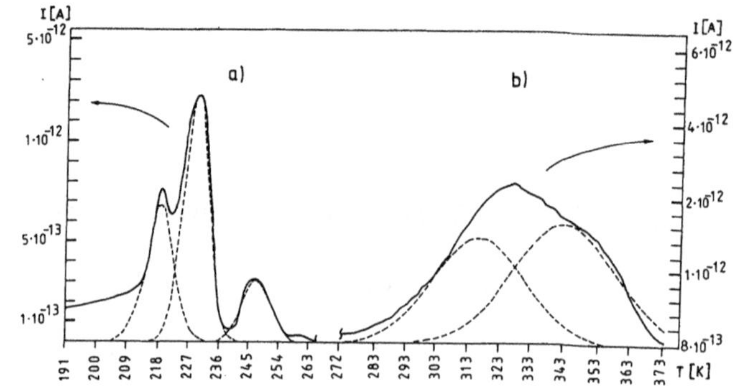

Fig. 2. TSC spectrum for 1,4-cis polyisoprene

voltage was removed, voltage of the order of 1 V at opposite polarization was applied, and the sample was heated at a constant rate. Current flowing through the electrometer was recorded as a function of temperature. At a temperature of about 380 K the polarizing field was reapplied, the sample re-cooled to the liquid nitrogen temperature, and the whole procedure was repeated with the same or different heating rate. A strict heating-cooling schedule was followed in order to obtain reproducible data. A scheme of this process is shown in Fig. 1.

3. Results and discussion

Typical TSC spectrum for polyisoprene films is shown in Fig. 2 a,b for low and high temperature ranges. The dashed curve represents a spectrum obtained after computer "cleaning" of the experimental data. This cleaning procedure was performed under the assumption that a TSC peak may be described by a Gauss distribution and it ensured more exact determination of the energy depth of localized levels. In the spectrum we can distinguish four TSC maxima, however, the broad maximum at 326 K is well-described by two Gauss distributions, so we suggest five groups of traps in polyisoprene. At temperatures below 300 K 1,4-cis polyisoprene contains both crystalline and amorphous regions whose relative proportions may be varied by a heat treatment. The temperature of maximum rate of crystallization for 1,4-cis polyisoprene varies in the range 245–248 K in dependence on the trans-fraction. In our case polyisoprene of average viscosic molecular weight $1.5 \cdot 10^6$ contains 98 % of cis fraction and the temperature of maximum rate of crystallization for this polymer is 247 K. We think that the TSC peak at 248 K is due to charges trapped at interfaces between crystallites and amorphous regions. However, the process of crystallization occures also at lower temperatures so we suppose that TSC maxima below 248 K are also connected with amorphous/crystalline boundaries. The temperature 295 K is known as the melting temperature of crystalline phase for this polymer [15], thus above 300 K 1,4-cis polyisoprene is completely in amorphous phase. We suggest that the high temperature peaks at 316 K and 342 K are due to charges trapped by field-induced dipoles or traps because a transition into a viscoelastic state is observed [16] in this region. Values of the energy depth for particular localized states starting from the lowest temperatures are: 0.43, 0.46, 0.50, 0.66 and 0.77 eV.

Acknowlegdement:

This work was sponsored by the Polish Ministry of Science and Schools of Academic Rank within a Central Project of Basic Research, CPBP 01.08.

References

1. Fritzsche H, Ibaraki N (1985) Phil Mag B 52:299
2. Agarwal SC (1974) Phys Rev B 10:4340
3. Agarwal SC, Fritzsche H (1974) Phys Res B 10:4351
4. Müller P (1974) Phys Stat Solidi (a) 23:165
5. Müller P (1975) Phys Stat Solidi (a) 28:521
6. Garlick FJ, Gibson AF (1948) Proc Phys Soc 60:574
7. Grossweiner LJ (1953) J Appl Phys 24:1306
8. Haering RR, Adams EN (1960) Phys Rev 117:451
9. Bucci C, Fieschi R (1964) Phys Rev Lett 12:16
10. Bucci C, Fieschi R, Guidi G (1966) Phys Rev 148:816
11. Simmons JG, Taylor GW, Tam MC (1973) Phys Rev B 7:3715
12. Nicholas KH, Woods J (1964) Brit J Appl Phys 15:783

13. Świątek J (1980) Trap States and Electrical Conductivity of Thin Polycrystalline p-Phenyl Layers. Scientific Bulletin of Łódź Technical University No 13
14. Tkaczyk S, Koćwin T, Kwiatkowska A, Świątek J (1985) Proc 5th Intern Symp Electrets, Heidelberg, p 305
15. Reff RAV, Doak KW (eds) (1965) Crystalline Olefin Polymers. Part I, Interscience Publ, New York London Sydney
16. Wintle HJ (1972) In: Dole H (ed) The Radiation Chemistry of Macromolecules. Academic Press, New York London

Received June 25, 1987;
accepted April 21, 1988

Authors' address:

J. Świątek
Institute of Physics
Pedagogical University
Al. Zawadzkiego 12/15
42-200 Częstochowa, Poland

Progress in Colloid & Polymer Science Progr Colloid Polym Sci 78:126–128 (1988)

Electronic properties of polyaniline

A. Kuczkowski

Faculty of Technical Physics and Applied Mathematics, Technical University, Gdansk, Poland

Abstract: We have investigated the electrical, thermoelectrical, and electrochemical properties of polyaniline (PANI) obtained by electrochemical oxidation of aniline hydrochloride. Van der Pauw's, two electrode and four-probe conductivity (σ) measurements were performed on compressed powder pellets of PANI equilibrated at different pH. PANI shows similar dependence of thermoelectric power (TEP) and $\log \frac{1}{\sigma}$ upon pH. At low pH the value of TEP was negative and low. At pH ~2 TEP changes its sign from negative to positive and increases with increasing pH. The temperature dependence of conductivity of PANI is closely matched by the expression for variable range hopping. A Hall mobility measurement test for PANI failed. The estimated value of the Hall mobility from this measurement was lower than 0.5×10^{-4} m^2/V. s. We have constructed a rechargeable Zn/aqueous $ZnCl_2$ + NH_4Cl/PANI battery showing promising characteristics. Using PANI obtained by anodic oxidation of aniline hydrochloride.

Key words: Polyaniline, conducting polymers, polymer electrochemical batteries.

Introduction

Polyaniline (PANI) is an attractive conducting polymer, not only of fundamental scientific interest [1] but also of potential technological importance. Polyaniline-based batteries and electrochromic devices seem very promising, mainly because of the excellent reversibility of their characteristic redox reactions and their stability [2, 3].

PANI can be prepared by chemical or electrochemical oxidation of aniline [4, 5]. The properties of PANI depend strongly on the preparation method [5].

The purpose of this work is to present the electrical properties of PANI prepared by electrochemical oxidation of aniline hydrochloride.

Experimental

Samples of PANI powder were prepared by anodic oxidation of aniline hydrochloride as described in [5], at a potential of 1.9 V. The samples were washed with water, and subsequently immersed for 12 h in solutions of given pH ($0 < pH < 10$).

The final pH value reached by the sample-solution system was assumed to be the equilibrium pH for the polymer. After drying the compressed powder, pellets were formed under a pressure of 6×10^8 Pa as tablets (diameter: 12 mm, thickness: from 0.7 mm to 2.4 mm).

As already reported [4] variation of pH has a strong effect on the conductivity of PANI. As seen from Fig. 1, treatment of PANI with strong alkalis (2M NaOH) decreases the conductivity from four to six orders of magnitude. The conductivity can be restored by treatment with a strong acid (5M HCl).

Van der Pauw's two electrode and four-probe conductivity measurements were performed in vacuum (10^{-2} Pa), as well as thermoelectric power measurements. In the latter case the sample were mounted between two copper blocks using pressure contacts.

Results and discussion

The temperature dependence of the thermoelectric power (TEP) of the two samples of PANI with different pH is shown in Fig. 2. At low pH, the value of TEP was low and negative. At pH ~ 2 TEP changes its sign from negative to positive and increases with increasing pH, indicating a transition from hole-dominated to electron-dominated transport with decreasing pH. The temperature dependence of the conductivity of

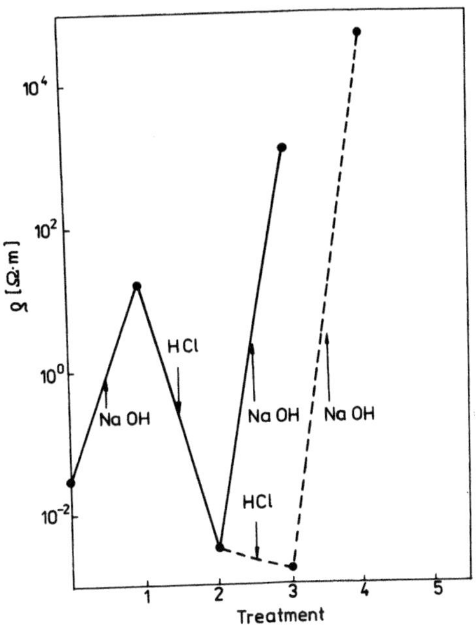

Fig. 1. The resistivity of 1.32 mm thick pelet of polyaniline subjected to series of alkali/acid treatments

PANI (Fig. 3) is closely matched by the expression for variable range hopping:

$$\sigma(T) = K_o T^{-1/2} \exp\left[-(T_o/T)^{1/4}\right].$$

The mean square least values for T_o and K_o for samples prepared at different pH are given in Table 1.

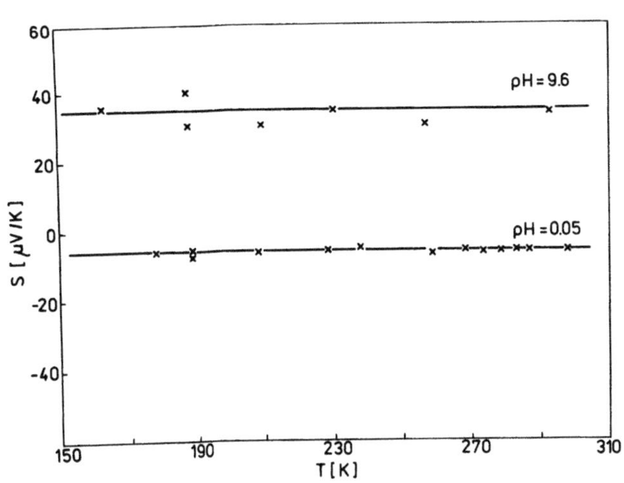

Fig. 2. Temperature dependence of thermopower for polyaniline samples equilibrated at indicated pH

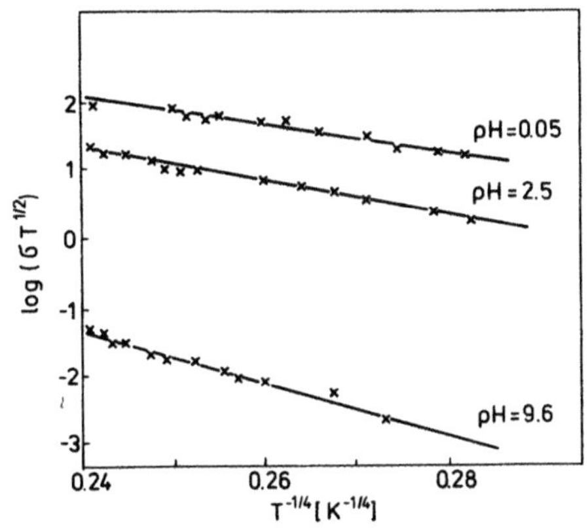

Fig. 3. Log ($\sigma \cdot T^{1/2}$) as a function of $T^{-1/4}$ for polyaniline samples equilibrated at different pH

The results in Table 1 are similar to the results obtained [4] for PANI prepared by chemical oxidation of aniline by $S_2O_8Na_2$ or $K_2Cr_2O_7$.

We have also carried out Hall mobility measurement tests for high conductivity samples of PANI. However, the value of the Hall voltage was below the sensitivity of our measurement set (10 μV). Estimated from the experimental conditions (thickness of sample $d = 0.02$ cm, resistivity $\varrho = 0.54\ \Omega \cdot \text{cm}$, $I = 25$ mA, $B = 1$ T) the value of the Hall mobility was lower than $0.5 \cdot 10^{-4}\ \text{m}^2/\text{V.s}$.

We have constructed a rechargeable Zn/aqueous $ZnCl_2 + NH_4Cl$/PANI battery (Fig. 4), using PANI obtained by anodic oxidation of aniline hydrochloride.

The charge-discharge characteristics of this battery are shown in Fig. 5. The properties of this battery are similar to the property of the battery described in [3].

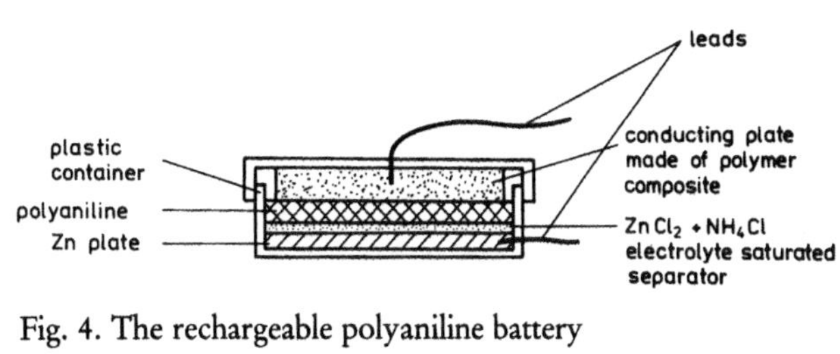

Fig. 4. The rechargeable polyaniline battery

Table 1.

pH	0.05	2.48	9.62
T_o [K]	$1.6 \cdot 10^6$	$8.87 \cdot 10^6$	$4.2 \cdot 10^7$
K_o [Ω^{-1} cm^{-1} K$^{1/2}$]	$4.9 \cdot 10^5$	$1.07 \cdot 10^7$	$1.2 \cdot 10^7$

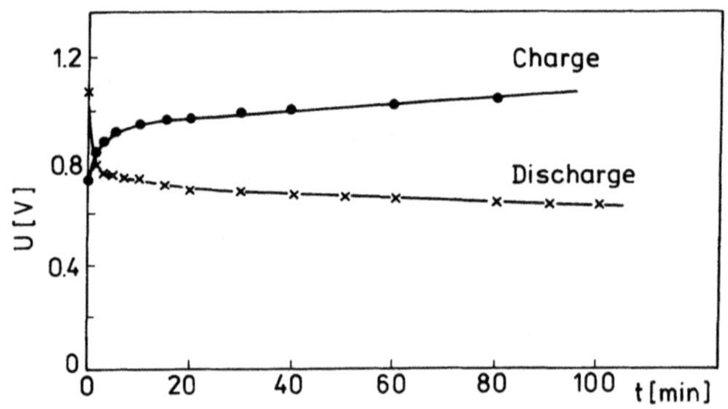

Fig. 5. Charge-discharge characteristics of the shown in Fig. 4 polyaniline battery, performed with a constant current equivalent to 4.5 mA/cm^2 for polymer electrode

References

1. Mac Diarmid AG, Chiang JC, Richter AF, Epstein AJ (1987) Synth Met, in press
2. Genies EM, Lapkowski M, Santier C, Vieil E (1987) Synth Met, in press
3. Mac Diarmid AG, Halpern M, Krawczyk J, Mammone RJ, Mu SL, Sommasiri NLD, Wu WQ (1984) In: ICSM '84 Abano Terme Conference Abstracts Bulletin, p 32
4. Travers JP, Chroboczek J, Devreux F, Genoud F, Nechtschein M, Syed A, Genies EM, Tsintavis C (1985) Molec Cryst Liq Cryst 121:198
5. Langer J (1984) Materials Sci X:173

Received June 25, 1987;
accepted April 21, 1988

Author's address:

A. Kuczkowski
Faculty of Technical Physics
Technical University
ul. Majakowskiego 11/12
Gdansk 80-952, Poland

 Progr Colloid Polym Sci 78:129–132 (1988)

The fluorescence of the concentrated solutions of N-vinylcarbazole and the structure of the second excimer sites of poly-N-vinylcarbazole

R. M. Siegoczyński and J. Jedrzejewski

Institute of Physics, Warsaw University of Technology, Warsaw, Poland

Abstract: The emission properties of N-vinylcarbazole have been investigated. Fluid solutions were observed covering a broad range of concentration to exhibit the structureless excimer emission bands at energies approximately 2 200 cm^{-1} and 4 100 cm^{-1} to the red of the 355 nm emission band of the monomer fluorescence of dilute solution. In high concentration range (1*M*) N-vinylcarbazole shows broad fluorescence composed of three components. The broad emission at the longer wavelength (λ_{max} = 415 nm) is assigned to the sandwich-like excimer fluorescence and the interesting, new emission in the short-wavelength region (λ_{max} = 385 nm) is assigned to the second excimer fluorescence in analogy to poly-N-vinylcarbazole.

Key words: Fluorescence, N-vinylcarbazole, excimer, poly-N-vinylcarbazole.

Introduction

It is well known that poly-N-vinylcarbazole (PVCz) exhibits a structureless broad emission band which peaks at about 420 nm and a structureless emission in the shorter-wavelength region (λ_{max} = 380 nm). Kloepffer has assigned the 420 nm emission band to excimer interaction between two adjacent side groups in polymer chain (sandwich-type excimer) [1]. Johnson proposed that the 380 nm emission band represents fluorescence from a second excimer state whose stabilization arises from configurational interaction of the lowest lying excitation and charge-resonance states with the departure from planarity of the two pendant groups [2]. Itaya et al. proposed that the structure of this second excimer seems to be the overlap of only one benzene ring in the neighboring chromophores in sandwich-type geometry [3]. Further insight into the question of the geometry of the second excimer is offered by investigations of excimer emissions of N-vinylcarbazole (N-VCz) and their decay times in the concentrated solutions.

PCPS HT 024

Experimental

Photostationary emission spectra were recorded on a spectrophotometer SF-1 Cobrabid and corrected for the detector response. The fluorescence-decay times were measured with a pulsed N_2 gas laser as the exciting light source.

Results

A. Photostationary results

Figure 1 shows the total emission spectra of dilute solutions of N-VCz in degassed benzene at room temperature in low concentration region (10^{-4}–10^{-3} M). The structured monomer fluorescence bands with (0.0) at about 355 nm do exhibit a slight dependence on concentration. As the concentration of solution is increased the intensity of the monomer band is observed to decrease slightly and the contribution due to excimer emission is seen as a weak, structureless tail (above 400 nm) extending to lower energies.

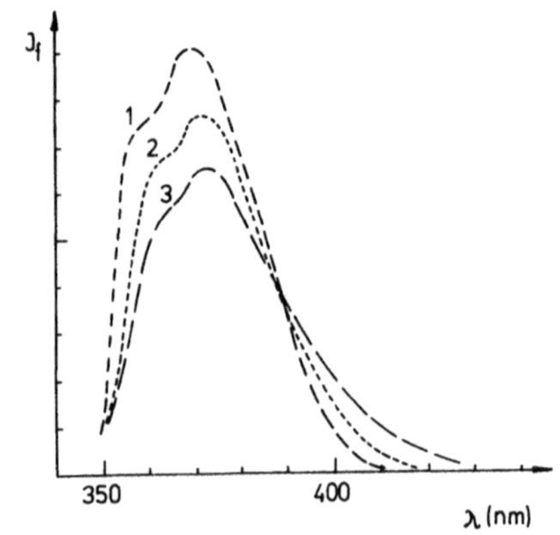

Fig. 1. Fluorescence spectra of N-VCz in benzene. Concentration: (1) 10^{-4} M, (2) $5 \cdot 10^{-4}$ M, (3) 10^{-3} M

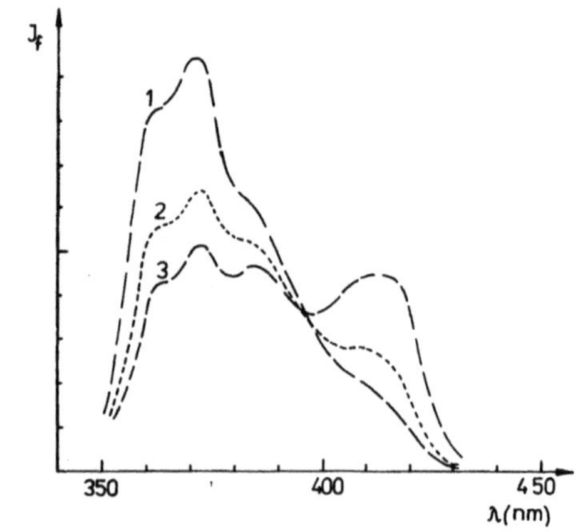

Fig. 2. Fluorescence spectra of N-VCz in benzene. Concentration: (1) 10^{-2} M, (2) 10^{-1} M, (3) 1 M

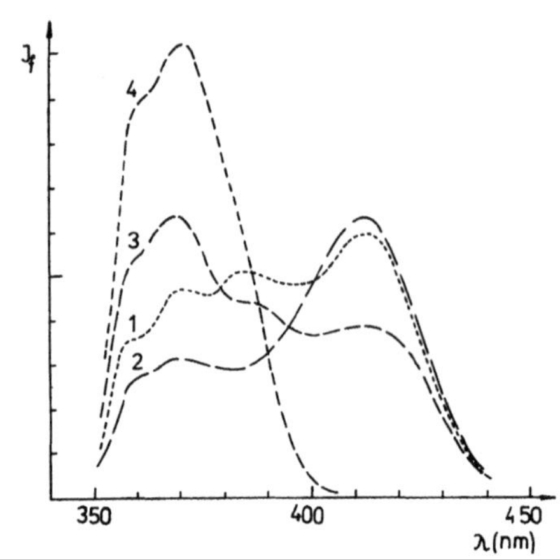

Fig. 3. Fluorescence spectra of N-VCz in benzene. Concentration: (1) 2 M, (2) 5 M, (3) $5 \cdot 10^{-1}$ M*, (4) 10^{-4} M*; (* solutions which were received from sample of 5 M by dilution)

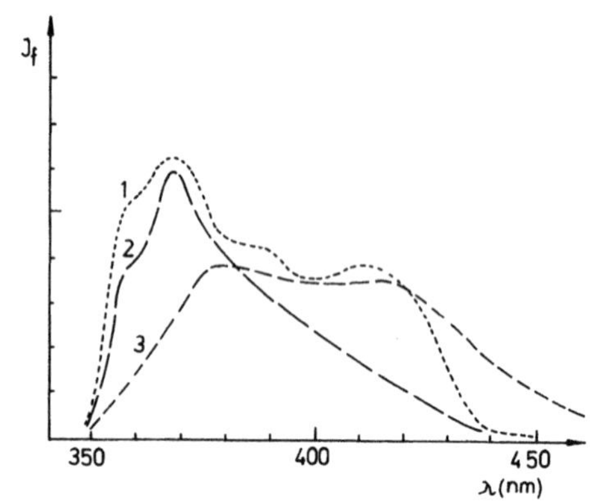

Fig. 4. (1) Fluorescence spectrum of N-VCz ($5 \cdot 10^{-1}$ M) in benzene at 293 K; (2) Fluorescence spectrum of N-VCz in MTHF at 108 K [2]; (3) Fluorescence spectrum of PVCz in benzene at 293 K [1]

Figure 2 shows the emission spectra in higher concentrations (10^{-2}–1 M). The high energy region of the total emission spectrum is characteristic of monomer-like fluorescence of N-VCz. In the low energy region the most important aspects of the emission, the appearance of the well-defined bands at 385 nm and 415 nm which increase in intensity with increasing concentration, are maintained.

The emission spectra shown in Fig. 3 were obtained for solutions of N-VCz in the highest concentration region (2–5 M). Two significant aspects of the emission as the concentrations of solutions are further increased, are evident. First, the 415 nm band is a dominant one in the fluroescence spectrum and second, the disappearance of the 385 nm band is observed. It must be emphasized that the extremely high concentrations of N-VCz used in these experiments do not lead to polymerization of N-VCz in fluid solution. When the highly concentrated solution was gradually diluted the changes in observed emission spectrum followed a reversible path to the complete reappearance of the low concentration monomer-like spectrum as seen in Fig. 3.

B. Fluorescence decay measurements

Emission decay curves for dilute solutions were detected at 355 nm and for concentrated solutions at 355 nm, 385 nm, 420 nm and 460 nm. The fluorescence decay observed at 355 nm displays concentration-dependent exponential decay. The fluorescence decays at 380 nm and 420 nm exhibit two-exponential decays with two distinctly separable components.

Discussion

It is clear that three concentration regions should be considered. Region I corresponds to the low concentration from 10^{-4} M to 10^{-3} M. As seen in Fig. 1, emission spectra exhibit the structured monomer fluorescence band with a (0.0) band at about 355 nm. The fluorescence decay curve of N-VCz in this region is clearly one-component with a decay constant of 10 ns.

The second concentration region, termed Region II, corresponds to high concentrations from 10^{-2} M to 1 M. As seen in Fig. 2, increasing concentration leads to a decrease in monomer fluorescence intensity, to a corresponding decrease in lifetime from 10 ns to 3.6 ns, to the appearance and the increase in overlapping 385 nm and 415 nm fluorescence intensities. This is the first observation of such a 385 nm band for N-VCz. The appearance of a well-defined, broad, structureless 415 nm emission band with the large Stokes' shift to the red is analogous to the concentration-dependent, diffusion-limited formation of excimer observed for many aromatics and heteroaromatics in solution. Furthermore, it is well known that solutions of the polymer of PVCz also show a structureless emission band similarly located and assigned earlier to the "sandwich-like excimer" [1]. On the basis of these results it is evident that the 415 nm emission follows from the excimer state [4, 5]. A similar interpretation of the nature of the state responsible for the 385 nm band is proposed. The structureless nature of the 385 nm band compared to the structured fluorescence of the parent monomer is suggestive of an emission to a repulsive ground state, i.e. an excimerlike fluorescence. Solution of the polymer PVCz also shows a structureless emission band similarly located [2] (assigned earlier to so-called second excimer).

Figure 4 presents, for example, a comparison of the emission spectrum of a concentrated solution of N-VCz in benzene at 293 K with that of PVCz. The long-wavelength part of the emission spectrum of N-VCz is similar to that of PVCz. The monomer fluorescence band of N-VCz (355 nm and 370 nm) is similar in location to that of the short-wavelength part of N-VCz at low temperature. A comparison of decay times of fluorescence found in this work with those previously obtained for PVCz and its useful model compound 1,3-bis(N-carbazolyl)propane (BCP) seems a further indication of a very close analogy between the physical origin of corresponding fluorescence bands in N-VCz and PVCz spectra. A short-decay time found for the short-wavelength region of N-VCz emission at 380 nm (4.6 ns) is comparable with the lifetime of 6 ns found for PVCz emission measured at 375 nm [6].

A long-decay time of N-VCz emission observed at 460 nm (15 ns) corresponds to the fluorescence lifetime 15 ns and 22 ns found for BCP [7–9] and PVCz [6], respectively. The above listed similarities in behavior of the fluorescence spectra of the highly concentrated solutions of N-VCz and of the fluorescence spectra of PVCz and its model compound BCP can be clearly considered as a proof that the nature of interaction and the origin of different fluorescence bands of N-VCz are the same as those in PVCz solutions.

The third concentration Region (III) occurs at 1 M and above. In this highest concentration region the most important aspects of the emission, the disappearance of the 385 nm band and a slight increase of intensity of the 415 nm band, are observed. These results are qualitatively like those observed for the pressure measurements of the fluorescence of PVCz [10, 11]. It has been found [11] that with a slight increase of pressure the intensity of the 420 nm band (sandwich-like excimer) is increased at the expense of intensity of the 380 nm band (second excimer). These results and similarities seem to provide evidence for a model of second excimer and a departure from planarity of the two pendant groups (Itaya model). It seems that pressure measurements of the emission of N-VCz solutions might be of crucial importance in the final choice between Johnson and Itaya models of "second excimer" (this is now being prepared in our laboratory).

Acknowledgment

We express our thanks to Dr. J. Dresner for helping us measure the decay times and to Professor J. Prochorow for his helpful discussions.

References

1. Kloepffer W (1969) J Chem Phys 50:2337
2. Johnson GE (1975) J Chem Phys 62:4697
3. Itaya A, Okamoto K, Kusabayashi S (1976) Bull Chem Soc Jpn 49:2082
4. Asai M, Matsui H, Takeda Y, Tazuke S (1972) International Symposium on Macromolecules, Helsinki, July

5. Siegoczyński RM, Jedrzejewski J, Kawski A (1975) Acta Phys Pol A47:707
6. Masuhara H, Tamai N, Inoue K, Mataga N (1982) Chem Phys Lett 91:109
7. Kloepffer W (1969) Chem Phys Lett 4:193
8. Kloepffer W, Liptay W (1970) Z Naturforsch 25a:1091
9. Kloepffer W (1970) Ber Bunsenges Physik Chem 74:693
10. Johnson PC, Offen HW (1971) J Chem Phys 55:2945
11. Chryssomallis G, Drickamer HG (1979) J Chem Phys 71:4817

Received June 25, 1987;
accepted April 21, 1988

Authors' address:

R. M. Siegoczyński
Institute of Physics
Warsaw University of Technology
ul. Koszykova 75
00 662 Warsaw, Poland

Low temperature spectroscopy of porphyrins in polymer hosts

J. Hala, I. Pelant, and M. Ambroz

Department of Chemical Physics, Faculty of Mathematics and Physics, Charles University, Prague, Czechoslovakia

Abstract: Low temperature spectroscopy of phthalocyanine and chlorophyll molecules embedded in polystyrene and polyethylene foils at temperatures below 10 K are reported. The role of polymer host is discussed in the frame of general theory of low temperature spectroscopy. The broad band and well resolved site selection fluorescence and fluorescence excitation spectra were measured. The vibrational analysis of these site selection spectra was performed and the frequencies of normal vibrations obtained in polymer hosts were compared to those in organic solvents.

Key words: Site selection spectra, porphyrins in polymer hosts.

Introduction

Absorption and fluorescence spectra of porphyrins in solutions as well as polymer matrices at room temperature consist of several broad spectral bands (FWHM ~ 100 cm^{-1}). Low temperature optical spectroscopy provides convenient experimental methods (site selection spectroscopy and hole burning technique) for extending the spectral resolution well above the factor of 100 [1–3]. The aim of this paper is to present site selection spectra obtained for both free base phthalocyanine (H_2-Phth) in a polyethylene host and chlorophyll-a (Chl-a) in a polystyrene host, including vibrational analysis (VA) and study of porphyrin-polymer interaction.

The Chl-a polystyrene samples were prepared by a slow evaporation, from the mixture of both the pigment and polymer chloroform solutions, from a mercury level. Concentration of Chl-a was ~ 10^{-5} M/l. The H_2-Phth polyethylene samples were prepared analogously by evaporation of the warm xylene solutions. The concentration of the H_2-Phth was $\lesssim 10^{-6}$ M/l. Thickness of measured samples ranged from 0,1 to 2,0 mm. Fluorescence spectra were excited by a pulsed nitrogen pumped dye laser ($\Delta\lambda$ = 0,02 nm, Δt = 2 ns), detected photoelectrically and processed by a boxcar integrator. Measurements were carried out at 5 K. More detailed description of the experimental set up can be found in [7].

H_2-Phth is a very stable symmetric porphyrin-like molecule, which has been widely studied using low temperature spectroscopy [3–6]. In n-alkane matrices especially both absorption and fluorescence spectra exhibit sharp line spectral structure under broad band excitation (Shpolskii effect). It has been found that H_2-Phth occupies two or several different positions (sites) in n-alkane single- or poly-crystals. The guest-host interaction is described by the site distribution function (SDF), which is a special function close to zero for all wavelengths, except for two or more very intense and narrow (~ 1 cm^{-1}) maxima at wavelengths corresponding to the O—O transitions in different sites. On the other hand no sharp lines were observed in broad band excited spectra of H_2-Phth in polyethylene (see the curve depicted 640–670 in the Fig. 1). This fact is due to the whole variety of guest-host interactions (many different sites) in the H_2-Phth polyethylene sample. The related SDF can be calculated from the theory of low temperature spectra [7]. It is a broad band gaussian-like function. The selection of particular sites needs a narrow laser excitation [1]. There are four site selected fluorescence spectra in Fig. 1 (curves 662, 660, 655, 643) corresponding to the different wavelengths of narrow laser excitation in nm. The observed sharp lines shift as the laser excitation is tuned, however the wavenumber differences between

PCPS HT 007

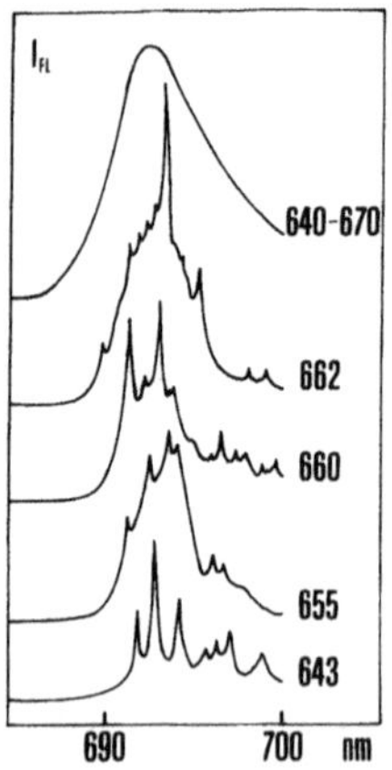

Fig. 1. Low temperature fluorescence spectra of H_2-Phth in polyethylene (T = 5 K, broad band excitation 640–670 nm, narrow band laser excitations at 662, 660, 655 and 643 nm)

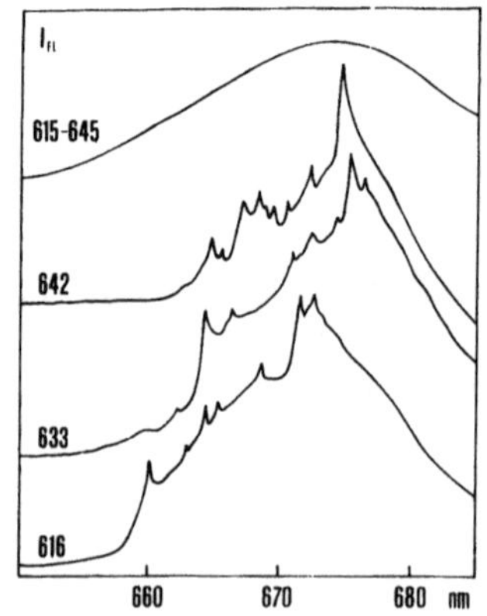

Fig. 2. Low temperature fluorescence spectrum of Chl-a in polystyrene (T = 5 K, broad band excitation 615–645 nm, narrow band laser excitations 642, 633 and 616 nm)

Table 1. The frequencies of normal vibrations (FNV) of free base phthalocyanine (H_2-Phth) in the first excited singlet state

polyethylene [This work] cm^{-1}	n-octane [4] cm^{-1}	n-dodecane [4] cm^{-1}	supersonic jet [5] cm^{-1}
685	677	676	685
730	719	717	728
805	796	793	795
830			
855			
905			
935			
1035	1028	1022	
1099	1081	1083	
1111	1110	1106	
1147			1142
1189	1190	1183	1185
1209	1200	1197	
1248	1235	1230	

Table 2. The frequencies of normal vibrations (FNV) of chlorophyll-a (Chl-a) in the first excited singlet state

polyethylene [This work] cm^{-1}	n-octane [1] cm^{-1}	n-hexadecane [1] cm^{-1}	n-eicosane [1] cm^{-1}
261	258	267	256
347	348	348	350
750	748	742	749
992	990	991	1000
1259	1253	1254	1256
1345	1353	1354	1356

laser line and O—O fluorescence ones remain unchanged and represent frequencies of normal vibrations (FNV) in the first excited singlet state. The most pronounced FNV are collected in Tab. 1 together with data obtained from literature for H_2-Phth in frozen solutions [5] and supersonic expansion jets [6].

Chl-a represents a biological active porphyrin molecule which does not exhibit any sharp line structure at all in broad band excited fluorescence. The SDF of Chl-a were found to be broad gaussian-like functions in an arbitrary matrix. The broad band excited Chl-a in polystyrene fluorescence spectrum is shown in Fig. 2 (the curve depicted 615–645 nm). Only narrow laser excited (site selection) fluorescence spectra provides vibrationally resolved sharp lines (see Fig. 2, the curves 642, 633, 616). Analogously to previous results the VA performed provides FNV in the first excited singlet state (see Tab. 2).

In conclusion we have presented a new application of low temperature site selection fluorescence spectroscopy of both H_2-Phth and Chl-a in polymer hosts. The VA of sharp line vibrationally resolved spectra brought FNV in the first excited singlet state, which are close to those measured in n-alkanes. Quite original results were obtained in the comparison of SDF for the pigments used reflecting a high number of different sites in both high and low symmetric porphyrin molecules embedded in polymer matrices.

References

1. Hala J, Pelant I, Parma L, Vacek K (1981) J Luminesc 24–25, 803, 26:117
2. Kador L, Schulte G, Haarer D (1986) J Phys Chem 90:1264

3. Hala J (1985) Proceed 8th Conference Czech Phys, Bratislava, p 349
4. Personov RI (1963) Opt & Spectrosk 15:61
5. Gorokhovkskii AA, Kaarli RK (1975) Izv Akad Nauk SSSR 39:2326
6. Bondybey VE, English JH (1979) J Amer Chem Soc 101:3446
7. Hala J, Pelant I, Ambroz M, Pancoska P, Vacek K (1985) Photochem Photobiol 41:643

Authors' address:

J. Hala
Department of Chemical Physics
Faculty of Mathematics and Physics
Charles University
Ke Karlovu 3
121 16 Prague, Czechoslovakia

Received June 24, 1987;
accepted April 21, 1988

Electrical properties of 1,4-cis polybutadiene

S. Tkaczyk, J. Świątek, and A. Kwiatkowska

Institute of Physics, Pedagogical University, Częstochowa, Poland

Abstract: Electrical properties of thin 1,4-cis polybutadiene films in the temperature range 150 K–390 K have been characterized by means of current-voltage characteristics, and also by thermally stimulated current (TSC) measurements. The TSC spectra have been discussed in terms of molecular relaxations connected with the glass transition and crystallization phenomena. Trap parameters have been determined using the relations proposed by Grossweiner and Zhdan.

Key words: 1,4-cis polybutadiene, current-voltage characteristics, thermostimulated currents, trap structure, glass transition, viscoelastic transition.

1. Introduction

The study of electrical properties of insulating polymers attracts considerable interest because the physical mechanisms responsible for the current in these materials have not been exactly determined. The purpose of this work is to characterize electrical properties of 1,4-cis polybutadiene on the basis of current-voltage characteristics and also using the method of thermally stimulated currents.

2. Current-voltage characteristics

The investigated polymer had average viscosic molecular weight $2.3 \cdot 10^5$. Samples had a sandwich configuration thickness of 0.7–4.5 µm with vacuum evaporated metal electrodes Au-Au and Au-Al. Measurement was performed in a vacuum of the order of 10^{-4} hPa in the temperature range 150 K–390 K. Typical current-voltage characteristics obtained for the investigated samples are shown in Figs. 1 and 2. One can distinquish two types of characteristics corresponding to two temperature regions below and above 330 K.

In the region below 330 K neither the type of applied electrodes nor their polarity influences the shape of I–V characteristics. They show an increase of a current up to the voltage about 1 V, a plateau up to 5–6 V, and finally an increase. We think that the increase at the beginning of characteristics may be explained by a release of earlier trapped charge carriers in the absence of carrier injection from electrodes. The plateau part is possible if the rate of carrier generation is smaller then their rate of removal by the field in the case when the number of carriers in the conduction band is smaller than the number of traps [1]. Further increase of the applied voltage improves the injecting properties of an electrode and the observed increase of a current is a result of injected carriers. It is known [2] that polybutadiene has the temperature of a transition into a viscoelastic state at 325 K. This explains the change in the shape of characteristics at temperatures above 330 K where one can observe a polarization component of a current added to the earlier described behavior. This component gives a reversed current because the gold electrode does not sufficiently inject and the internal carriers move in a strong internal field. Only at higher voltages does the gold electrode inject sufficiently to cause a further increase of a current.

3. TSC measurements

The TSC method is an experimental technique which allows us to obtain parameters that describe trapping levels of semiconductors or insulators. The

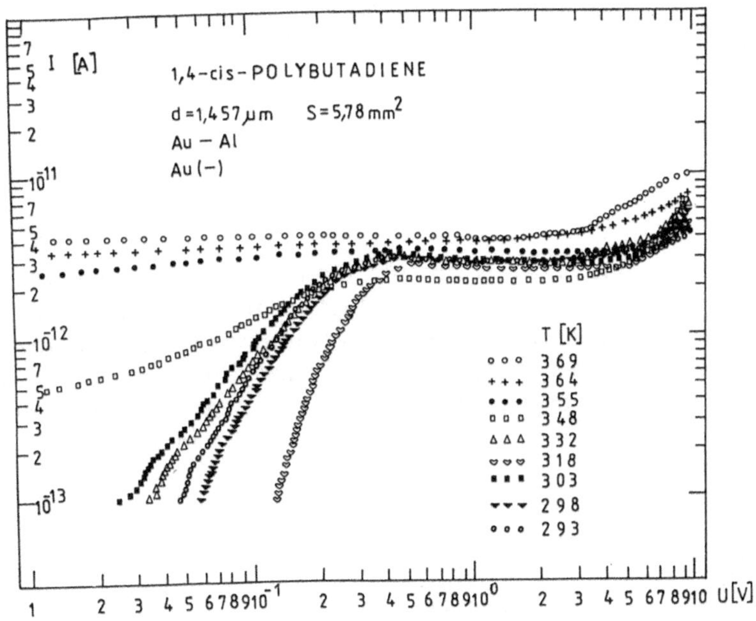

Fig. 1. Current-voltage characteristics for 1,4-cis polybutadiene, electrode polarity Au (–), Al (+)

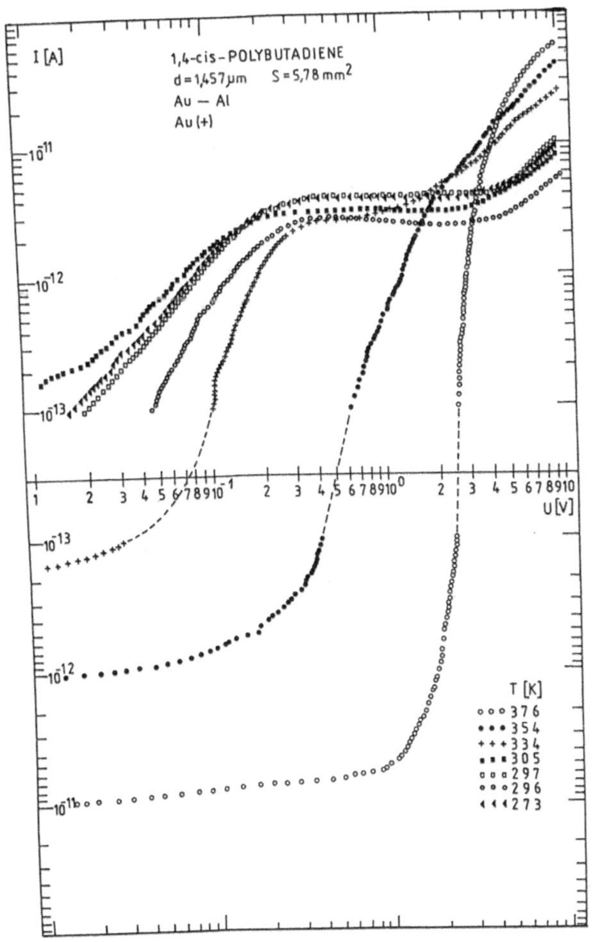

Fig. 2. Current-voltage characteristics for 1,4-cis polybutadiene, electrode polarity Au (+), Al (–)

investigated material is polarized in a static electric field for a certain time interval at a temperature T_p, then cooled down to a temperature $T_0 \ll T_p$ where the external field is taken off and the solid is warmed up at a constant rate b. The discharge current registered as a function of temperature is called a TSC current. Peaks appearing in TSC can be analyzed to provide information about the trap depth in energy E_t, the capture cross-section of the trap S_t, and the attempt-to-escape frequency factor f. We calculated E_t from a dependence proposed by Grossweiner [3] and modified by Zhdan et al. [4], based on a rising part of a TSC peak

$$E_t = kT_m\left(\frac{1{,}455 \cdot T_1}{T_m - T_1} - \frac{0{,}790 \cdot T_1}{T_m}\right) \tag{1}$$

where T_m = temperature at which a current maximum appears, T_1 = temperature corresponding to the half of maximum value on the rising part of a TSC peak, k = Boltzman constant. The relation in Eq. (1) may be used under condition that $1{,}012 \leq \frac{T_m}{T_1} \leq 2$.

The capture cross-section, S_t, and the attempt-to-escape frequency factor f were evaluated from the following relations reported by Grossweiner [3]

$$f = \frac{3 \cdot T_1 \cdot b}{2\,T_m\,(T_m - T_1)} \cdot \exp\left(\frac{E_t}{kT_m}\right)[s^{-1}] \qquad (2)$$

$$S_t = \frac{f}{2{,}9 \cdot 10^{24}\,T_m^2}\,[m^2]\,.$$

The obtained TSC spectrum for 1,4-cis polybutadiene is shown in Fig. 3. One can distinquish three peaks in this curve. The low temperature peak at 187 K we associate with molecular relaxations connected with the glass transition, which for this polymer occurs at a temperature of about 180 K. The calculated activation energy E_t for this peak is 0,84 eV. The peak at 272 K we relate to a release of charges trapped at interfaces between crystallites and amorphous regions since 273 K is the temperature of crystallization of 1,4-cis polybutadiene. The high temperature peak at 365 K is probably due to charges trapped by field-induced dipoles or traps and released above the temperature of a viscoelastic transition. The calculated trap parameters for the last two peaks are given in Table 1.

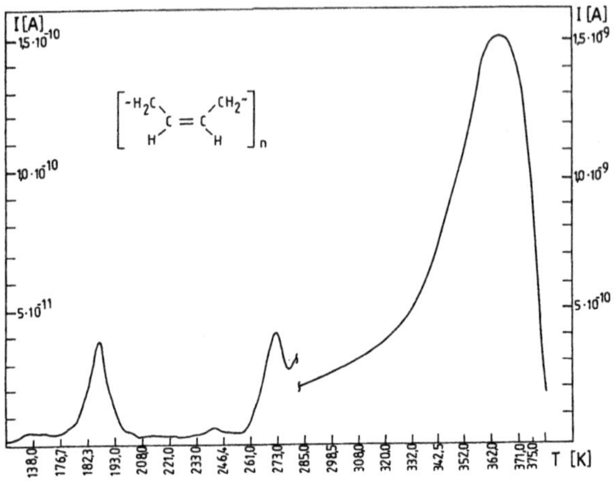

Fig. 3. TSC spectrum for 1,4-cis polybutadiene. Polarizing field 10^7 V/m, time of polarization 30 min, heating rate $b = 4$ K/min

Table 1. Trap parameters for 1.4-cis polybutadiene

T_m [K]	E_t [eV]	f [s^{-1}]	S_t [m^2]
272	0,67	$1{,}7 \cdot 10^{11}$	$7{,}9 \cdot 10^{-19}$
365	0,69	$7{,}0 \cdot 10^{7}$	$1{,}8 \cdot 10^{-22}$

Acknowledgement

This work was sponsored by the Poland Ministry of Science and Schools of Academic Rank within a Central Project of Basic Research, CPBP 01.08.

References

1. Monteith LK, Hauser JR (1967) J Appl Phys 38:5355
2. Wintle HJ (1972) In: Dole M (ed) The Radiation Chemistry of Macromolecules. Academic Press, New York, London
3. Grossweiner LJ (1953) J Appl Phys 24:1306
4. Zhdan AG, Sandomirsky VB, Ozheredov AD, Yakovleva CN (1969) Phys Techn Semicond 3:1755

Received June 25, 1987;
accepted April 21, 1988

Authors' address:

S. Tkaczyk
Institute of Physics
Pedagogical University
Al. Zawadzkiego 13/15
42-200 Częstochowa, Poland

Physical processes during the long-time decay of the surface potential of thin films

I. Chudáček, M. Marvan, and D. Slavínská

Charles University, Department of Polymer Physics, Prague, Czechoslovakia

Abstract: The time decay of the surface potential U_s of thin polymer layers was measured. It was found that for time $t > t_d$ (t_d is the characteristic quantity of the sample) the decay of U_s can be described by the formula $U_s \sim \exp[-(t-t_d)/\tau_d]^{1/2}$ where τ_d is the material constant. We explain this behaviour as a diffusion process of the space charge. The value of the constant τ_d is interpreted in terms of the microscopic material parameters.

Key words: Surface potential, polymer film, diffusion, decay of potential.

1. Introduction

Photoconductive polymer thin films are used in the electrophotography [1] as dielectric coatings deposited on a metal substrate. These films are charged to relatively high surface potentials U_s, usually by applying a corona discharge. After the charging is finished the decay of potential U_s is observed. This decay is described by one or more exponential curves [1, 2]. Nevertheless after some time $t > t_d$ the time decay of U_s is experimentally given by Eq.

$$U_s = U_{sd} \exp\left(-\sqrt{\frac{t-t_d}{\tau_d}}\right). \quad (1)$$

The material constant τ_d can be considered to be the "relaxation time" of the process. Relaxation (1) has been verified by many authors and also by us for very different materials (e. g. amorphous selenium, polyvinylcarbazole (PVK), zinc oxide, electrophotographic lithographic layer of Elfasol type).

Our results for different materials are shown in Fig. 1. We have measured $U_s(t)$ using a non-contact electrometer. The parameters t_d and τ_d are given in Table 1. The parameters t_d and τ_d depend on the chemical composition and on the manner of preparation of the samples.

2. Theoretical model

For the explanation of the long time behaviour of the decay we suppose that for $t > t_d$ the self diffusion process is superior to the influence of the electric field $\vec{E}$:

$$|e\, D \operatorname{grad} \varrho| \gg \sigma\, |\vec{E}| \quad (2)$$

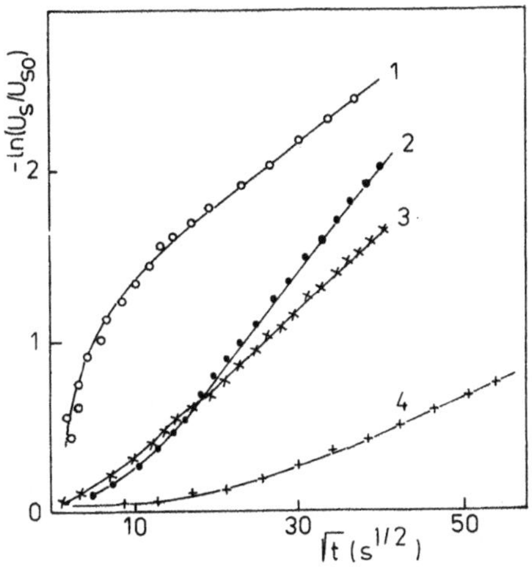

Fig. 1. The time decay of the relative surface potential U_s/U_{so} for different layers (1: amorphous selene, 2: ZnO in a polymer matrix, 3: PVK : NVK = 1 : 1, 4: Elfasol multilayer system)

Table 1. The decay characteristics

Material	$t_d(s)$	$\tau_d(s) \cdot 10^{-3}$
Selenium	240 ± 40	2,4
Zinc oxide	270 ± 50	0,85
PVK	44 ± 20	2,6
PVK:NVK = 8:1	94 ± 30	18,6
PVK:NVK = 2:1	540 ± 100	18,0
PVK:NVK = 1:1	780 ± 150	8,4
PVK:NVK = 1:2	920 ± 200	7,6

(e-charge on moving particles, D-diffusion coefficient, ϱ-density of space distribution of moving particles, σ-conductivity). We suppose that samples contain very deep traps which are remote from one another so interference between them can be neglected. Let l_i be the x-coordinate of the i-trap (the axis x is considered to be perpendicular to the film, the linear geometry is sufficient to describe the problem) then the diffusion equation

$$\frac{\partial \varrho_i}{\partial t} = -\lambda \frac{\partial^2 \varrho_i}{\partial x^2} \tag{3a}$$

$$(\lambda = e D) \tag{3b}$$

is fulfilled in the neighboorhod of l_i.

The boundary condition at the region of the i-th trap is:

$$\begin{aligned} &\text{for } t = t_d,\ x < l_i && \varrho_i = \varrho_{oi} = \text{const.} \\ &\text{for } t > t_d,\ x = l_i && \varrho_i = 0. \end{aligned} \tag{4}$$

The solution of Eq. (3) with the conditions (4) is

$$\varrho_i(x,t) = \varrho_{oi} \frac{2}{\sqrt{\pi}} \int_0^{[(l_i - x)/2\sqrt{\lambda(t-t_d)}]} c^{-y^2}\, dy. \tag{5}$$

Our model is formally similar to the model of energy transfer given by J. Heber [4]. Now we take into account the average influence of many traps. The "mean death rate" ($d\varrho/dt$) of the particles is given by the two following factors: (a) the average number of free traps (their concentration is $N(t)$ at the time t). (b) the mean current $\bar{j}$ of particles at the points $x = l_i$. This gives:

$$\left(\frac{d\varrho}{dt}\right) = -K\bar{j}\,N(t). \tag{6}$$

where K is the rate constant ($\varrho, \bar{j}$ are the space averages of $\varrho_{oi}(M)$, $j(l_i)$, respectively). Under supposition (2) the current is given by the following equation

$$j = -e D \frac{\partial \varrho}{\partial x}, \tag{7}$$

and if (5) is used we obtain

$$\bar{j} = \bar{\varrho}_{io} \sqrt{\lambda/[\pi(t-t_d)]}\,. \tag{8}$$

The concentration of free traps $N(t)$ on density ϱ dependence is given by

$$N(t) = N_d - (\varrho_d - \varrho(t)) = \tilde{N}_d + \varrho(t) \tag{9}$$

where N_d, ϱ_d are the concentrations of free traps and moving particles at the beginning of the diffusion process, respectively, and $\tilde{N}_d$ is defined by the relation

$$\tilde{N}_d = N_d - \varrho_d. \tag{10}$$

From the Eqs. (6), (8), (9) follows:

$$\frac{d\varrho}{dt} = \frac{\alpha}{\sqrt{t-t_d}} (\tilde{N}_d + \varrho) \tag{11}$$

where (see also (3b))

$$\alpha = \bar{\varrho}_{io} K \sqrt{\lambda/\pi} = K \bar{\varrho}_{io} \sqrt{eD/\pi}. \tag{12}$$

The solution of Eq. (11) for $\tilde{N}_d > 0$ (and $t > t_d$) is

$$\varrho = \frac{\tilde{N}_d}{(N_d/\varrho_d)\left[\exp\left(2\tilde{N}_d \alpha \sqrt{t-t_d}\right) - 1\right]} \tag{13}$$

and the approximation of this expression for $t \gg t_d$ gives

$$\varrho = \frac{\varrho_d}{N_d} \tilde{N}_d \exp\left(-2\tilde{N}_d \alpha \sqrt{t-t_d}\right). \tag{14a}$$

In the case of ($\tilde{N}_d = 0$, ($N_d = \varrho_d$), the solution of Eq. (11) is $\varrho = (2\alpha\sqrt{t})^{-1}$. Supposing that the decrease of the potential is proportional to the decrease of space charge, then Eq. (14a) is in agreement with the experi-

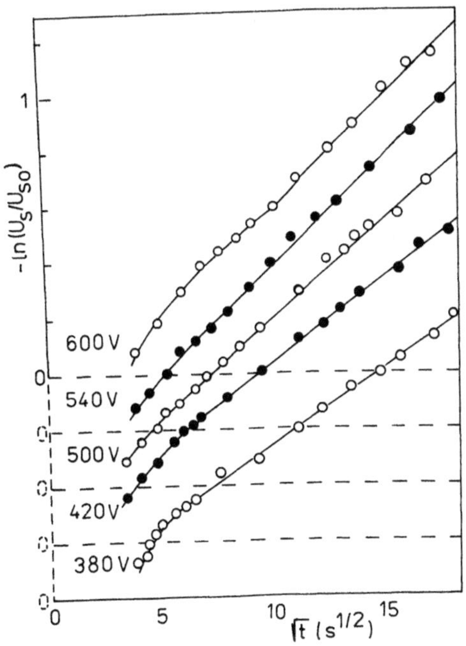

Fig. 2. The time decay of U_s/U_{so} for a thin layer of PVK:NVK = 1:1 for different initial potential U_{so}, indicated at the left side of each curve. The zero levels of the curves are shifted

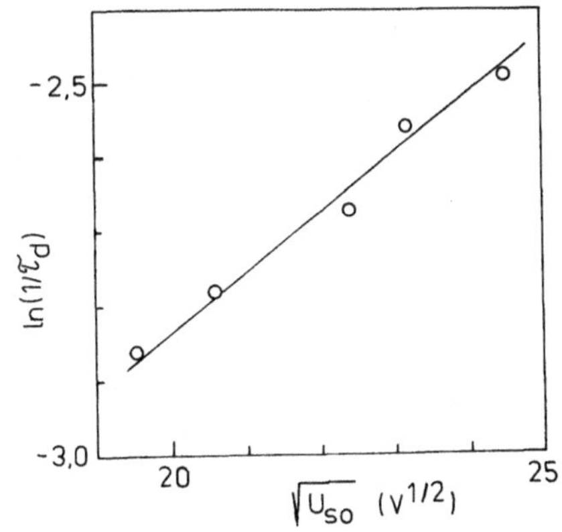

Fig. 3. The field dependence of τ_d^{-1} for a thin layer of PVK:NVK = 1:1

mental result (1). Comparing (14a) with (1) we obtain the microscopic interpretation of the "relaxation time" τ_d:

$$\tau_d = \frac{1}{(2\alpha N_d)^2}\,. \tag{14b}$$

3. Dependence of the "relaxation time" τ_d on the initial surface potential U_{so}

In this paper we also present our results concerning the surface potential time decay $U_s(t)$ of a thin layer of PVK:NVK = 1:1, for different initial surface potentials U_{so} (NVK is N-vinylcarbazole). The results are given in Fig. 2 from which the dependence of τ_d on U_{so} may be reconstructed. This has been done in Fig. 3 from which the relaxation

$$(\tau_d)^{-1} \sim \exp\sqrt{\gamma U_{so}} \tag{15}$$

follows. In Eq. (15) γ is a constant.

Comparing (15) and (14b) we obtain

$$4\tilde{N}_d^2\,\alpha \sim \exp\sqrt{\gamma U_{so}}\,. \tag{16}$$

Because the quantity α is independent of U_{so}, we can write

$$\tilde{N}_d \sim \exp\sqrt{\gamma U_{so}}\,. \tag{17}$$

If we consider both the physical interpretation of $\tilde{N}_d$ and the dependence of the carrier mobility on the field, in a form similar to (17) (Pool-Frenkel effect), then the Eq. (17) is easily understood. Indeed in the case where the number of traps and the number of moving particles was the same at time $t = 0$, $\tilde{N}_d$ is equal to the concentration ϱ_c of the particles which reached the electrode without being trapped ($\tilde{N}_d$ is the growing function of ϱ_c in each case). On the other hand, the above mentioned dependence of the mobility on the field implicates that more particles reached the electrode in a strong field than in a weak field at a time $0 < t < t_d$. From these facts we obtain (17).

4. Conclusion

The long time behaviour of the decay of the surface potential has been studied on photoconductive amorphous films. The main experimental results are expressed in Eqs. (1), (15). The explanation of the relations is done rather schematically. We devide the period of decay into two parts: the carrier transport by drift under the influence of the field in the first part of the period ($0 < t < t_d$) and the diffusion process with trapping in the second part of the period ($t > t_d$). The process of trapping is controlled by the diffusion that is reflected in the factor j (instead of ϱ, as in classical chemical kinetics) on the right-hand side of Eq. (6). Although our model is a considerable simplification of reality, we believe that it describes the main feature of the studied effects.

References

1. Schaffert RM (1975) Electrophotography, Focal Press, London
2. Chudáček I, Slavínská D (1981) J Signal AM 9:433
3. Gill WD (1972) J Appl Phys 43:5033
4. Heber J (1971) phys stat sol 48:319

Received November 13, 1987;
accepted April 21, 1988

Authors' address:

Prof. I. Chudáček
Department of Polymer Physics
Charles University
V Holešovičkách 2
180 00 Praha, CSSR

Progress in Colloid & Polymer Science Progr Colloid Polym Sci 78:143–151 (1988)

Carbonaceous thin films

H. Pagnia

Institut für Angewandte Physik, Technische Hochschule Darmstadt, Darmstadt, F.R.G.

Abstract: Carbonaceous thin films prepared by ionizing radiation from organic contamination or pure monomers reveal various compositions and physical properties. An attempt was made to review development in this field, and to show how its improved technologies now allow controlled tailoring of particular features of the films.

Key words: Carbonaceous thin films, plasma polymerization, carbon films, carbon phases.

1. Introduction

In the field of physics of carbonaceous films a substantial number of papers during the last decade reveal growing interest in the application point-of-view, as well as new insights in fundamental. Interest peaked in this development in the 1950s and early 1960s. It is the aim of this paper to link these results and to point out lines of progress in investigating substances with various compositions between organic polymers and pure carbon phases. For simplicity the view shall be limited mainly to CH_x (x: integer or fractional number) and C; further components or doping atoms are not considered.

2. The graphite-like film period

It has been well known for more than a century that solid films grow under the influence of irradiation with ionizing particles or electromagnetic waves on contaminated surfaces in vacuum vessels in the absence of an oxydizing atmosphere. In the 1930s it was established that those films are of a carbonaceous nature [1, 2] and that they are created from adsorbed, cracked, organic molecules whose origin are, e.g., vacuum grease, rubber gaskets, and rotary pump oils.

Much effort was made to determine growth of such contaminating films, especially by electron microscopy, which unfortunately obscured object details, e.g., by reducing contrast. But a positive effect was discovered which led to utilizing these films as a stabilizing envelope for object structures volatile or destructed under the electron beam influence [3]. It should be mentioned that in 1948, König [4] had proved by electron diffraction that they are made up of "a powder of carbon six-rings." Half-width diffraction ring intensities revealed an amorphous, or at least, a fine crystalline material.

Handling these "coal" covers (which today frequently called "contamination") certainly stimulated exeriments in order to characterize these films, and also to develop preparation methods, e.g., for the application in replica methods. König and Helwig [5] first described a dc glow discharge apparatus for the preparation of carbonaceous polymeric films from the monomer benzene on the anode (Fig. 1). Widely used were dc glow discharges from a few hundred to some thousend volts between plane (cooled or heated) electrodes in small distance causing a restrained discharge with the anode located within the negative glow regime, and hence, therefore mostly the whole applied voltage dropped near the cathode. Positive ions originating from the quasi-neutral plasma were accelerated to impinge at the cathode erasing electrons, most of which reach the anode with equal energy. Taking into account the different momentums of the particles, transferred energy is extremely different, revealing the strongest effects of the ions on adsorbed molecules.

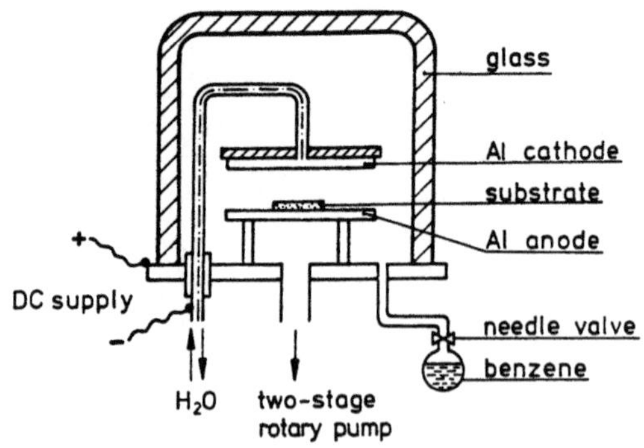

Fig. 1. "Coaling" apparatus [5]

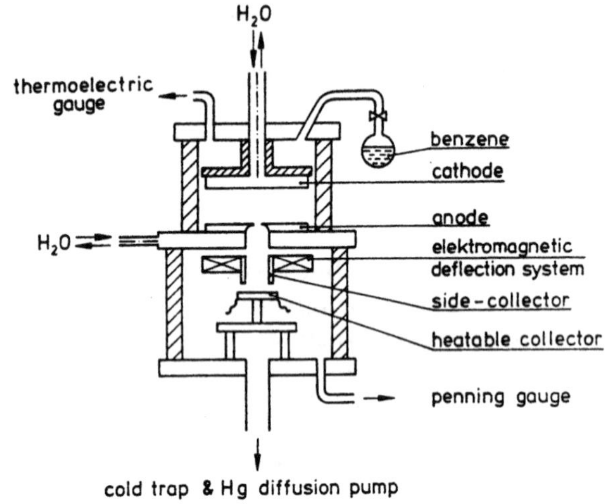

Fig. 2. Pressure-stage glow discharge apparatus [6]

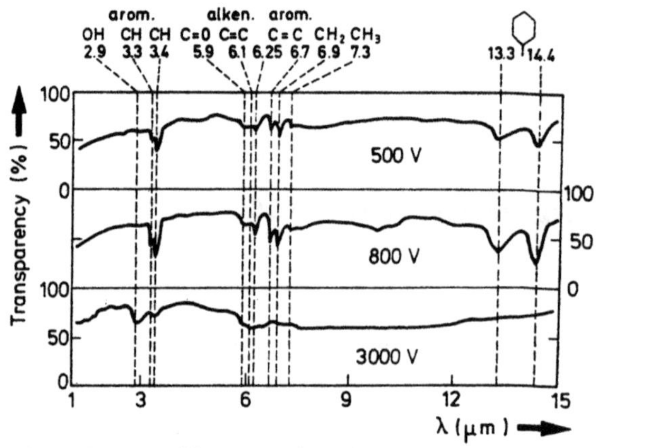

Fig. 3. Spectra of benzene glow discharge polymer films [10]

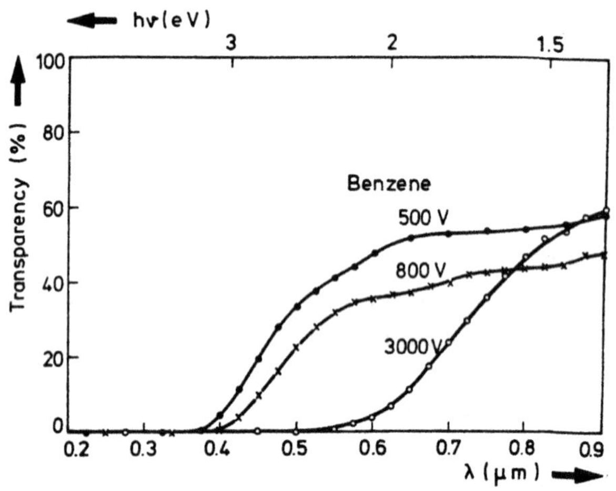

Fig. 4. Spectra of glow discharge polymer films [10]

Even when provided equal doses the growth rate ratios were found (by Heisen [6]) to be about 6 (at 1 keV) (Fig. 2). In this investigation negative ion effects were also proved and separated from those of the electrons by deviating the electron beam. The author came to the conclusion that these ions were low in number and energy, in accordance with the potential profile of the discharge. Most attention was paid to anode grown films which were found to deliver (regarding the provided applications) some advantages over cathode grown ones. Clearly, electrons are strongly scattered at the substrate and other surfaces and secondaries produce carbon film on shadowed object surfaces, while ions are only weakly scattered, revealing film growth at surfaces adjacent to the discharge space.

Films on anodes were produced with different organic gases (e.g., benzene [5], methane [7]) at dc voltages between 0.4 and 10 kV [5, 7, 8]. The adhesion problem was also discovered. Thicker films are likely to peel off, especially in a humid atmosphere. Infrared (*ir*) investigations exhibited spectra very similar to electron irradiated polymers. The alkenic and aromatic vibrations, including those of the monomer benzene, are to be seen (Fig. 3), while films grown in methane only show alkenic ones; therefore it is surely justified to speak of these films as a kind of polymer. At a somewhat diffuse threshold of the applied voltage (sometimes obscured by the not well known substrate temperature) all these absorptions cease and spectra deliver no further information. They exhibit a weakly monotoneous increasing transparency (50–60 %), similar to evaporated carbon films [9]. Obviously, with increasing voltage and temperature during growth, films change from a composition near CH_x ($x = 1$ in benzene) to mainly carbon ($x \cong 0$).

It can be observed that low voltage samples are yellow in color and transparent, while those at higher voltages are brownish, or even black. Quantitatively this

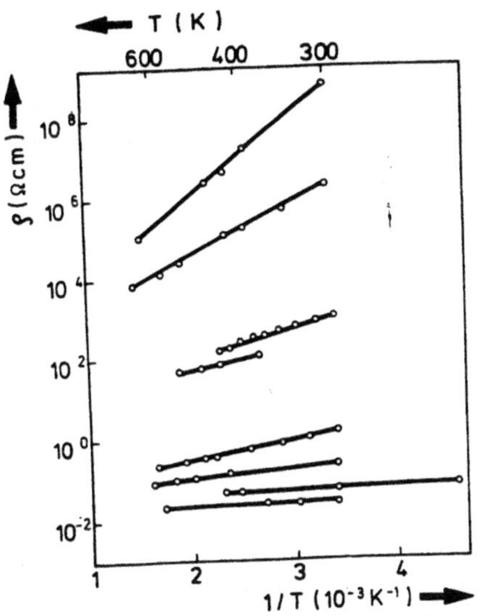

Fig. 5. Resistivity of benzene-glow discharge polymer films [8]

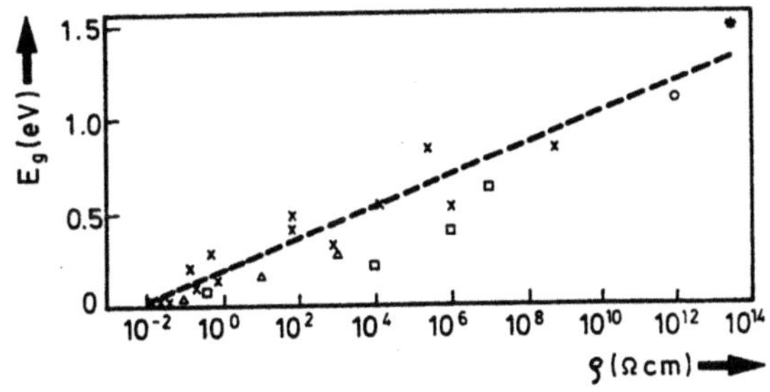

Fig. 6. Activation energies of different carbonaceous material: × [8]; □ [13]; △ [14]; ✱ [11]; ○ [31]

behavior is evinced by the absorption coefficient α, and a shift of the electronic absorption edge E_g (Fig. 4) [10]. It should be mentioned that within the framework of a more general view about the behavior of carbonaceous films, the same tendency is revealed for organic foils under electron irradiation influence. In the case of polystyrene [11], e.g., after impinging a dose of 0.5 As/cm^2, the gap shifted from 4 eV to 1.5 eV. Heat treated cellophane films [12] behave similarly.

This carbonizing by different methods, as well as preparation of differently carbonized matter, was extensively studied. The most obvious property which could be easily measured and which well characterizes the state of carbonizing is the electrical resistivity at a fixed temperature. Its decrease (at *rt*), related to the decrease of the activation energy E_g, derived from the slopes of log ϱ vs $1/T$ plots (Fig. 5), and being very nearly linear, at least in the high resistivity cases, was always found (Fig. 6). Such behavior is characteristic of intrinsic semiconductors, and it was concluded that the π electrons fill a low band built up from overlapping p-orbitals (while the hybridized three sp^2-orbitals are engaged in σ bonds) separated E_g form an empty conduction band [15]. The above mentioned irradiated polystyrene with $\varrho \cong 5 \cdot 10^{13}$ Ω cm fit the discussed rule.

It was concluded that this class of materials which are formed by heat treatment or irradiation of organic films or synthesized in glow discharges with increasing applied voltages (and/or substrate temperatures) tend to a graphitelike configuration. Irrespective of whether the starting material (or the monomer in the case of glow discharge forming) was composed of long chain molecules or aromatic molecules, the result of polymerization was assumed to be a system of cross-linked planar-condensed benzene-ring molecules or in more developed state stacks of (in the c-axis direction), uncorrelated graphite planes [9]. Low resistance films (10^4 Ω cm) show six rings with no regularities in the c-direction of a hexagonal unit cell, and a turbostratic type at about 1 Ω cm [8, 9] that exhibits distances between planes 0.37 nm, instead of 0.34 nm, in graphite. That is in agreement with similar results gained from pyrolytic carbons (e.g., by [15]). Absence of mixed reflexes ($h, k, l \neq 0$) in even the best conducting glow discharge films exhibits that the material consists of pseudocrystals with linear dimensions estimated to be less than 1 nm. Nearneighbor distances were always found to be reduced to the lowest value 0.138 nm in the case of polymer films and evaporated carbon films. Possibly the weakened bonding perpendicular to the planes enhances the in-plane bond strength (via *p* orbitals). Changes of both *c* and *a* are always found to be parallel to a crystal size reduction.

Carbon films with diamond structure were first identified by Schmellenmeier [17]. Also Kakinoki et al. [18] (from analysis of observed eight rings of diffraction patterns from evaporated carbon films) came to the conclusion that the structure is a three-dimensional network composed of usual diamond-like C-C distance, and graphite-like shorter distance.

3. New aspects of carbon phases

It had seemed to be well established that pure carbon can only exist in the graphite and diamond phases and in certain compositions, all based on sp^2 and sp^3

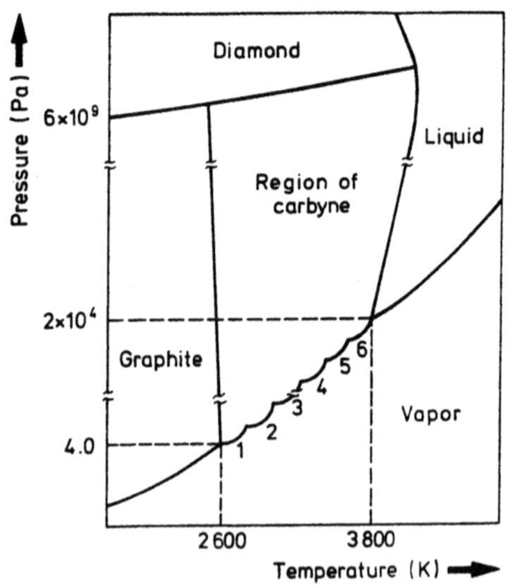

Fig. 7. Carbon phase diagram [21]

hybridization. But new carbon forms were discovered in 1968 by El Goresy and Donnay [19] and Sladkov and Kudryavtsev [20], who opening a new field of research on the stable carbon phases. Ultimately, a new phase diagram was put forth by Whittaker [21]. Though somewhat speculative, it was able to remove difficulties in the comprehension of known facts fitting the low and high pressure data of carbon. This was solved by introducing a low pressure, high temperature phase of the carbynes (Fig. 7). Because of the linear σ bonding between –C≡C– units, chains (polyyne), rather than crystalline networks, are expected. But it was shown that there exist at least 10 linear polymorphs containing these units [22] which build up hexagonal crystalline material with a density between that of graphite and diamond. The chains (6–12 carbon atoms) are thought to be arranged parallel to the *c*-axis, and bonding is achieved by intermolecular linkage via interaction of unoccupied *p*-orbitals. Some of the carbynes were transparent, but electrical properties are unknown. Therefore, it is not yet settled whether the alternating bonds deliver semiconduction like, e.g., in trans-polyacethylene, or if there also exists a cumulene-type metallic phase. Of course cumulene has been found besides polyyne in *ir* adsorption spectra so that Sladkov et al. [23] came to the conclusion that carbyne consists of these two linear polymeric forms.

But before discussing consequences, one should bear in mind some properties of carbon atoms, listed in Table 1.

Although bond length undoubtedly indicates stronger in-plane bond strength in the graphite case, diamonds are effectively harder. This is caused by the weaker Van der Waals bond of π-electrons interacting between the planes of graphite. From the thermodynamic point-of-view, graphite is the more stable configuration, but the transformation energy from one structure into the other is small compared with the bond energies. The in-chain bond energies are stronger in the carbynes, as indicated by the small bond length. This was also deduced from Sladkov et al. [23] based on calorimetric determination of the heat of combustion and on specific heat measurements. They came to the conclusion of a maximum thermodynamic stability in the equal sequence: diamond < graphite < carbyne. The transition from, e.g., carbyne to graphite, was supposed to proceed rapidly and exothermically, while the reverse transition via single bond breaking should be the stronger process. Nevertheless, carbynes may be frozen-in by rapid cooling, and then exist under normal conditions.

But while the stable elements are in graphite planes, here the basic elements are chains which possibly interact via π-electrons building the crystals. There exist flaws at the chain ends where they are kinked, and therefore indicate higher hybridization. But there exist, as was already mentioned, various carbynes due to a different number of carbons in the unit cell, and also systems of polyyne or cumulene forms and mix-

Table 1. Characterization of the carbon-carbon bond

hybridization (n = 2 shell orbitals)	number and type of bond	C-C bond length (nm)	unit	bond angle (degrees)		phase
sp^3	4 σ	0.154	–C–C–	109		diamond
sp^2	3 σ 1 π	0.142	=C=C=	120		graphite
sp^1	2 σ 2 π	0.138	≡C–C≡	0	carbyne	polyyne
		0.121	–C≡C–	0	carbyne	
		0.128	=C=C=	0	carbyne	cumulene

Table 2. Features of the solid state carbon species

	lattice structure	a (nm)	c (nm)	near-neighbor distances (nm) (see Table 1)	atomic density (nm^{-3})	mass density (g/cm^3)
diamond	two shifted fcc	0.356		0.154	177	3.5
graphite	hexagonal AB layers	0.246	0.67	0.142	114	2.3
hypothetical polyyne compound	primitive hexagonal	0.275	0.448	0.121–0.138	140	2.8
β-carbyne [27]	enlarged hexagonal cell	0.824	0.768	similar	159	3.2

tures; soft and hard carbynes are observed, colorless and transparent, as well as opaque ones, and concerning electronic features, even superconduction was expected [24]. The origin of the natural carbynes is assumed to lie in a low temperature (about 400 K) chemical $CO + H_2$ reaction [25], but in the meantime, other processes became known, such as quenching carbon gas on a metal surface in an argon atmosphere at 13.3 Pa [21]. This links the three types of possible carbon-carbon bonding into the framework under consideration. The question is under which conditions either is present in thin film growth, and which physical features of relevance for application may be tailored by suitable processes.

Some further details as to structure and density of the carbon phases are listed in Table 2. The carbyne polymorphs are caused by the variety of different chain lengths stacked hexagonally. The exact structural details are still unknown, but a good impression is given by a discussed hypothetical polyyne compound [26] which $n = 4$ carbon atoms per chain, the last kinked by 60°, and then delivering, (including the different triple and single bond distances), an average carbon-carbon distance in the c direction of $\bar{c} \equiv 0.112$ nm. Then, with lowest interchain distance a, densities lie between diamond and graphite (Table 2). Investigations of real carbynes usually reveal large primitive unit cells with $a = 3\,a_0$ in the case of β-carbyne and $z = 72$ atoms/cell (nine atoms in the base layer and eight atoms/chain (including the kinked)) delivering densities that seem to be the highest well-established value of these species.

4. The diamond-like carbon film period

While in the first period, besides the discussed electron microscopy applications, high conductivity of films and hardness were the main goals, essentially realizable in graphite-like pyrolytic films; the development of techniques like ion beam and plasma deposition revealed novel results. The different methods and apparatus shall not be reviewed here (see, e.g., [28–30]). Cleaner environmental conditions were revealed than in the early *dc* apparatus by using either separate ion guns or magnetron-type cathodes, (e.g., in the hf plasma case), and, naturally, improved vacuum facilities. Condensation of the organic molecules and bombarding with particles (e.g., noble gas ions and electrons) could be independently controlled in some equipment. Substrate temperature (heating or cooling) and potential (relative to ground, gun, or, e.g., plasma sheath) were further parameters used to tailor special film properties.

Parameters controlling film growth and composition are particle beam fluxes f_{po}, energy E_{po}, and direction δ_{po} of organic species, respectively, to the amount of adsorbed ones (Fig. 8). In the latter case and, additionally, noble gas ion or electron flux f_p (and E_p, δ_p), are essential. The overall gas pressure p, and composition x, must be mentioned, as well as substrate temperature T_s, and, possibly, an electrical field F (and direction ϕ) on the substrate surface. Organic-to-noble gas ratio influence the number of impinging valuable species, T_s the sticking coefficient and surface diffusion,

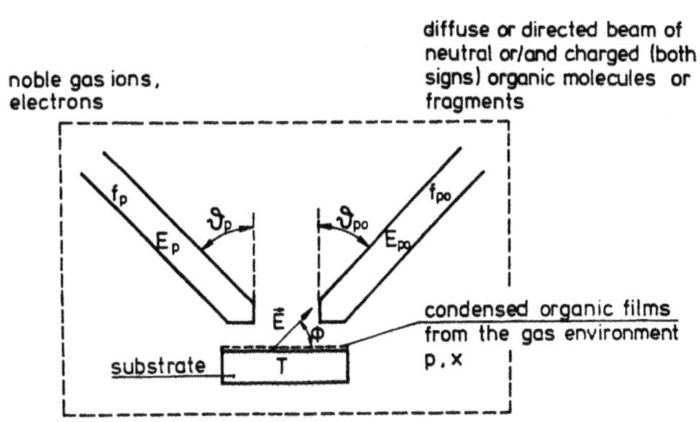

Fig. 8. Parameters controlling carbonaceous film growth

Table 3. Parameters influencing growth of various carbonaceous thin film material; for comparision, three solid state carbon species (scale: mass density)

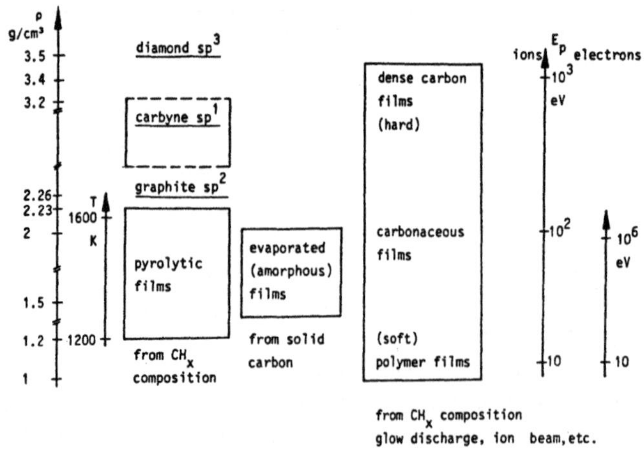

and a local field F, possibly growth orientations. Impinging particles transfer energy, momentum, and charge. They may induce excitation, ionization, breaking of bonds (and by these, cross-linking and polymerization), atom displacement (radiation damage), decomposition (e.g., dehydration), and sputtering (also desorption), the yield of course increases with particle energy.

Secondary effects of particles as well as of electromagnetic waves are neglected for simplicity, in spite of the non-neglectable role they frequently play. Evidently, there exist besides a threshold, an upper limit for effective film growth, and as was already mentioned, polymer films are the first and pure carbon materials (see Table 3) are achieved below preponderance of the destructive mechanisms. Variations of the parameters led to films with a very broad spectrum of properties like soft or hard, conducting or insulating, opaque or transparent, with different refraction indices, etc., from which feature based on simple physical reasons, one excludes the others so that compromises are required for practical applications.

Diamond-like films which were empirically tailored to be hard (hard carbon) contain a high amount of sp^3 bonded carbon. Consequently they are good insulators, and offer such other features as high-mass density, chemical inertness, and optical transparence (optimal in the *ir*, weak in the visible). But changing the parameters supplies various proportions of the different bonded carbon. The physical properties, as expected, were found to depend strongly on the sp^2/sp^3 proportion. The sp^1 bonded carbon was detected by Dischler et al. [31], but only to be incorporated at a low level.

In order to provide an appreciable conductivity, as was stated by Gambino and Thompson [32], at least half the carbon atoms must be sp^2 bonded. That is qualitatively true for arguing the percolation theory, predicting an interconnected network of p-orbitals extending through the film, and first bridging electrodes at a critical proportion. Those investigations also revealed that an increased substrate temperature during deposition and a simultaneously decreasing hydrogen content (starting with CH_x monomers) both favor sp^2 over sp^3 bonding. This is of course in accordance with the early findings on the condensation temperature or post-treatment influences on conductivity (see also [33], results on ion-deposited diamond-like films).

It is not an easy task to find a quantity that changes uniformly in one direction over the whole range. Here, we will try to discuss this spectrum on the basis of the macroscopic mass density ranging from 1 g/cm³ to 3.5 g/cm³. In such an attempt there are surely uncertainties, because after growth and treatment the material is often inhomogeneous. We shall neglect those effects in Table 3.

The data are extremely rough, and, at least concerning the particle energies, speculative. Results on pyrolysis are well established. Since growth conditions are purely thermal at about equilibrium, the low pressure, low temperature phase features are revealed, e.g., like black color, hardness, and high conductivity (basal planes are frequently oriented parallel to the substrate). Concerning evaporated films, particle energies in the vapor have a broad velocity distribution, but of a comparatively low average. Therefore, film properties are stongly dependent on substrate temperatures. Those films are either amorphous or fine crystalline, have only a medium conductivity, (the latter increasing with higher substrate temperature), or after post-annealing they approach graphite values.

Films grown form CH_x compounds under particle beam influence reveal these new features. With increasing particle energy they change from soft to hard. But since there are the other parameters, there exists no straight forward correlation. There are definite trends. If, for example, a carbonaceous film grows under distinct conditions revealing a hard species with proportion sp^3:sp^2:sp^1 = 68:30:2 (from *ir* analysis of *a*-C:H, [31]) a post-treatment up to 873 K converts it to material with 100% sp^2 with reduced optical gap

(1.2 → 0.2 eV). No doubt, it can no longer be a good insulator. A temperature too high during the growth process obviously has a similar influence. Nyaiesh and Holland [29] reported that a temperature of about 450 K should not be exceeded (due to the excessive *rf* power input) for gaining hard films; above these values graphitizing was obtained. Diamond-like films have been produced even at about 570 K (150 V negative bias voltage [34]). But as was stated by the authors, similarity of ring pattern with textured graphite (*c* axis normal to the film surface) possibly points to another explanation.

An acceleration voltage of at least 50–100 V is evidently required to gain hard films. But the negative bias voltage ($-V_B$) is not a direct measure for the ion energy. In the *dc* glow discharges, ion energies are uncertain. Data from electron and ion gun experiments are rare, in spite of being naturally the best methods to study impact energy processes. The density of impinging particles also affects film growth in the direction of increasing hardness with increasing flux, at constant energy [33].

One further reason for boundaries between the different species being very broad and diffuse is possibly the different particle mass and the transferred energy. Masses of the noble gas atoms and the organic fragments are at least of the same order of magnitude. That is definitely not the case for electrons. Of these, no diamond-like products are known. But since the transferred energy is provided an elastic collision and non-relativistic particle energy, $T_m = \gamma \cdot E_p$ with $\gamma = (4 m_1 m_2)/(m_1 + m_2)$, normalizing eU into a T_m scale would shift electron values gained from the *dc* glow discharge experiments into the proper regime.

That leads to the question of what the analogies are, if any, between the integral quantity pressure required for growth of diamond in thermodynamic equilibrium, and locally transferred energy and momentum in the case of production of diamond-like films. One would anticipate that there exists a flux threshold for broken bonds because some compulsory effects on neighboring atoms must be achieved in order to build diamond precursors. The minimum transferred energy should then be about identical to the displacement energy, known from radiation-damage investigations to lie between 25 eV (graphite [35]) and 80 eV (diamond; the high value possibly due to multiple processes [35]). The usual particle energies of the ions are between 50 and 500 eV, and T_m lie (considering their masses), in all cases above those values. Accordingly, electrons with more than about 10^6 eV would be required. Evidently those energies are also higher than thresholds for sputtering (e.g., graphite: 15 eV [36]), and growth can only take place as long as the sputtering yield is below unity. But in this region of transferred energies, all discussed effects occur simultaneously and a stationary growth rate must be established. Spencer et al. [37] proposed the hypothesis that sp^3 bonded species are the most resistant, and that preferential sputtering of others should be the determining factor for building hard films.

But, there is also an explanation based on pressure arguments. Impinging particles transfer momentum due to their mass and velocity. Provided those particles already deposited and built in the growing film reveal no motion (or only a weak one compared to the impinging ones), the process may be approximated by elastic particle impacts on a wall. For charged (q) particles with mass m accelerated in a potential U one may derive the pressure $p = i(8\ Um/q)^{1/2}$, with i current density. For frequently used parameters of *hf* plasma hard carbon deposition, e.g., $i = 25\ \text{A/m}^2$ and $U = 500\ V$, and provided Ar^+ ions ($m = 30$ mu), this rough estimation delivers 1 Pa – far below the graphite-diamond transition pressure (see Fig. 7). Possibly, charge transfer reveals up to an order higher particle densities of neutrals [38]. But even this cannot repair the quantitative discrepancy and particle beam pressure is therefore only a model that reveals, qualitatively, the right tendencies. But, it is a mechanism involving additive particle density, and also acceleration voltage with another exponent other than in a transferred energy model. Experiments yielding all relevant parameters can thus be performed in order of find out a proper explanation.

Growth of diamond-like films from hydrocarbons is, concluding all facts and discussed models, thermodynamically a non-equilibrium process. But it may be, nevertheless, a quasi-equilibrium mechanism put forth by useful boundary conditions. Hydrocarbon molecules must be transferred to a reactive state by particle impacts. But it seems necessary that the decomposed elements C and H in proper proportion be confined for a long enough time within a narrow sheath at the growing film surface in order to predominantly form sp^3 precursers which ultimately make up the film.

Such conditions are favored by a high particle beam energy and density (the pressure argument) or a high hydrogen density on the surface. Requirements to maintain the tetrahedral configuration are not as strong provided presence of free negative charges (e.g., electrons) on the growing surface for it is even preserved in the carbanion (CH_3^-) species. A temperature too high on the other hand, disturbs such conditions of local confinement as well as reducing the stay-times in a local volume. The latter arguments seem contradicted by the observation of a hot filament chemical vapor deposition (HFCVD, e.g., [40]) method, where a minimum temperature (873 K [39]) was required to form diamond species. This, and even typical conditions of electron assisted HFCVD via biasing the hot filament to substrate (positive) by, e.g., 100 V, 773–1023 K [39], are far above the range known from plasma polymerization (e.g., upper limit 450 K [29]) and it is also true that an 873 K post-treatment temperature was found high enough to transform sp^3parts of a hard film to sp^2 [31].

But HFCVD is usually practiced in an environment of extremely high hydrogen content (2 vol% CH_4 in H_2 [39]), and it seems possible that the residual carbon atoms frequently stripped off from their hydrogens, and caused sp^2, sp^1, or even atomic configurations, even though they, in a dynamic equilibrium, tetrahedrally bond to hydrogen. Diamond-like films can also be produced directly from carbon in form of C_n^+ ions (inclusively $n=1$) [41]. Although ion energies and current densities were of course similar to those in the CH_x cases, growth rates observed were much lower. Taking into consideration the positive charge, repulsive forces should yield closed-packed atomic arrangements under the confinement due to the beam pressure. One can imagine that the growing material is the densest form.

5. Concluding remarks

Now it is well established that there exist three forms of pure carbon due to the different hybridization: graphite, carbyne, and diamond. The question if it is possible to produce thin films of either form, and what physical properties may be achieved, is not yet fully answered.

Since graphite is the low-pressure, low-temperature phase of carbon, films of it are easily to gain under near-equilibrium conditions. Applying a high enough temperature during deposition (or post-treatment) of films suffices. Such films have a medium to good conductivity (due to the in-plane component) dependent upon crystal size and orientation as well as composition, as to, e.g., amorphous inclusions. Other properties (like mechanical or optical ones) approach bulk graphite values.

Diamond crystals are known to be practically stable at room temperature and atmospheric pressure. But it should remembered that, from a thermodynamic point-of-view, they are not in equilibrium under those conditions, though the potential barrier is high enough to preserve this phase under reduced pressure. During film growth it is the primary problem to stabilize sp^3 hybridization of C atoms in close proximity and over a long enough lifetime, in order to allow ultimate reaction toward the diamond phase.

It is also possible to produce thin films of the linear carbon form. Because, thermodynamically, carbynes are in a high temperature phase, direct transference from the vapor is only possible by quenching; during a synthesis by chemical reactions in all stages the sp^1 state must be preserved [23]. Reactions on surfaces are often very different, e.g., catalytic, and require less drastic conditions as to the reaction or quenching temperature. But there may possibly exist quite different methods to grow chain-like material. Strong anisotropic conditions may be produced by high electrical fields in a microstructured system, forcing preferential growth in a destined direction. This is still an open question and much effort is necessary to find out if this is viable, but there are supporting indications, e.g., switching carbon filaments [42]. Linear carbon species strongly related to chain molecules like $(CH)_x$ with their promising electrical features, should also be an interesting material regarding applications.

References

1. Harkins WD (1934) Trans Faraday Soc 30:221
2. Steward RL (1934) Phys Rev 45:488
3. Leisegang S (1956) Handbuch der Physik. Springer, Berlin 33:512
4. König H (1956) Naturwiss 35:261
5. König H, Helwig G (1951) Z Phys 129:491
6. Heisen A (1958) Ann Phys 2:23
7. Haefer R (1954) Acta Physica Austriaca 9:1
8. Pagnia H (1961) Phys Stat Sol 1:499
9. Pagnia H (1962) Phys Stat Sol 2:336
10. Brockes A, König H (1958) Z Phys 152:75
11. Brockes A (1957) Z Phys 149:353
12. Pinnick HT (1956) Proceedings Conf Carbon, Buffalo
13. Kmetko EA (1951) Phys Rev 82:456
14. Toyoda H, Nagashima M (1959) J Phys Soc Jpn 28:274
15. Loebner EE (1956) 102:46
16. Hofmann U, Wilm D (1936) Z Elektrochem 42:504
17. Schmellenmeier H (1955/56) Z Phys Chem 205:349
18. Kakinoki J, Ino T, Katada K, Hanwa T (1957) Act Cryst 10:829
19. El Gorsy A, Donnay G (1968) 161:363
20. Sladkov AM, Kudryavtsev YuP (1969) Priroda 58:37
21. Whittaker AG (1978) Science 200:763

22. Heimann RB, Kleiman J, Salansky NM (1983) Nature 306:164
23. Sladkov AM, Kasatochin VI, Kudryavtsev YuP, Korshak VV (1969) Bull Acad Sci USSR, Chem Dev Sci 2560
24. Little WA (1964) Phys Rev 1416:134
25. Hayatsu R, Scott RG, Studier MH, Lewis RS, Anders E (1980) Science 209:1515
26. Heimann RB, Kleiman J, Salansky NM (1984) 22:147
27. Kasatochin VI, Korshak VV, Kudryavtsev YuP, Sladkov AM, Sterenberg LE (1973) Carbon 11:70
28. Anderson L-P (1981) Thin Solid Films 86:193
29. Nyaiesh AR, Holland L (1984) Vacuum 34:519
30. Weissmantel (1981) J Vac Technol 18:179
31. Dischler B, Bubenzer A, Koidl P (1983) Solid State Comm 48:105
32. Gambino RJ, Thompson JA (1980) Solid State Comm 34:15
33. Mori T, Namba Y (1983) J Vac Sci Technol A1:23
34. Mori T, Namba Y (1984) J Appl Phys 55:3276
35. Corbett JW (1966) Electron Radiation Damage in Semiconductors and Metal. Academic Press, New York
36. Carter G, Colligon JS (1968) Ion Bombardment of Solids. Elsevier, New York
37. Spencer EG, Schmidt PH, Joy DC, Sansalone FJ (1976) Appl Phys Lett 29:118
38. Davis WD, Vanderslice TA (1963) Phys Rev 131:219
39. Sawabe A, Inuzuka T (1986) Thin Solid Films 137:89
40. Matsumoto S, Sato Y, Tsutsumi M, Setaka N (1982) J Mater Sci 17:3106
41. Miyazawa T, Misawa S, Yoshida S, Gonda S (1984) J Appl Phys 55:188
42. Bach Th, Blessing R, Pagnia H, Sotnik N (1983) Thin Solid Films 103:283

Received June 25, 1987;
accepted April 21, 1988

Author's address:

H. Pagnia
Institut für Angewandte Physik
Technische Hochschule Darmstadt
Schloßgartenstr. 7
D-6100 Darmstadt, F.R.G.

Semidilute polymer solution near the wall

W. Korneta[1]) and Z. Pytel[2])

[1]) Department of Physics, Technical University, Radom, Poland
[2]) Institute of Physics of the Polish Academy of Sciences, Warsaw, Poland

Abstract: The behaviour of long flexible polymer chains in good solution near an adsorbing wall, a nonadsorbing wall and a wall at the critical point of adsorption is studied. Only the case of a semidilute polymer solution is considered. It is assumed that the monomer-monomer correlation function satisfies the generalized Edwards equation. The second moments of the monomer-monomer correlation function have been calculated. These moments determine shapes of polymer subunits. The dependence of the shape of polymer subunit on the distance to the wall, on the total monomer concentration and on the interaction between the wall and monomers is discussed.

Key words: Polymer solution, subunits, adsorption, wall, mean field.

For a long time only macroscopic parameters of polymer solutions were measured. In recent years it has become possible to study one single-labeled chain among other chains which are chemically identical, but not labeled. We discuss the behaviour of long flexible polymer chains in good solvents near a wall. We consider both the attractive wall and the repulsive wall for monomers. We assume that polymer chains are not attached to the wall. In the discussion we concentrate on semidilute polymer solutions in which polymer chains interpenetrate one another. Polymer chain in semidilute solution may be considered as a succesion of subunits. The shapes of subunits are determined by the second moments of the monomer-monomer correlation function. The effective interactions between subunits of the same chain are weak and their behaviour is expected to be gaussian. The idea of polymer subunits was introduced by Edwards [1] in an infinite semidilute polymer solution. The idea was presented most clearly and simply by De Gennes [2]. We generalize this idea for semidilute polymer solutions having different local monomer concentration in different regions.

Consider a polymer solution bounded by two infinite parallel flat walls. The z-axis is perpendicular to the walls. The position in the direction parallel to the walls is described by a two-dimensional vector $\bar{\varrho} = (x, y)$. The surfaces are at $z = 1$ and $z = L$. We assume that the polymerization index N satisfies the inequality $1 \ll N \ll L$. The equation for the monomer-monomer correlation function in an infinite semidilute polymer solution was proposed by Edwards [1, 2]. This equation generalized for a semidilute polymer solution in which the local monomer concentration $c(\bar{r})$ is different in different regions of the solution, has the form

$$\nabla^2 G(\bar{r}, \bar{r}') - 3(6vc(\bar{r}) - 2\varepsilon)\, G(\bar{r}, \bar{r}') = -\,\delta(\bar{r}, \bar{r}') \quad (1)$$

where v is the coefficient for the monomer-monomer interaction, and ε is a constant. In good solvents $v > 0$. The profile of the local monomer concentration $c(\bar{r})$ was obtained using the mean-field theory similar to that in Ref. [2]. The constant ε is the Lagrange coefficient in the mean-field equation for the local concentration $c(\bar{r})$. This constant is determined from the total monomer concentration. Assuming that between the walls and the monomers there are only the short-ranged interactions, the effect of the walls on the monomers have the following boundary conditions [3]

$$\left.\frac{\delta G(\bar{r}, \bar{r}')}{\delta z}\right|_{z=1,L} = \pm\,(1-4D)\, G(\bar{r}, \bar{r}')\Big|_{z=1,L}. \quad (2)$$

The parameter D may be varied independently in experiments e.g. by adding different constituents to the walls [4] or by varying the temperature [3]. For adsorbing walls $D > 1/4$, for nonadsorbing walls $D < 1/4$, whereas for walls at the critical point of adsorption $D = 1/4$. Let $\mathcal{J}^{\parallel}(\bar{r})$, $\mathcal{J}^{\perp l}(\bar{r})$ and $\mathcal{J}^{\perp r}(\bar{r})$ be the second moments of the monomer-monomer correlation function in the directions parallel and perpendicular to the walls. These moments are defined as

$$[\mathcal{J}^{\parallel}(\bar{r})]^2 = \frac{1}{2}\frac{\int_1^L\int[(x'-x)^2+(y'-y)^2]\,G(\bar{r},\bar{r}')\,d\bar{\varrho}'\,dz'}{\int_1^L\int G(\bar{r},\bar{r}')\,d\bar{\varrho}'\,dz'},$$

$$[\mathcal{J}^{\perp l}(\bar{r})]^2 = \frac{\int_1^z\int(z'-z)^2\,G(\bar{r},\bar{r}')\,d\bar{\varrho}'\,dz'}{\int_1^z\int G(\bar{r},\bar{r}')\,d\bar{\varrho}'\,dz'},$$

$$[\mathcal{J}^{\perp r}(\bar{r})]^2 = \frac{\int_z^L\int(z'-z)^2\,G(\bar{r},\bar{r}')\,d\bar{\varrho}'\,dz'}{\int_z^L\int G(\bar{r},\bar{r}')\,d\bar{\varrho}'\,dz'}. \quad (3)$$

By expansion of $G(\bar{r},\bar{r}')$ into eigenfunctions and converting all the above equations from continuous to discrete form one, can calculate $\mathcal{J}^{\parallel}(\bar{r})$, $\mathcal{J}^{\perp l}(\bar{r})$ and $\mathcal{J}^{\perp r}(\bar{r})$ numerically. Because of translational invariance in the plane parallel to the walls, the second moments of $G(\bar{r},\bar{r}')$ do not depend upon x and y. The dependence of $\mathcal{J}^{\parallel}(\bar{r})$, $\mathcal{J}^{\perp l}(\bar{r})$ and $\mathcal{J}^{\perp r}(\bar{r})$ on z near the adsorbing wall ($D = 1/2$), near the nonadsorbing wall ($D = O$) and near the wall at the critical point of adsorption ($D = 1/4$) is shown in Fig. 1. The quantities $\mathcal{J}^{\parallel}(\bar{r})$, $\mathcal{J}^{\perp L}(\bar{r})$ and $\mathcal{J}^{\perp r}(\bar{r})$ determine shapes of polymer subunits. The shapes of subunits of one polymer chain near the adsorbing wall, the nonadsorbing wall and the wall at the critical point of adsorption are shown in Figs. 2 and 3.

Near the adsorbing wall, when the total monomer concentration c is small, almost all polymers are adsorbed onto the wall. Moreover almost all monomers of each polymer chain adhere to the wall and polymer subunits are extended parallel to the wall. The size of subunits quickly decreases further from the wall. When c increases, the polymer adheres less for the wall and the fraction of its monomers close to a the wall reduces. Subunits near the wall decrease, whereas subunits further from the wall increase. Near the wall at the critical point of adsorption the second moments of $G(\bar{r},\bar{r}')$ in the direction parallel and oposite to the wall are equal and they are the same in any position. Near the nonadsorbing wall the local monomer concentration decreases and the polymer subunits increase. Near the nonadsorbing wall and near the wall at the critical point of adsorption all polymer subunits increase when the total monomer concentration decreases.

We conclude that the behaviour of polymers in semidilute solution near the wall and shapes of polymer subunits strongly depend on the parameter D, which characterizes the wall adsorption quantities and on the total monomer concentration.

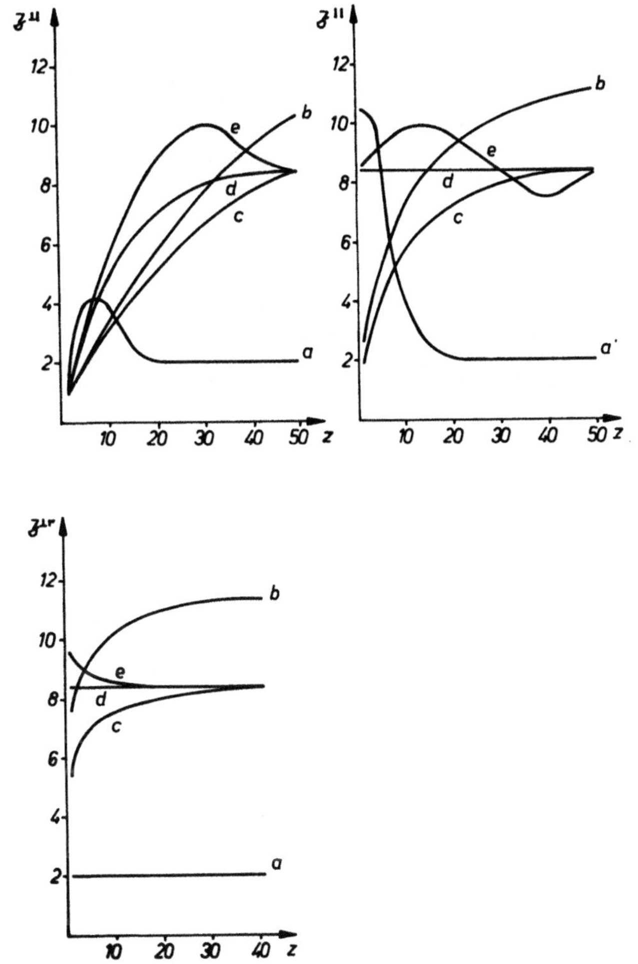

Fig. 1. The second moments of the monomer-monomer correlation function in the direction parallel ($\mathcal{J}^{\parallel}$) and perpendicular ($\mathcal{J}^{\perp l}$, $\mathcal{J}^{\perp r}$) to the wall. z denotes the distance to the wall. The parameter D and the local monomer concentration close to the wall (c_1) and far from the wall (c_b) are the following: (a) $D = 1/2$, $c_1 = 0.0017$, $c_b \sim 0$; (b) $D = 1/2$, $c_1 = 0.1867$, $c_b \sim 0$; (c) $D = 1/2$, $c_1 = 0.1900$, $c_b = 0.0024$; (d) $D = 1/4$, $c_1 = 0.0024$, $c_b = 0.0024$; (e) $D = 0$, $c_1 = 0.00002$, $c_b = 0.0024$. The coefficient $v = 1$

Fig. 2. The shapes of subunits of one polymer chain in semidilute solution near the adsorbing wall. The total monomer concentration increases from A to C

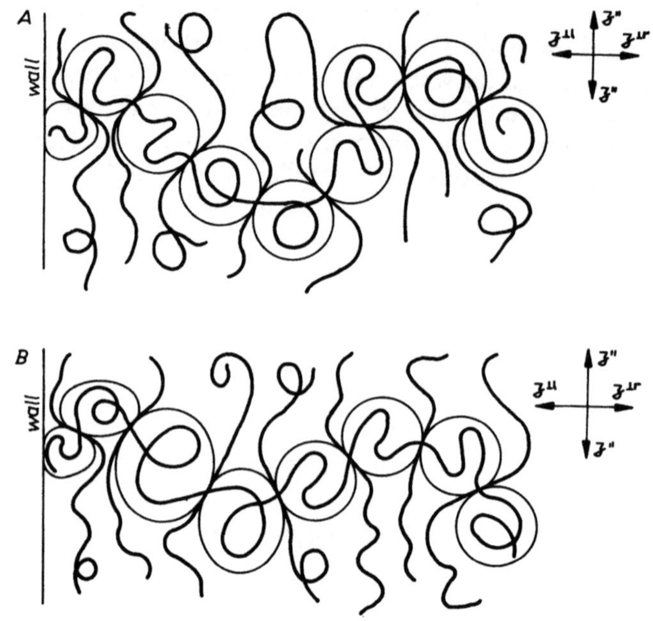

Fig. 3. The shapes of subunits of one polymer chain in semidilute solution near the wall at the critical point of adsorption (A) and near the nonadsorbing wall (B)

References

1. Edwards SF (1966) Proc Phys Soc 88:265
2. De Gennes PG (1979) Scaling Concept in Polymer Physics. Cornell University, Ithaca
3. De Gennes PG (1969) Rep Prog Phys 32:187
4. Eisenriegler E (1984) J Chem Phys 81:4666

Received June 24, 1987;
accepted April 21, 1988

Authors' address:

Z. Pytel
Institute of Physics
Polish Academy of Sciences
Al. Lotników 32/46
Warsaw, Poland

This work was supported by the Ministry of Higher Education in Poland under Grants CPBP - 01.08.

Progress in Colloid & Polymer Science

Progr Colloid Polym Sci 78:155–157 (1988)

Phase diagram of polymer solution bounded by walls

W. Korneta[1]) and Z. Pytel[2])

[1]) Department of Physics, Technical University, Random, Poland
[2]) Institute of Physics of the Polish Academy of Sciences, Warsaw, Poland

Abstract: A polymer solution that is bounded by two infinite parallel flat walls is studied. It is assumed that the polymer is in good solution. The dependence of properties of the polymer solution on the total monomer concentration, on interactions between the walls and the monomers, and on the relation between the polymerization index and the distance between walls is considered. The phase diagram of the solution was obtained.

Key words: Polymer solution, wall, mean-field, polymer subunits.

The behaviour of polymer solutions near walls is of considerable scientific and technological interest. In this paper we study the effect of walls on the critical concentration of semidilute polymer solution and on the profile of the local monomer concentration. The critical concentration of a semidilute polymer solution is defined as the total monomer concentration below which separated polymer chains appear in the solution. We consider both adsorbing walls and nonadsorbing walls for monomers. Because the mean-field theory provides a useful picture of semidilute polymer solution [1], we use this theory in this paper.

Consider a semidilute polymer solution confined in the space bounded by two infinite parallel flat walls. The z axis is perpendicular to the walls. The walls are at $z = 1$ and $z = L$. In the mean field approximation, the free energy of a semidilute polymer solution can be written as [1]

$$F = \frac{1}{V} \int_V \left\{ \frac{1}{6} (\nabla\psi(\vec{r}))^2 + \frac{1}{2} v\psi^4(\vec{r}) \right\} dV, \tag{1}$$

where V is the volume of polymer solution, $\psi(\vec{r})$ is the eigenfunction for the ground state of polymers in semidilute solution and v is the coefficient for the monomer-monomer interaction. In good solvents $v > 0$. The local monomer concentration $c(\vec{r}) = \psi^2(\vec{r})$. The total monomer concentration c is defined as

$$c = \frac{1}{V} \int_V c(\vec{r})\, dV. \tag{2}$$

Taking only short-ranged interactions between walls and monomers into account, the boundary conditions for the function $\psi(\vec{r})$ can be written in the form [2]

$$\left.\frac{\partial\psi(\vec{r})}{\partial z}\right|_{z=1,L} = \pm (1 - 4D)\, \psi(\vec{r})\Big|_{z=1,L}. \tag{3}$$

The parameter D characterizes interactions between walls and monomers. For adsorbing walls $D > 1/4$, for nonadsorbing walls $D < 1/4$, whereas for walls at the critical point of adsorption $D = 1/4$. Minimizing F with respect to $\psi(\vec{r})$ assuming that c is a constant one obtains

$$\frac{\partial^2\psi(\vec{r})}{\partial z^2} = -6\varepsilon\psi(\vec{r}) + 6v\psi^3(\vec{r}), \tag{4}$$

where the Lagrange'a coefficient ε is connected with c by Eq. (2). The above equation together with the boundary conditions Eq. (3) can be solved both analytically and numerically. Knowing $\psi(\vec{r})$ we have obtained profiles of $c(\vec{r})$ for different values of c, D, and L.

The above approach to polymer solutions fails when separated polymer chains appear in the solution. Edwards [3] defined the criterion which indicates whether an infinite polymer solution is semidilute or not. According to this criterion, an infinite polmyer solution is semidilute when the diameter of the polymer subunit is smaller than the diameter of the separated chain. Assuming that the configuration of separated chain is that of a random walk, the diameter of a space occupied by a separated chain is $\sqrt{N}$. We generalize the above criterion to polymer solutions having different monomer concentration in different regions. We define the critical concentration c^* as the total monomer concentration at which the length of any subunit in any direction is equal to $\sqrt{N}$. At c^* separated chains appear in the solution. We have calculated numerically the second moments of the monomer-monomer correlation function. From this we obtained the dependence of c^* on L, D, and N. The equation for the monomer-monomer correlation function can be derived from the free energy given by Eq. (1) by a standard method. The boundary conditions for the correlation function follow from the boundary conditions Eq. (3) and they are given in [2]. To simplify further discussion we define two parameters τ_* and c_s and the function $f(t)$. The parameters τ_* and c_s are defined as

$$\tau_* = \frac{2}{3}\frac{1}{L}\left|\frac{4D-1}{4D+1}\right| e^{-(4D-1)^2L^2} + \frac{\pi^2}{24\left(\frac{L+1}{2}+\lambda\right)^2} e^{-\frac{1}{(4D-1)^2L^2}}, \tag{5}$$

$$c_s = \frac{4}{3}\frac{1}{L}\frac{4D-1}{4D+1}, \tag{6}$$

where

$$\lambda = \begin{cases} \frac{4D}{1-4D} & D < 1/4 \\ \frac{1}{4D-1} & D > 1/4 \end{cases} \tag{7}$$

The function $f(t)$ is defined as

$$f(t) = \frac{t}{v}\left\{ e^{-(4D-1)^2L^2} + e^{-\frac{1}{(4D-1)^2L^2}} \times \left[\frac{5}{6} + \left(-\frac{1}{6} + \frac{\frac{5}{18}\frac{t}{\tau_*}}{1+\frac{5\pi}{36\sqrt{2}}\left(\frac{t}{\tau_*}\right)^{3/2}}\right) \times \frac{1-\left(\frac{t}{4\sqrt{2}\tau_*}\right)^2}{1+\left(\frac{t}{4\sqrt{2}\tau_*}\right)^2}\right]\right\}. \tag{8}$$

For $D < 1/4$ the critical concentration $c_1^* = f(t = 1/N)$. In the case when c approaches c_1^*, polymer chains situated far from the walls become separated. For $D > 1/4$, when c is small, polymer subunits near the walls are the biggest and their shapes determine the critical concentration. The formula for the critical concentration has the form

$$c_2^* = \frac{1}{Nv}\left[e^{-(4D-1)^2L^2} + e^{-\frac{1}{(4D-1)^2L^2}}\frac{2}{L}\frac{\ln(4D)}{sh^2(\ln(4D))}\right]. \tag{9}$$

For $D > 1/4$ and $N < L^2/9$ there is another range of the total monomer concentration in which separated chains appear in the solution. In this case when c is bigger than c_2^* and increasing, polymer subunits far from the walls increase and become the biggest. This means that for c between c_3^* and c_4^*, where

$$c_3^* = c_s - \frac{1}{Nv}\left[e^{-(4D-1)^2L^2} + e^{-\frac{1}{(4D-1)^2L^2}}\frac{2}{3}\frac{1+\frac{1}{12N\tau_*}}{1+\frac{\pi^2}{72}\left(\frac{1}{\tau_* N}\right)^{3/2}}\right] \tag{10}$$

and $c_4^* = c_s + f(t = 1/N)$, separated chains are present in the solution.

The profile of the local monomer concentration depends on the parameter D, the total monomer concentration and the relation between the polymerization index N and the distance L between walls. For $N \ll L$ one wall has a negligible effect on the behaviour of polymer chains near the opposite wall. The profile of the local monomer concentration near one of the walls may closely be approximated by a hyperbolic function [1]*. This profile may be obtained assuming that one of the walls tends to infinity. For $N \gg L$ it is possible to observe the effect of both walls on the behavior of polymer chains in a bounded polymer solution. Characteristic concentrations can be defined: for $D < 1/4$, $c_1 = f(t = \tau_*)$ and $c_1^h = 5f(t = \tau_*)$; whereas for $D > 1/4$, $c_2' = c_s - f(t = \tau_*)$, $c_2'' = c_s + f(t = \tau_*)$ and $c_2^h = c_s + 5f(t = \tau_*)$. For $c > c_1^h$, $c_2^* < c < c_2'$ and $c_2^h < c < 1$ the profile of $c(\vec{r})$ and the shapes of polymer subunits are the same as for $N \ll L$. For $c_1^* < c < c_1$ each polymer subunit extends from one wall to the other. For $c_1^* < c < c_1$ and $c_2' < c < c_2''$ the profile of $c(\vec{r})$ is closely approximated by trigonometric functions [1].

The results obtained in this paper are summarized in the Fig. 1.

A

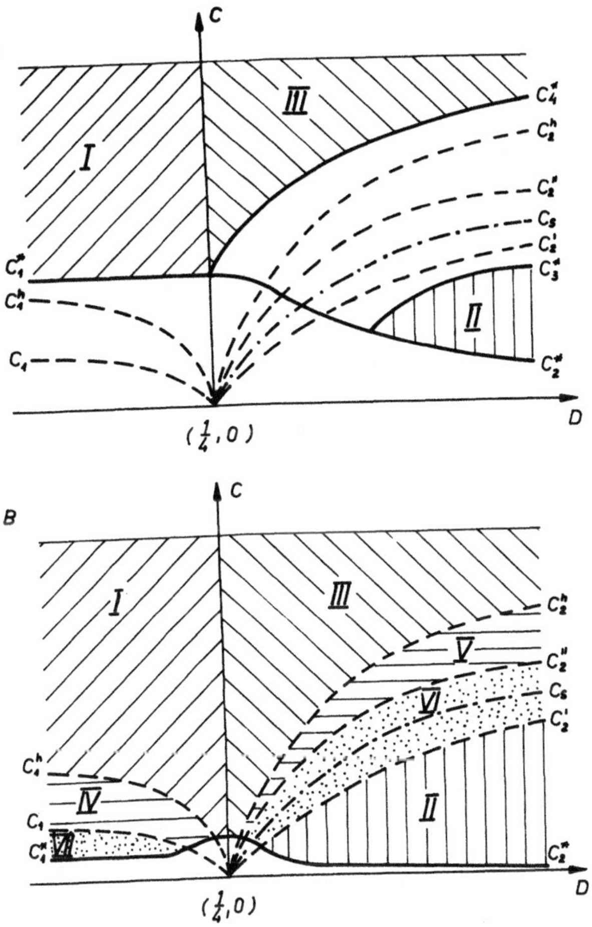

Fig. 1. The phase diagram of polymer solution bounded by two infinite parallel flat walls for $N \ll L(A)$ and $N \gg L(B)$. The lines of critical concentrations are denoted by c_1^*, c_2^*, c_3^*, and c_4^*. The profile of the local monomer concentration is given in regions I, II, and III by hyperbolic functions, in regions VI and VII by trigonometric functions, and in regions IV and V by elliptic functions. In region VII polymer subunits extend from one wall to the other

Acknowledgement

This work was supported by the Ministry of Higher Education in Poland under Grants CPBP – 01.08.

References

1. De Gennes PG (1979) Scaling Concept in Polymer Physics. Cornell University, Ithaca
2. De Gennes PG (1969) Rep Prog Phys 32:187
3. Edwards SF (1966) Proc Phys Soc 88:265

Received June 25, 1987;
accepted April 21, 1988

Authors' address:

Z. Pytel
Institute of Physics
Polish Academy of Sciences
Al. Lotników 32/46
Warsaw, Poland

Progress in Colloid & Polymer Science Progr Colloid Polym Sci 78:158–160 (1988)

Calorimetric investigations of phase transitions in amyl-ethoxy-cyanstilbene (AECS)

O. Valentová[1]), J. Biroš[2]), L. Valenta[1])

[1]) Faculty of Mathematics and Physics, Charles University, Prague, Czechoslovakia
[2]) Institute of Macromolecular Chemistry, Czechoslovak Academy of Sciences, Prague, Czechoslovakia

Abstract: Caloric measurements of AECS enabled estimation of the phase transition temperatures and evaluation of the corresponding latent heats. Combination of these results with those of [1] leads to complete determination of the Gibbs potential for AECS.

Key words: Liquid crystals, phase transitions, latent heat, Gibbs potential.

Introduction

Amyl-ethoxy-cyanstilbene (AECS)

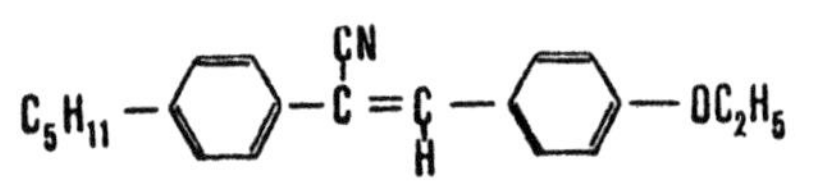

reveals the properties of a nematic liquid crystal in the temperature range of about 27.9–38 °C. Its temperature dependence of the order parameter S has been studied by means of the birefringence measurements in [1] where it has been demonstrated that the experimental results may successfully be interpreted using the Landau theory of phase transitions of first order. This theory is based on the discussion of the Gibbs potential/Gibbs free energy/$G(T,P,S)$, T being the absolute temperature, P pressure and S the long-range-order parameter. G may approximately be written for uniaxial crystals like AECS in form of a polynomial in S

$$G = G_o + \frac{1}{3} A S^2 - \frac{2}{27} B S^3 + \frac{1}{9} C S^4 . \quad (1)$$

The experiments described in [1] enabled the evaluation of B/A and B/C only. It was found

$$\frac{B}{C} = 1{,}5627; \frac{B}{A'} = 48{,}3470 \quad (2)$$

where A' is the coefficient in

$$A = A' (T - T^*), A' > 0 . \quad (3)$$

Full information on the coefficients A,B,C requires more experimentation that gives a third equation including A',B,C in addition to (2). Such a possibility is given by the measurement of the latent heat of a phase transition nematic → isotropic using the formula for the latent heat per unit volume [2] or the enthalpy change ΔH of the phase transition

$$\Delta H = \frac{1}{27} \frac{A' T_{NI} B^2}{C^2} \quad (4)$$

where T_{NI} is the temperature of the phase transition nematic → isotropic liquid. The measurement of ΔH is described in the next section.

Measurements of the latent heat

The calorimetric measurements were realized by means of the DSC-2 Perkin Elmer dynamic calorimeter. Its signals were digitalized taking into account the nonlinearity of the instrument and thermal retardation. The obtained data were then expressed in terms of the specific heat capacity $C_p(T) Jg^{-1}K^{-1}$ as a func-

PCPS HT 029

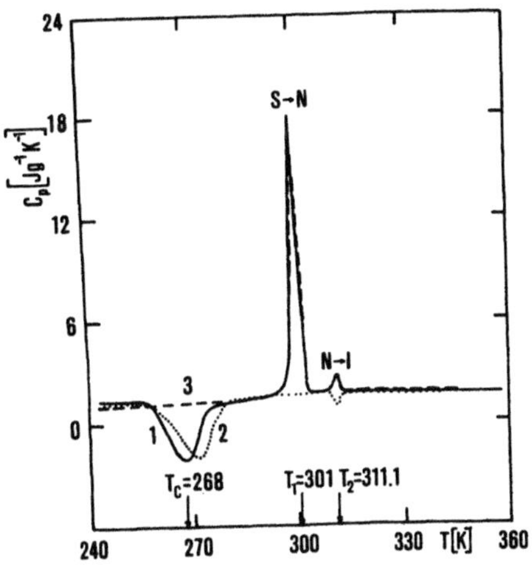

Fig. 1. Temperature dependence of specific heat of AECS: 1. Solid line——: heating of the supercooled specimen from 260 K to 360 K (first step); the rapidly supercooled specimen was unable to crystallize; it crystallizes after heating at $T = 268$K; 2. Dotted line·····: cooling from 360 K to 240 K (second step); the supercooled nematic crystallizes at $T = 270$ K; 3. Dashed line - - -: heating from 240 K to 360 K (third step); the properly crystallized specimen does not need to reveal a peak at $T \simeq 268$ K

tion of temperature T and integrated in order to obtain the change of the specific enthalpy $\Delta H(T) Jg^{-1}$

$$\Delta H(T) = \int_{T_1}^{T_2} C_p(T')\, dT .$$

The obtained data were tabled and graphs of $C_p(T)$ and $H(T)$ were obtained from a plotter Wang 2200 C.

The phase transitions were studied in the temperature interval 240 K – 360 K / solid ↔ liquid and nematic ↔ isotropic / and, moreover, in the interval 290 K–340 K for the phase transition nematic liquid ↔ isotropic liquid. The heating or cooling rate was 2.5 K/min. The weight of all specimens used was within the limits 12–13 mg.

The melting point of the crystal going over to nematic liquid (see Fig. 1) was ~ 30.1 ± 0.5 K. The transition nematic → isotropic liquid took place at ~ 311.1 ± 0.3 K.

Figure 1. illustrates that the investigated specimen was rapidly cooled to 240 K at the beginning. The supercooled specimen did not crystallize completely; that was only possible at higher temperatures (see the negative peak of the curve No. 1). The specimen was then gradually heated revealing the phase transitions solid → nematic and nematic → isotropic. Going in the opposite direction to lower temperatures (see curve No. 2) it again registered the transition point isotropic → nematic (see the negative peak), practically at the same temperature; the shape of the curve was about the same as the inverse of curve No. 1. Nevertheless, the nematic phase remained supercooled, passed the phase transition temperature nematic → solid, and, finally, the supercooled nematic crystallized at a little higher temperature (by ~ 2 K) than in the first step indicated by curve No. 1. Now in a crystallized phase and going again from 240 K to 360 K (see curve No. 3) there was no longer a necessity for a peak at $T \simeq 268$ K. However, it is desirable that curves Nos. 1 and 3 coincide in what follows and as may be seen in Fig. 1., they do indeed coincide.

We were primarily interested in the latent heat of the transition nematic ↔ isotropic which is much smaller than the total enthalpy change of the transition crystalline phase → isotropic liquid, which in dependence on the thermic history of the specimen, was within the limits of 40–55.2 J/g. The enthalpy difference ΔH of the transition nematic ↔ isotropic was approximately $\Delta H = 2.50 \pm 0.1$ J/g.

Inserting this numerical value of ΔH in Eq. (4) and combining it with Eq. (2), we obtain

$$A' = 0.091 \pm 0.004\ \mathrm{Jg^{-1}K^{-1}}$$

$$B = 4.374 \pm 0.175\ \mathrm{Jg^{-1}}$$

$$C = 2.799 \pm 0.112\ \mathrm{Jg^{-1}} . \quad (5)$$

Discussion

The numerical values A', B, C given in Eq. (5) enable the evaluation of various thermodynamical "bulk" properties by means of the Gibbs potential Eq. (1). Unfortunately, Eq. (1) is not sufficient to involve the surface properties, for which more information than Eqs. (1) and (5) would be necessary.

A typical problem of experiments with liquid crystals is their dependence on a history, leading, in general, to difficulties with the reproducibility of results. We frequently discover supercooled or superheated phases, hysteresis phenomena, etc. It is, therefore, satisfying if there are some independent experiments leading to the same conclusions or statements supporting the interpretation of our results.

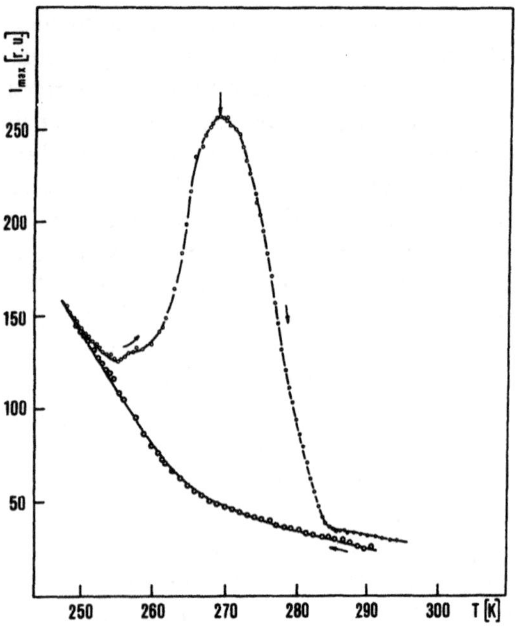

Fig. 2. Temperature dependence of the maximum intensity of luminescence of AECS near the phase transition solid → nematic

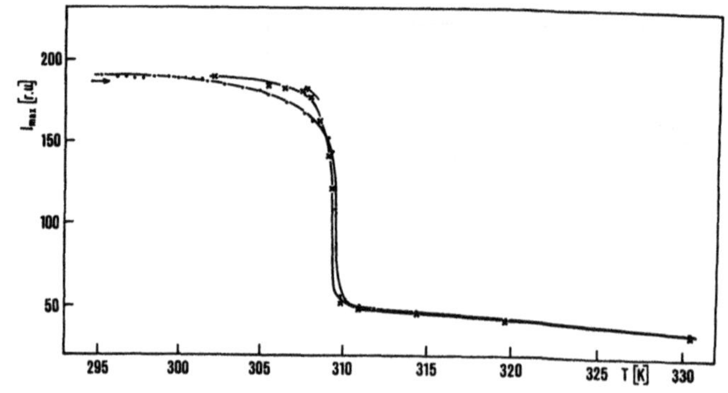

Fig. 3. Temperature dependence of the maximum intensity of luminescence of AECS near the phase transition nematic ↔ isotropic liquid

Examples of such experiments are given in Fig. 2. and Fig. 3. [3]. In these experiments the spectral distribution of the intensity of the luminescence of AECS was observed and the wave length λ_{max} of the maximum intensity was found. It was equal $\lambda_{max} = 462$ nm throughout the whole interval of temperatures. The maximum luminescence intensity was then plotted against temperature.

In Fig. 2 we can see no apparent behavior in the neighborhood of the phase transition nematic ↔ solid when cooling the specimen. However, coming back to higher temperatures we can observe a pronounced peak at the temperature 268 K, in good agreement with the position of the peaks of specific heat capacity in Fig. 1. Similar examples may be seen in Fig. 3. The temperature dependence of the maximum luminescence of AECS reveals an evident change of its behavior near the temperature 309.16 K, evidently corresponding to the phase transition nematic ↔ isotropic. The change in the temperatures $T = 311.1$ K and $T = 309.16$ K need not be surprising, since measurements were done on different specimens possibly prepared under slightly variable conditions.

References

1. Valentová O, Valenta L, Skála L (1985) Czech J Phys B 35:734
2. Stephen MJ, Straley JP (1974) Rev Mod Phys 46:617
3. Valentová O, Koňák Č, Sedláček B, Šorm M (1981) in Proceedings of IV. International Seminar on Energy Transfer, Prague, edited by Pantoflíček J, Zachoval L, Union of Czechoslovak Mathematicians and Physicists, p 238

Received April 24, 1987;
accepted April 21, 1988

Authors' address:

Dr. O. Valentová
Faculty of Mathematics and Physics
Charles University
Ke Karlovu 3
121 16 Prague 2, Czechoslovakia

Progress in Colloid & Polymer Science

Progr Colloid Polym Sci 78:161–176 (1988)

Crystallization in oligomer and copolymer systems comprising lattice incompatible units

H.-G. Kilian

Abteilung Experimentelle Physik, Universität Ulm, Ulm, F.R.G.

Abstract: The presumptions are given on which the thermodynamics of eutectoid oligomer- and copolymer systems is based. It is typical for both systems that crystallization segregation of non-crystallizable chain units occurs on "molecular scales". Solid solution is therefore possible only by forming crystal lamellae with diffuse basal defect layers. The existence of these inhomogeneous micro-phases is demonstrated. The solid solution is limited to a maximum number of components for thermodynamic reasons. The understanding of isobaric state diagrams and of melting in eutectoid oligomers and copolymers with the aid of thermodynamics includes that chain and colloid structure are related in a fundamental manner. That this holds true in eutectoid oligomers and copolymers is verified by a combined description of caloric melting curves and WAXS- and SAXS-measurements. Some fundamental aspects and problems are discussed.

Key words: Oligomers, copolymers, multi-component systems, thermodynamics, melting, superstructure.

Introduction

Polymers and oligomers are capable of undergoing crystallization if the molecules are stereoregular [1–24]. The crystallizability is therefore affected by the presence of structural irregularities which are lattice-incompatible (*ic*-units); *ic*-units are assumed not to be incorporated in the lattice formed by crystallizable sequences (*c*-sequences). Lattice-incompatible chain units must therefore in any case be segregated during crystallization. Underlying processes are therefore controlled by the chemical structure of the chains.

Chain ends mostly operate as lattice incompatible units [2,4,25–32]. Lamella-shaped units are most favorable for squeezing the chain ends out of the *c*-sequence crystal core. The superstructure in crystallized oligomers is for this reason characterized by densily packed stacks of lamellae. This also holds true for solid solutions whereby mixed extended chain crystals are in addition characterized by basal defect layers. This is illustrated in Fig. 1.

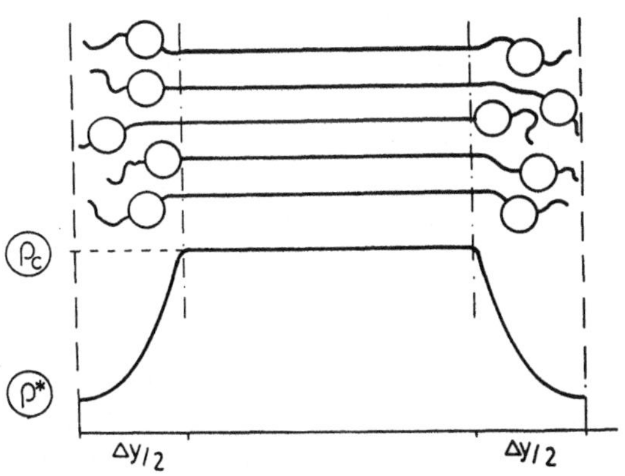

Fig. 1. Two dimensional sketch of an inhomogeneous microphase. Longitudinal defect layers are originated with the disparity in lengths of the crystallized sequences. Within the defect boundaries the density drops from ϱ_c continuously down to the value of ϱ^* which is de facto independent on the mean thickness of the crystal core. Depending on the system ϱ^* is found to be equal or lower than the density of the melt (ϱ_m). Δy characterizes the thickness of the defect layers

PCPS HT 042

The degree of order in the boundaries decreases continuously with increasing distance from the crystal core. An oligomer solid solution represents an "inhomogeneous microphase" (IMPs) [31,32]. Due to the excess-energy in the defect layers, the possible number of components within an IMP is limited to a maximum value ("defect-saturated" IMP ≡ IMP*): Oligomer multi-component systems are therefore eutectoid systems. According to the thermodynamics of eutectoid oligomers [27–32, 40–43] crystallization is characterized by two processes: segregation of the chain ends and micro-phase separation for chains with a large disparity in length.

One of the simplest non-stereoregular polymers is a copolymer comprising two lattice-incompatible units. If the concentration of one of these units is small (*nc* units ≡ "noncrystallizable units") and if sequences comprising the other units are stereoregular, these sequences are, on principle, crystallizable (*c*-sequences composed of *c*-units) [3, 4, 31, 32]. Having a broad *c*-sequence-length distribution and considering *c*-sequences of different length as components, these copolymers must be treated as eutectoid multi-component systems. In copolymers with a broad *c*-sequence-length distribution solid solutions are formed with a maximum number of components. The crystallized lamellae represent defect-saturated inhomogeneous micro-phases ("inhomogeneous *c*-sequence micro-phases" (ISMP*)).

For oligomers or copolymers with a broad distribution of the components, the thickness distribution of IMP*s and ISMP*s is thus related to the chain- or *c*-sequence-length distribution. On "primary and secondary levels", the superstructure is therefore controlled by the chain structure. We would like to call this the "chain-super-structure equivalence principle".

Methods of describing melting in eutectoid oligomers and copolymers are reviewed in this paper.

Chain-ends

The simplest molecule with *ic*-units is a stereoregular linear oligomer with lattice incompatible chain-ends. For *n*-alkanes chain ends are in any case excluded from the crystal core. Chain-end couples are segregated in crystallographic layers (Fig. 2). If we consider the lamellae as subunits of a superstructure, the chain-superstructure equivalence principle has as its simplest form: the thickness of the lamellae within the *n*-alkane single crystal is uniquely determined by the chain-length parameter *y*.

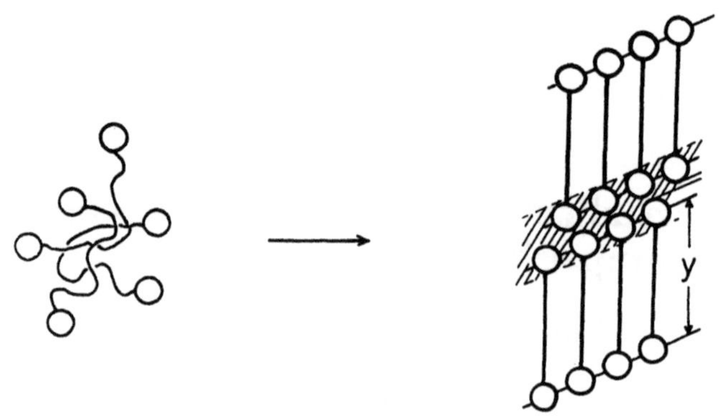

Fig. 2. Illustration of the chain-end segregation running off during crystallization of oligomers. The hatched area indicates the crystallographic layers formed by the chain ends in *n*-alkane single crystals

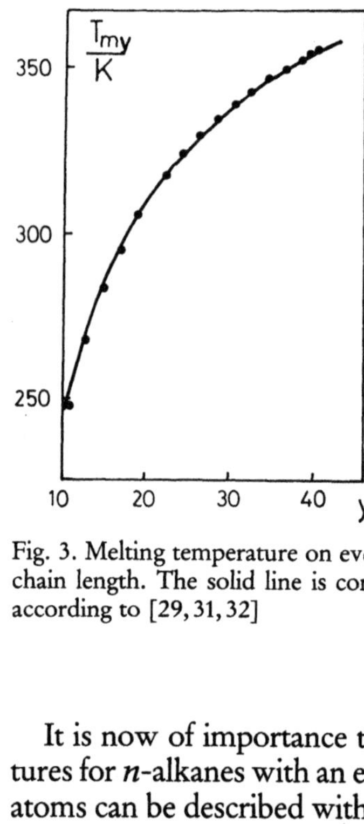

Fig. 3. Melting temperature on even-numbered *n*-alkanes against chain length. The solid line is computed with the aid of Eq. (1) according to [29, 31, 32]

It is now of importance that the melting temperatures for *n*-alkanes with an even or odd number of C-atoms can be described with the aid of the "Flory Frij equation" (Fig. 3) [25–32]

$$T_{my} = T_m(1 - 2\sigma_e/y\Delta h)/\{1 + (RT_m/y\Delta h)\ln y\} \quad (1)$$

where T_m is the melting temperature of polyethylene single crystal of "infinite thickness", $\Delta h(T)$ is the molar melting enthalpy per the *c*-unit. R is the gas constant. Δh and σ_s do not depend on the chain length. It is therefore a small and invariant number of chain units (the chain ends included) which contribute to the surface free-enthalpy correction $2\sigma_e$.

According to Eq. (1) the "primary superstructure" parameter y enters now into thermodynamics implicating the discussion of the chain-end excess properties. It is therefore possible to consider each single lamella within the macroscopic single crystal as a thermodynamically autonomous subunit. These identical lamellae are thermodynamically equivalent so that they altogether melt at the same temperature. The model proposed is fully equivalent to the classical treatment of melting of single crystals in single component systems.

Binary oligomer mixtures

In binary oligomer solid solutions the chain ends must also be squeezed out of the c-unit lattice. Extended chain lamellae with defect layers are now formed (Fig. 4). The structure in the layers is, of course, controlled by the disparity in chain length and the concentration of the components.

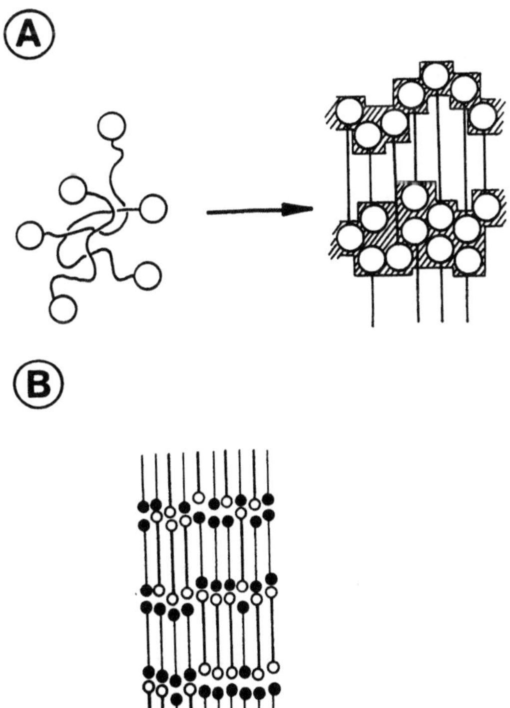

Fig. 4. (A) Chain-ends segregation which characterizes the structure of oligomer binary extended-chain crystals. The hatched area indicates the defect layers with a distorted atomistic structure; (B) Model of the binary mixed extended-chain crystal according to [27]

For the thermodynamic description let us define the chemical potential of the chains y_i in the IMP

$$\mu_i^{(c)} = \mu_{io}^{(c)} + RT\ln(x_i^{(c)}) + g_{i.\mathrm{excess}}^{(c)} + 2\sigma_e;\; i = 1, 2\,. \tag{2}$$

$2\sigma_e$ is the surface free enthalpy per chain, $\mu_{io}^{(c)}$ is the standard potential of the component i, T the absolute temperature, $x_i^{(c)}$ the molar fraction of the $i-th$ component and $g_{i.\mathrm{excess}}^{(c)}$ its excess energy. If controlled by pair-interactions, the excess-free-enthalpy can be formulated by the use of the second virial coefficients $A_{12}^{(c)}$.

$$g_{\mathrm{excess}}^{(c)} = A_{12}^{(c)}\, x_1^{(c)}\, x_2^{(c)})\, (n_1^{(c)} + n_2^{(c)})\,. \tag{3}$$

n_ν^c is the molar number of the chains ν in the solid solution.

Constraints which are expected to arise within the defect layers [27–32] should not severely restrict lateral exchange of chains within the IMPs. By introducing Eq. (2) into the conditions of coexistence, isobaric binary state diagrams can be described fairly well (Fig. 5) [29–32]. In Eq. (2) ideal mixing is supposed to exist.

To find the second virial coefficients related to the relative disparity in chain length [29–32] Eq. (4)

$$A_{12}^{(c)} = A^{(c)}\; |\Delta y_{12}|/\langle y_{12}\rangle$$

$$\langle y_{12}\rangle = (y_1 + y_2)/2;\, \Delta y_{12} = y_1 - y_2 \tag{4}$$

gives evidence that the defect energies are really controlled by pair interactions. Defects are located in the longitudinal layers of IMPs [28–32] (Fig. 4). Their inhomogeneous distribution among the micro-phase must not be accounted for in thermodynamics because the IMP is defined as a cooperative unit: the defect layers are non-autonomous parts of the IMP.

"Boundary-crystal core effects" due to the very small thickness of the IMPs can in fact be observed: the mean density in binary IMPs plotted against the molar fraction of one of the components reveals a non-linear dependence (Fig. 6 a) [27]. Lateral lattice distances behave accordingly (Fig. 6 b) [28–32]. These cooperative effects are in the present case not at all originated with ic-units which are incorporated in the crystal lattice.

Because the excess properties in IMPs are linked with structure parameters like the disparity of lengths and the mean thickness of the IMPs it suggests to

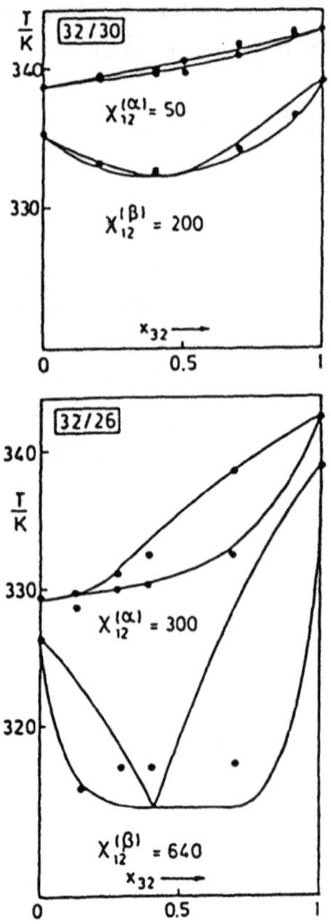

Fig. 5. Isobaric state diagrams of binary *n*-alkane systems. The chain lengths y_i are indicated for each of the drawings. Two distinct solid phases, of different crystal structure (α) and (β), exist which undergo a phase transition. $X_{ik}^{(\alpha)}$ ($A_{ik} = RT\,X_{ik}$) is the enthalpy component of the second virial coefficient in the crystal modification (α) which is defined by $X_{ik}^{(\alpha)} = (X_{ik}^{(s)} + X_{ik}^{(h)}/T)^{(\alpha)}$. The solid lines are theoretical [29, 31, 32]. $X_{ik}^{(\beta)}$ is the second virial coefficient *m* the modification β

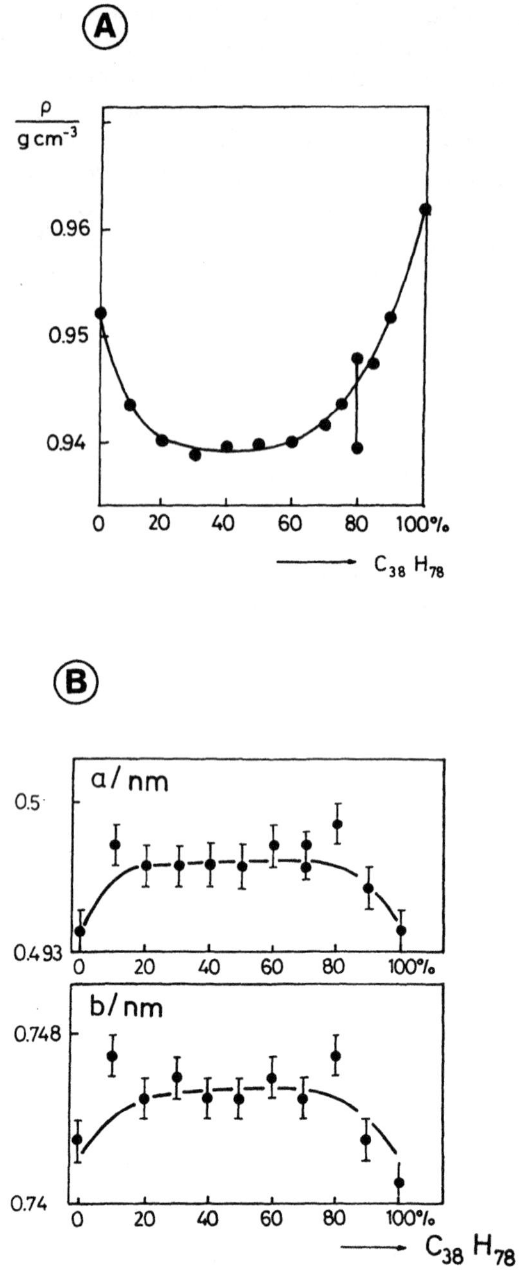

Fig. 6. (A) The density of oligomer binary extended chain crystal ($y_1 = 34$; $y_2 = 38$) over the molar fraction of C38 ($C_{38}H_{78}$); (B) the lattice parameters (a- and the b-axis) against composition

apply this treatment to oligomer multi-component systems.

Eutectoid oligomer multi-component systems

To describe isobaric state diagrams of oligomer multi-component systems the free enthalpy excess within IMPs must only be considered. Defects which are produced or annihilated do not change free enthalpy ($\Delta\mu^{(i)} = 0$).

With the use of second virial coefficients ($A_{ik} \equiv A_{ik}^{(c)}$) which have been inter- or extrapolated from the data deduced from the description of binary state diagrams [29–32], we are led to the representation of the liquidus in *n*-alkane mixtures which are shown in Figs. 7 and 8 [30, 31]. For *n*-alkane multi-component systems [30] the set of solid solutions and their liquidus temperatures are fairly well predicted (Figs. 8 and 9). This is an impressive demonstration that chain-ends are excluded from the core of IMPs. The excess free-enthalpy is mainly controlled by the disparities in chain lengths.

SAXS-pattern displays sets of diffraction peaks. More than three orders can be identified in each set

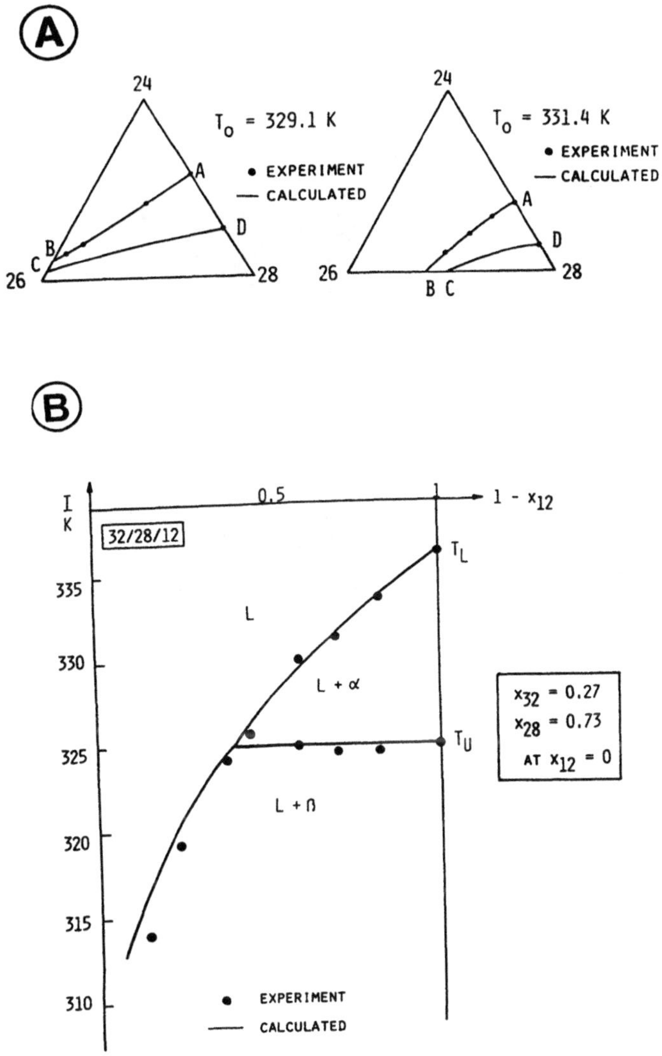

Fig. 7. (A) Isothermal cuts in the ternary n-alkane phase-diagram (C24, C26, C28) at temperatures as indicated [29, 31, 32]. AB and CD are the traces of the two-dimensional liquidus- and solidus surfaces; (B) quasi-binary cut of the eutectoid ternary phase diagram of n-alkanes C12, C28, C32 according to [29, 31, 32]. L is the region of the ternary melt. The regions are to be seen where both of the binary crystal modifications (α) and (β) are in coexistence with the ternary melt

(Fig. 9) [30, 32]. Lamella cluster must be formed, each one comprising, on the average, more than eight identical IMPs. Such a segregation must be the consequence of the minimum interlamellar contact-energies between identical IMPs.

Because we know the amounts of stable IMPs in oligomer multi-component systems the chain-superstructure equivalence principle is verified e.g., (Figs. 8 and 9). It is also demonstrated that thermodynamic equilibrium is fairly well approximated.

The defect-saturated inhomogeneous micro-phase

In oligomer multi-component systems with a broad chain-length distribution IMPs should be formed which are for reasons of thermodynamic stability "defect saturated": Each of these IMP*s contains the maximum number of components.

To prove this hypothesis let us consider mixed microphases with equimolar amounts αn of each of the N components such that the molar fraction is given by

$$x_i^c = n_i / \sum n_i = \alpha n / \alpha N = 1/N; \; \alpha = \text{const} \tag{5}$$

$$n = \sum n_i \, . \tag{6}$$

The excess free-enthalpy should be proportional to the total number of "mismatch increments" between chain pairs of different lengths, $\sum x_i x_k \Delta y_{ik}$. Defining a defect energy increment per unit, the virial coefficient can be written as $A_{ik} = \langle A \rangle (y_k - y_i)$. Summing up the excess contributions and introducing $y_i = y_0 + i$ and $y_k = y_0 + k$, we are led to the excess free-enthalpy in the Porter version

$$G_{ex} = \sum_i^N \sum_k^N A_{ik} x_i x_k n = \langle A \rangle \, n(1/2) \sum_{i=1}^N \sum_{k=1}^N (i-k)$$

$$= \langle A \rangle \, n(N-1)/6 \, . \tag{7}$$

The total free enthalpy of the micro-phase given by

$$G^c = \sum n_i y_i G_0 + RT \sum n_i \ln x_i + G_{ex} \tag{8}$$

the molar free enthalpy reads then

$$g^c = G^c/n = \langle y \rangle G_0 - RT \ln N + (N-1) \langle A \rangle / 6;$$

$$\langle y \rangle = \sum y_i n_i / n \, . \tag{9}$$

By seeking the minimum free enthalpy in dependence of N

$$dg^c/dN = -RT/N + \langle A \rangle / 6 = 0 \tag{10}$$

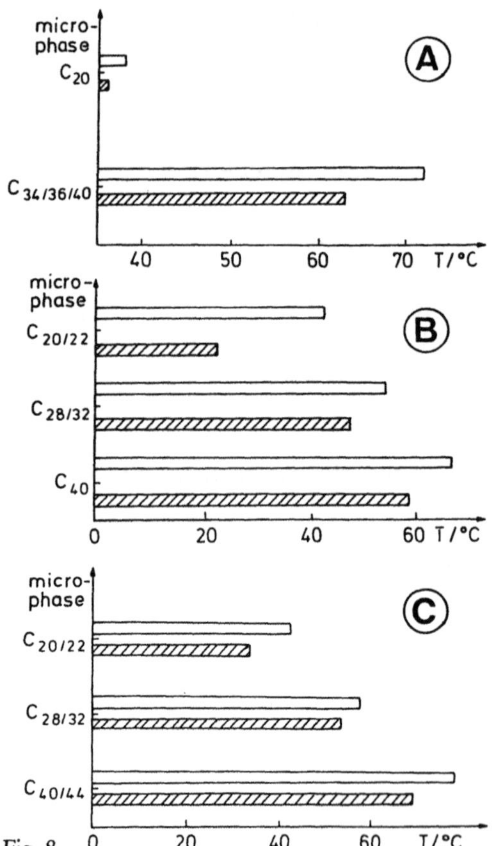

Fig. 8

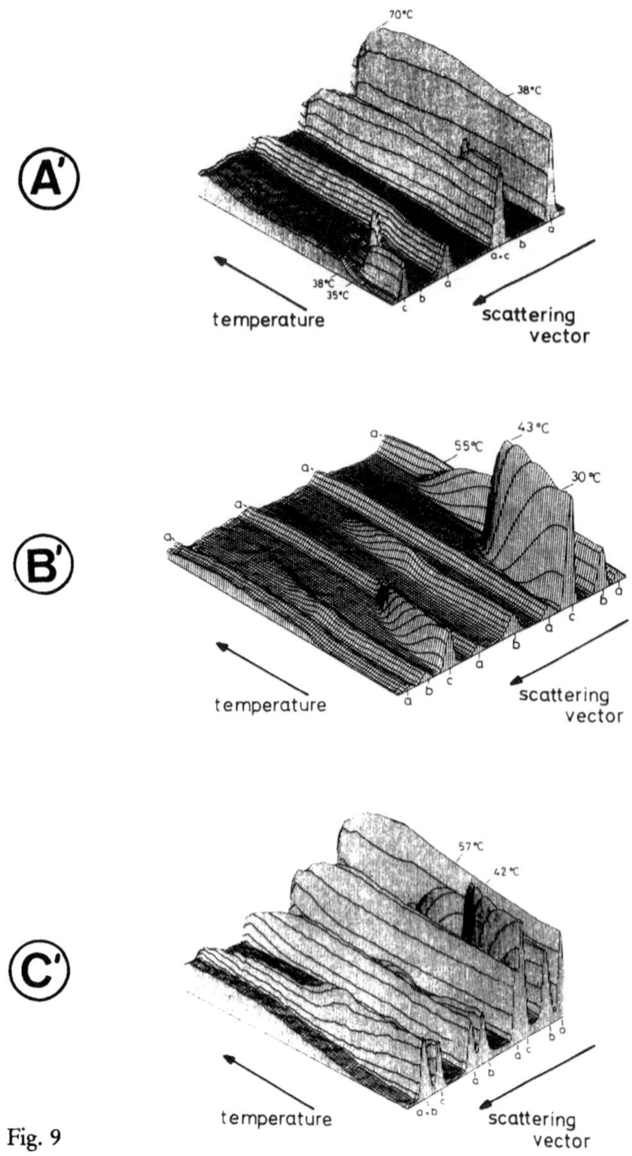

Fig. 9

Fig. 8. and 9. (A) The height of the bars is assigned to indicate the liquidus temperatures in the four component *n*-alkane system (equal weight fractions g_i = .25) (C20, C36, C38, C40). The hatched bars are theoretical. The components mixed in these crystals are indicated with each of the bars [30–32]; (A′) Synchrotron SAXS-pattern against temperature of the four component *n*-alkane system [30] (A): (a) ternary mixed crystal clusters C36, C38, C40, (b) hexagonal phase of the single component crystal clusters C20 the crystal lattice of which is transformed into a triclinic lattice a few degrees below (c); (B) Liquidus in the five component *n*-alkane systems (C20, C22, C28, C32, C40) each of the component with the same weight fraction of g_i = .2 [30]; (B′) Synchrotron SAXS-pattern against the temperature of the system [30] (B): (a) single component crystal clusters C40; (b) binary mixed crystal clusters C28, C32; (c) binary mixed crystal clusters C20, C22; (C) Liquidus of a six component *n*-alkane system [30]: C20: $g_{20} = 0.08$; C22: $g_{22} = 0.11$; C28: $g_{28} = 0.13$; C32: $g_{32} = 0.19$; C40: $g_{40} = 0.23$; C44: $g_{44} = 0.26$; (C′) Synchrotron SAXS-pattern against temperature of system (C) [30]: (a) binary mixed crystal clusters C40, C44; (b) binary mixed crystal clusters C28, C32; (c) binary mixed crystal clusters C20, C22

a maximum number of components N^* results

$$N^* = 6\,RT/\langle A\rangle\,. \tag{11}$$

If homologues in the solid solution vary by just one *c*-unit in length, N^* is identical with the maximum disparity in chain length, Δy. Setting $\langle A\rangle$ inversely proportional to the mean chain length $\langle y^{c*}\rangle$ ($\langle A\rangle = A/\langle y^{c*}\rangle$) we are led to

$$N^* = \Delta y = \varkappa\,\langle y^{c*}\rangle;\ \varkappa = 6RT/A\,. \tag{12}$$

This is exactly the relationship which has been deduced from the description of the state diagrams of *n*-alkane multi-component systems [29–32]: the relative maximum miscibility ($\Delta y / \langle y^{c*} \rangle$) is invariant, the width of the defect layers in IMP*s grows in proportion to $\langle y^{c*} \rangle$. The excess energy in IMP*s depends on the mean thickness of the crystal core as a result of "boundary-core effects": non-autonomity of defect layers of oligomer IMP*s is evidenced.

This result implies an interesting suggestion: according to

$$\lim_{\langle y^{c*} \rangle \to \infty} \langle A \rangle = 0 \qquad (13)$$

the defect energy should dissappear only in IMP*s of "infinite thickness". A nematic solid solution of *n*-alkanes is not stable under realistic circumstances.

The superstructure-molecular weight-distribution equivalence relationship

It is typical for an oligomer multi-component system with a broad chain length distribution $\Phi(y)$ (molecular weight distribution) that IMP*s are formed. In the fully crystallized state, the crystal thickness distribution $C(\langle y^{c*} \rangle)$ must uniquely be related with the molecular weight distribution $\Phi(y)$ [31, 32]

$$\Phi(y) = \gamma \, C(\langle y^{c*} \rangle); \; \gamma\text{: normalization factor}. \qquad (14)$$

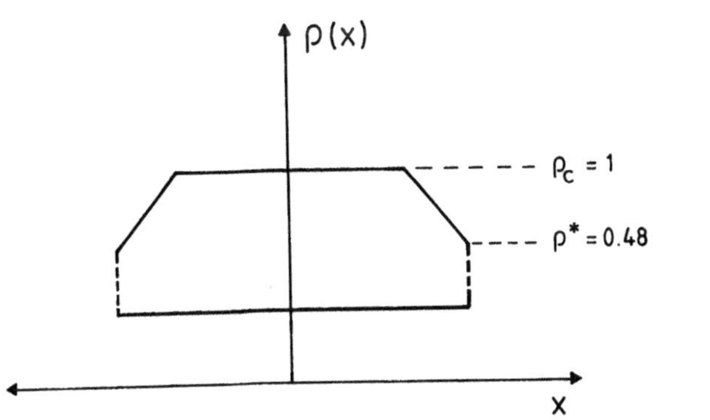

Fig. 10. Normalized ($\varrho_c = 1$) longitudinal electron density profile of a defect-saturated eight component IMC* (C21: $g_{21} = 0.058$; C22: $g_{22} = 0.077$; C23: $g_{23} = 0.096$; C24: $g_{24} = 0.116$; C25: $g_{25} = 0.135$; C26: $g_{26} = 0.154$; C27: $g_{27} = 0.173$; C28: $g_{28} = 0.192$) deduced from SAXS-pattern according to [26]. The density drops down to 48% of the crystal core density, a value which is below the mean density of the multi-component melt

Fig. 11. Electron microscopic picture of a fraction of PE according to Anderson [48]

The "primary elements" in the colloid structure (the mean thickness and the thickness distribution of the IMP*s) are determined by the molecular weight distribution.

The superstructure

The utility of the inhomogeneous microphase model comes now in evidence: as deduced from SAXS-investigations the mean electron-density falls across the boundary IMP*s down to exceedingly low values reaching in the outer layer values of about $.5\varrho_c$ where $\varrho_c = 1$ is normalized density in the crystal lattice (Fig. 10) [27]. The chains should, on the average, have defined positions so as to yield the observed density profile.

It should be noticed that the density in the outer layers of the boundaries arrives at nearly the same value for IMP*s of different thickness (Fig. 10). Interlamellar contacts between IMP*s are therefore identical. There are no longer any energetical reasons for forming clusters comprised of identical IMP*s. Every spatial distribution of IMP*s is iso-energetical so that it is more likely to have them nearly randomly distributed among clusters.

This apparently holds true for fractions of polyethylene with extended chain crystals: according to the electron micrographic picture shown in Fig. 11, lamellae of different thickness are nearly randomly scattered over the whole region. The thickness distribution is observed to be a mirror of the molecular weight distribution. The structure-molecular weight-distribution relationship (Eq. (14)) again is verified within the limits of confidence.

The colloid-structure in eutectoid oligomers is then characterized by some features which are altogether related with eutectoid segregation of c-sequences:

1. The lamella-shaped defect-saturated inhomogeneous micro-phase is the prototype of primary structure units;
2. the thickness distribution of the IMP*s is determined by the molecular weight distribution;
3. clusters are formed as statistically well defined constituents of a secondary level of colloid structure;
4. IMP*s of different thickness are nearly randomly distributed among clusters. Interlamellar contacts between IMP*s are, on the average, isoenergetical;
5. due to this randomness there is a broad scale of different clusters which display "cluster-gas" interreference behavior [39].

Melting process

IMP*s are autonomous. Idenical IMP*s are thermodynamically equivalent. In coexistence with the same melt they have therefore the same melting temperature independent of where they are located. When homogeneity in this sense is guaranteed, it is possible to understand the macroscopical behavior by computing the amount and composition of each of the coexisting IMP*s (therewith the degree of crystallinity) with the aid of "micro-phase thermodynamics" [31, 32].

Based on Eq. (14) thermodynamics of eutectoid multi-component systems has been developed. If the maximum disparity in chain lengths within the IMP*s obeys relation (12) it is possible to describe the melting process [31, 32]; the quality of a fit to experimental data for fractions of polyethylene can be seen by evidence in Fig. 12.

We arrive at the conclusions:

1. IMP*s are formed. They can be treated as thermodynamically autonomous microphases;

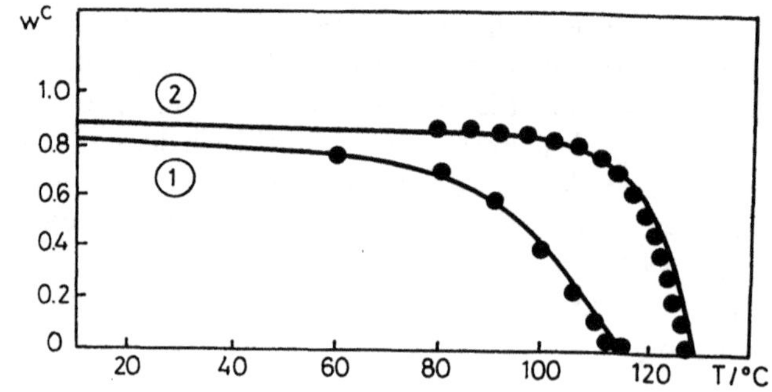

Fig. 12. The molar degree of crystallinity w^c against temperature for fractions of linear polyethylene with different mean chain lengths (1) $\langle y \rangle = 206$ CH_2-units, and (2) $\langle y \rangle = 232$ CH_2-units according to [31, 32]

2. on heating melting starts with IMP*s of smallest thickness at the lowest temperature. Next, fractions of IMP*s are then transferred into the melt selectively and consecutively in respect to their mean thicknesses (see Fig. 16);
3. local concentration fluctuations in the molten regions must be small.

It is in principle possible to deduce the molecular weight distribution from the melting process of eutectoid oligomers provided that folded chain crystallization does not occur.

The thermodynamically equivalent-subsystems

According to thermodynamics composition and amount of each of the coexisting IMP*s must continuously change during the melting process. On heating, these processes were found to run off within a comparatively short period of time (for n-alkane systems below 10–30 s [30]). An interesting example of synchrotron SAXS-measurements is shown in Fig. 13. Below $T = 43\,°C$ one set of small-angle peaks splits into two. Even this "solid state transformation" is finished within a period of time shorter than 15 s, the time needed for registering a single SAXS-diagram. According to theory [39] each of the new clusters must comprise more than eight parallel IMP*s to produce more than three interference peaks [39]. Regions with a minimum diameter of about 20–30 nm must continuously reshuffle.

This is understandable only if the system is comprised of thermodynamically equivalent subsystems comprising a certain number of clusters (EQS). This number ought to be so large that each of the EQSs shows identical properties. Every reorganization of

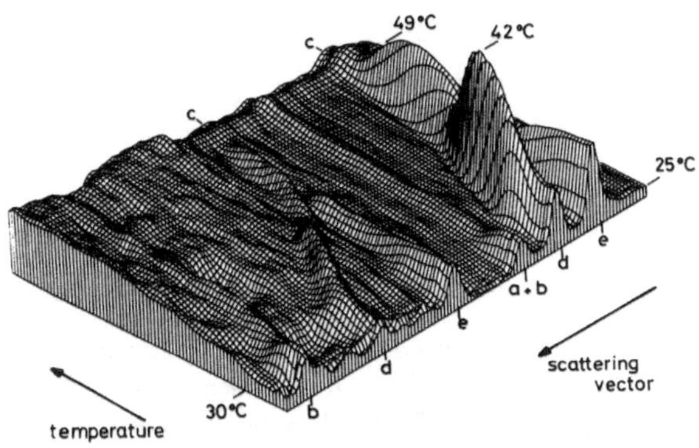

Fig. 13. Synchrotron SAXS-pattern of a five component n-alkane system against temperature [30] C20: $g_{20} = 0.3$; C26: $g_{26} = 0.25$; C28: $g_{28} = 0.2$; C32: $g_{32} = 0.15$; C40: $g_{40} = 0.1$. The ternary mixed crystals C26, C28, C30 become unstable below 42 °C: A fraction of C26 is squeezed out of this IMP, forming pure crystals C26. On further cooling the structure of these crystals C26 passes continuously from the hexagonal phase into a monoclinic symmetry. The long period is diminished accordingly

the colloid structure runs therefore on very local scales, at least within EQS-subsystems. That chains can quickly be exchanged between neighboring IMP*s ("membrane effect") is possibly the reason why thermodynamic equilibrium is approximated even in eutectoid oligomer multi-component systems.

Eutectoid bi-polymers

The crucial role of "lattice incompatible chemical units" in copolymers was first recognized by Flory [2–4]. With a small number of one of the co-units, sequences of the other co-units are generated which are crystallizable if they are stereoregular (c-sequence). It results from Flory's theory that the maximum melting temperature of copolymers is depressed with increasing numbers of "non-crystallizing co-units" (ic-units $\equiv$ nc-units). This effect is accompanied with typical broadening of the melting range. Despite of qualitatively correct predictions, this theory fails to give a full understanding of melting. This might be one of the reasons behind suggestions that "nc-units" are built into crystals [4–24]. Observations that unit-cell dimensions and lattice distortions increase with increasing numbers of nc-units were taken as evidence for incorporating nc-units [11–13].

There are observations according to which evidence is produced for the existence of "diffuse" longitudinal "melt-to-crystal" interlayers [35–37]. Yet, an interpretation of combined SAXS- and SANS-measurements leads to the hypothesis that the interlayers are nevertheless very small [22].

A weak point in most of the above arguments is the use in thermodynamics of the two-phase model as deduced from x-ray or neutron-scattering experiments. In view of the oligomer-multicomponent systems it appears to be imperative to postulate the existence of defect-saturated extended c-sequence crystals (ISMP*s, (Fig. 14)) with extended and diffuse defect boundaries. Size and structure of these "thermodynamic boundaries" cannot straightforwardly be deduced from SAXS-investigations.

Thermodynamics

Bi-copolymers with nc-units may be considered as a special type of a eutectoid multi-component system under the following assumptions:

1. Stereoregular c-sequences of different length are considered as components despite their being linked in the chains [31, 32, 34];
2. lamella-shaped inhomogeneous, defect-saturated microphases (ISMP*) are formed;
3. c-sequences can be exchanged between microphases, at least on local scales;
4. crystallization segregation processes are confined to thermodynamically equivalent subsystems (IQS).

In view of the last points it is of importance that the maximum disparity of c-sequence lengths in the ISMP*s of eutectoid copolymers is much larger than in eutectoid oligomers ($\Delta y = A\langle y^{c*}\rangle$): the maximum disparity in chain length within an ISMP* has been defined by

$$\Delta y = A\langle y^{c*}\rangle + B; B = B_1 x_c/(1 - x_c) + B_2;$$

$$A = 0.15\text{–}0.3\,. \tag{15}$$

x_c is the molar fraction of the c-units. Because of having B_2 in the range of 30–60 (Table 1) crystallization segregation of c-sequences is relevant for ISMP*s with a mean thickness of $y > B$. This situation is met for copolymers with not too large a number of nc-units ($0.2 < x_{nc} < 0.5$) at high enough temperatures where first ISMP*s crystallize. For copolymers with a larger x_{nc} or at the end of crystallization, the thickness of the ISMP*s becomes extremely small until there is no longer any crystal core (Fig. 15); these microphases might be considered as "continuous", compris-

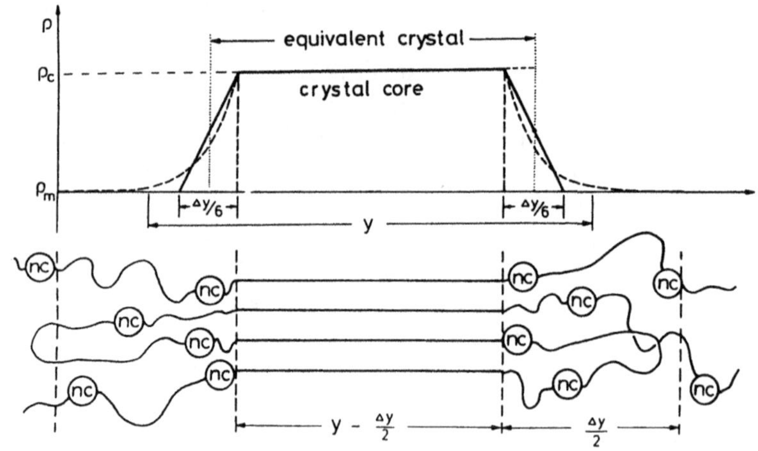

Fig. 14. Scheme of a defect saturated extended c-sequence mixed crystal (ISMP*). Δy is the maximum disparity in c-sequence length tolerated. The mean density drops within the defect layers from ϱ_c rapidly down to the mean density as in the melt (ϱ_m). It is to be seen that the mean chain length characterizes fairly well the mean thickness of the ISMP*

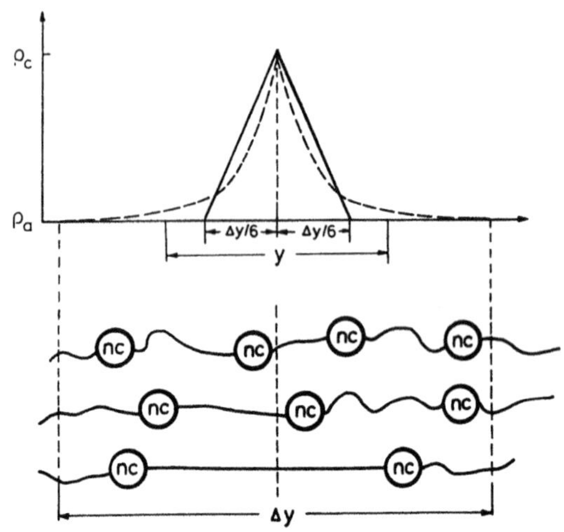

Fig. 15. Schematic of a degenerated ISMP* no longer showing a perfectly ordered lattice core

ing defect layers only. "Meso-phases" can be formed by small translational shifts of the c-sequences involved – this is indeed one of the reasons why eutectoid crystallization is possible even in the presence of crosslinks (permanent chemical crosslinks, crystals as multi-functional crosslinks).

According to Eq. (15) "defect-compatibility" of ISMP*s increases with decreasing thickness whereby the mean intrinsic degree of disorder grows. Smallest ISMP*s have no ordered crystal core at all. It should be emphasized here that this broad scale of defect structures is thermodynamically described with the aid of the non-homogeneous micro-phase model. The fundamental role of interlayers comes into evidence. Their existence is necessary for thermodynamic reasons. One might take this as the most appropriate identification of eutectoid copolymers as colloid systems.

The primary colloid-structure

Because ISMC*s are defect-saturated, the thickness distribution of the ISMC*s $C(\langle y^{c*}\rangle)$ must strictly be related with the c-sequence-length distribution $\psi(y)$ ("c-sequence-length-colloid-structure) (equivalence principle") [31, 32, 37, 38].

$$C(\langle y^{c*}\rangle)\,|_{y_{\min}} = \gamma\psi_{(y)}\,|_{y_{\min}}\,. \qquad (16)$$

$y_{\min}$ gives the smallest ISMP* which is thermodynamically completely stable. γ is a normalization factor. Crystallized from the melt under optimum conditions [4], folded c-sequence crystallization does not occur to measureable extents if the molar fraction of nc-untis is at least larger than about two percentage [40].

It is possible to characterize the defect situation in the layers of ISMP*s by defining the parameter y_k

$$y_k = B/3/(1 - A/3)\,. \qquad (17)$$

$\Delta y/3 = Ay^{c*}/3 + (1 - A/3)\,y_k$ gives the number of "non-crystallizable" c- units located in defect layers.

Table 1. Thermodynamic parameters for PE networks [37, 38, 40]

$2\sigma_e/\Delta h\ (T_m) = 2.11$
$\Delta h\ (T_m) = 970$ cal mol^{-1} pro CH_2-unit
$\Delta C = 1.4$ cal mol^{-1} K^{-1} pro CH_2-unit
$T_m = 415$ K
$B_1 = 0.9$; $B_2 = 50$; $X^{(m)} = 0$

The molar fractions x_c are indicated for each of the drawings.

With the aid of this definition the molar fraction of crystals is written as [34, 37]

$$w^c = (1 - A/3) \sum_{y = y(T)}^{\infty} (y - y_k)\, \psi(y) / \sum_{y=1}^{\infty} y\psi(y) \quad (18)$$

with $y(T)$ characterizing the ISMC*s of smallest thickness which are just stable at the temperature T. On heating $y(T)$ increases so that w^c decreases. $w^c(T)$ is apparantly controlled by the c-sequence-length distribution function $\psi(y)$ and $y(T)$.

It is possible to derive various "primary structure parameters" [31, 32, 37, 38, 40]. The mean thickness of the ISMC*s is for example given by

$$\langle y_c \rangle = (1 - A/3) \sum_{y(T)}^{\infty} (y - y_k)\, \psi(y) / \sum_{y(T)}^{\infty} \psi(y) \,. \quad (19)$$

We find the "primary level" of the superstructure in eutectoid copolymers in unique dependence on temperature clearly related with concentration and distribution of nc-units.

Thermodynamics

An analogue for the thermodynamics of eutectoid copolymers has been developed for oligomer multicomponent systems [31, 32, 34, 37, 38, 40]. The "liquidus temperature" of ISMC* of the thickness $y(T)$ has been explicitely formulated (see appendix).

One of the most important outcomes is then that the ISMP*s should melt selectively and consecutively in respect to their thicknesses accompanied with a conjugated transformation of the colloid-structure. Due to the selective melting the crystal thickness distribution narrows increasingly as it is to be seen from Fig. 16. Melting, observed for example, by measuring the heat of fusion is therefore a mirror of the c-sequence-length distribution in eutectoid copolymer.

Networks as eutectoid copolymers

Thermodynamics of eutectoid copolymers has been applied to networks (electron-crosslinked polyethylene) [31, 32, 34, 37, 38, 44]. Crosslinks are nc-units the distribution of which defines the structure of the network. Chains of different length are components of a eutectoid system. Because of having the chains linked in a network, eutectoid crystallization is

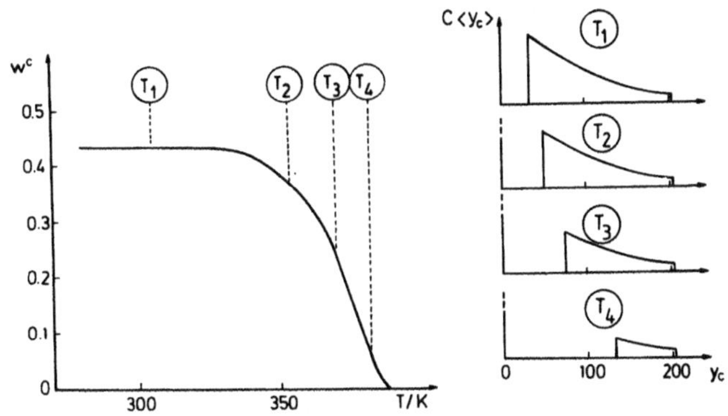

Fig. 16. Molar degree of crystallinity against temperature. The ISMP* thickness distribution $\Phi(\langle y^{c*} \rangle)$ is schematically shown for distinct temperatures T_i

expected to be constrained. It may thus be questioned whether ISMP*s can be formed as demanded by thermodynamics.

If the network is assumed to show a random c-sequence-length distribution we succeed without difficulties in fitting c_p-curves of differently crosslinked polyethylene (Fig. 17) [38]. The calculations have been carried out with the parameters listed in Table 1.

Typical phenomena which arise when the density of crosslinks is increased (like depression of the maximum melting temperature or reduction of the maximum crystallinity) are fairly well described. The

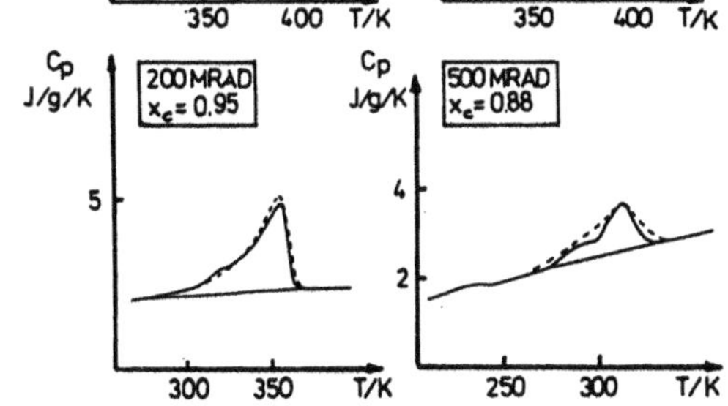

Fig. 17. DSC-measurements of PE networks (radiation crosslinked in the melt). The dosis and x_{nc} are indicated with each of the drawings. The dotted lines are computed [38]

Table 2. PE-networks [38]

Dosis Mrad	x_{nc} mole %	Δh_{exp} J/g	Δh_{theo}	L_{exp} nm	L_{theo}	y_{exp} nm	y^{theo}
400	0.1	35	31	6.5	16.5	0.9	1.1
200	0.05	58	53	10.0	9.3	2.5	2.6
110	0.028	73	79	11.0	9.0	4.8	4.4
55	0.01	113.5	114	13.5	10.4	5.1	6.3

L = long period; y = mean thickness of x-ray coherent blocks

whole melting process is understood including its broadening in highly crosslinked PEs.

It is a matter of consistency that the theoretical thickness of the x-ray coherent core of ISMP*s is in agreement with the experimental data (Table 2). From the cluster structure observed in electron microscope pictures (Fig. 18) [37, 40] the same statistical parameters are deduced which were introduced in the calculation of SAXS-pattern (Fig. 19). It is important that SAXS-pattern at elevated temperatures are fairly well computed by only accounting for the selective melting of ISMP*s. The experimental changes of intensity are obtained correctly without any parameter adjustments. The calculations were carried out by using Hosemann's model of linear paracrystals [37, 39, 40].

It is apparantly typical for eutectoid copolymers that segregation on *nc*-units occurs on "molecular scales". This implicates that the lateral width of the ISMP*s is also decreased with decreasing thickness of the ISMP*s (Fig. 18). A discussion of this phenomenon is needed.

For eutectoid copolymers the mean size of EQS-subsystems should be relatively small: the mean number of ISMP*s per x-ray coherent clusters ranges between 1 and 5. Introducing reasonable dimensions of the constitutive units the minimum diameter of the

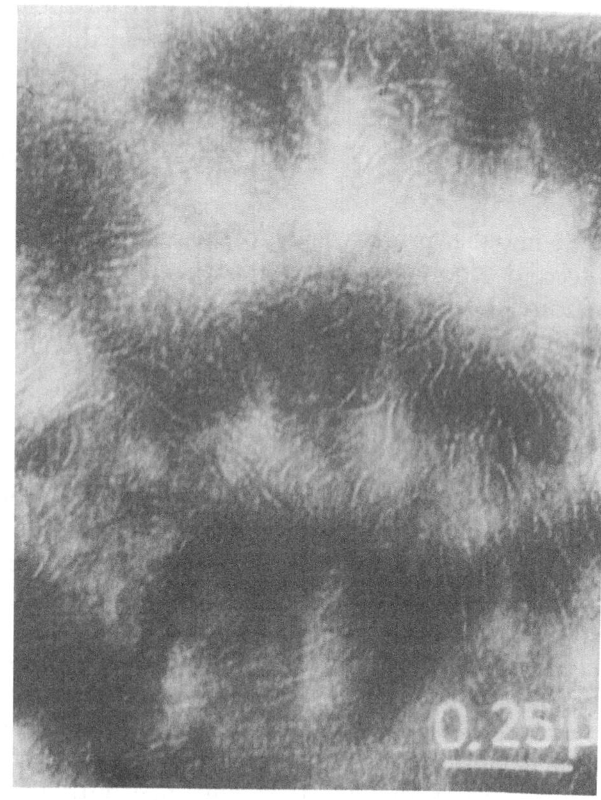

Fig. 18. Electron micrograph of 80 Mrad PE according to [27]

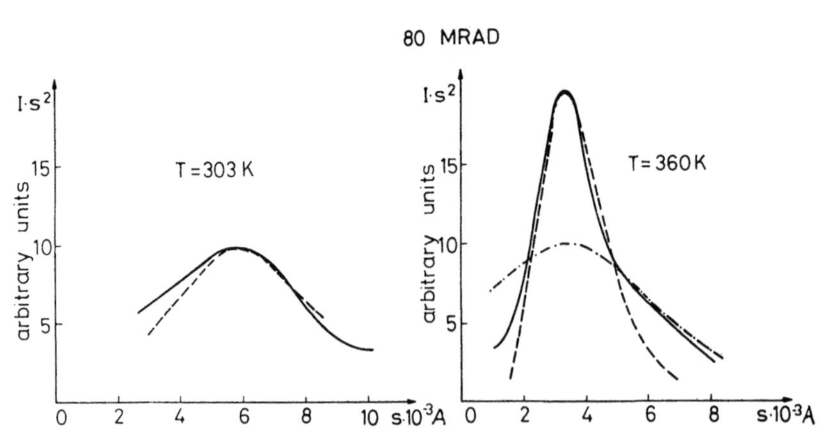

Fig. 19. Is^2 of radiation crosslinked (80 Mrad) at the temperatures as indicated ($s = 2\sin(\theta)/\lambda$; θ scattering angle, λ wave length). The broken lines are computed with the concept of the one-dimensional paracrystal-cluster [37, 39, 40]. The colloid structure at 360 K is assumed to be represented by a random distribution of *c*-sequence-lengths and therewith of the thickness distribution of the ISMP*s: on heating the ISMP*s within clusters have to reshuffle accordingly. The dash-dotted line is theoretical as well, computed under the assumption that ISMP*s melt selectively without any rearrangements of stable ISMP*s

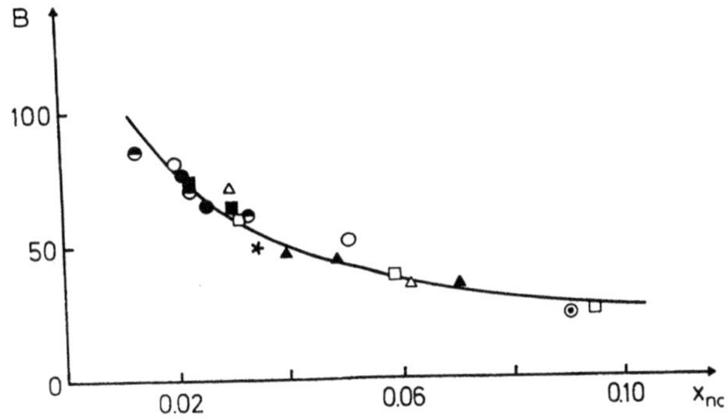

Fig. 20. $B(x_{nc})$: radiation crosslinked PE; random copolymers; ethylene-fluorethylene copolymers; * LDPE; polyoxi-methylene copolymers; vicinal clorinated PE

EQS-systems is estimated to be larger than 10 nm–200 nm [32, 37, 38, 40]. This is in the same order of magnitude as in oligomer multi-component systems.

By these results it is verified that crosslinks in polyethylene generated by high energy electrons [37, 38, 40] are indeed randomly distributed.

General aspects

The melting process of various eutectoid copolymers (for example short-chain branched PE [37, 40, 44], ethylene-tetrafluor-copolymers [40], ethylene copolymers with sidestanding vicinal clorines as *nc*-units [40]) has been described with success including a correct characterization of the primary (mean crystal thickness, ISMP*-thickness distribution) and the secondary colloid-structure (long periods, SAXS-pattern).

Parameter B depends of the molar fraction of *nc*-units in a universal manner (Fig. 20) [40]. The chemical structure of the *nc*-units is of minor importance. Each *nc*-unit produces extreme defect energies; larger sequences are incommensurable to the lattice so that these units must be squeezed out. This unique behavior must, in addition, originate with the way in which chain sequences pass through the boundaries of the ISMP*s. It is a future task to find a solid interpretation.

Analysis of the c-sequence-length distribution in "non-random copolymers"

In view of the above results it suggests itself to use a computer-aided analysis of the melting curves of eutectoid copolymers ($x_{nc} > .02$) as a method for characterizing the chemical chain structure in the *c*-sequence-length range of $y > y_{min}$ (y_{min} in the range of 16–30).

We confine ourselves to the representative example of $-O-(CH_2)-$ containing copolymers with $-O(CH_2)_n$-groups as *nc*-groups as *nc*-units ($n > 1$) [40]. A bimodal *c*-sequence-length distribution is obtained in any case (Fig. 21) which is the consequence of topochemical processes during copolymerization.

The theoretical mean thickness of the x-ray coherent core of the ISMP*s is in good accord with the experimental data including their typical dependence on composition (Fig. 22). SAXS-pattern can be computed fairly well in dependence of temperature with the aid of Hosemann's linear theory of paracrystalline structures [39]. It is significant that this agreement is

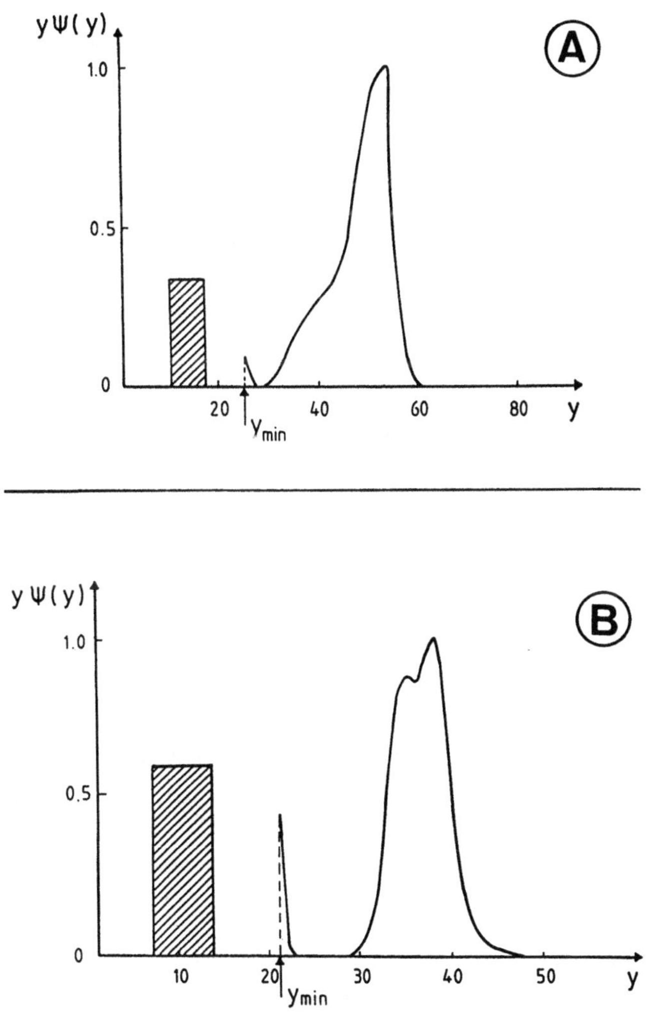

Fig. 21. Weight distribution function of poly-oxi-methylene y^{Ψ}: (A) n_c-unit: $(CH_2)_4$–O, $x_{nc} = 0.031$; (B) n_c-unit: $(CH_2)_6$–O, $x_{nc} = 0.03$

dependent on the use of bimodal sequence-length distribution which has been obtained by analyzing the melting process [37, 40] (Fig. 23).

If *nc*-units make chain sequences incommensurable to the crystal lattice, it is understandable that their lattice mismatch is of importance. Only few lattice compatible co-units are known [1, 3]. It is therefore in most cases possible to deduce the *c*-sequence-length distribution of copolymers within the range of $y > y_{min} \simeq$ 18–30 by means of the method presented. This cannot be achieved with the aid of other analytical methods in use [46].

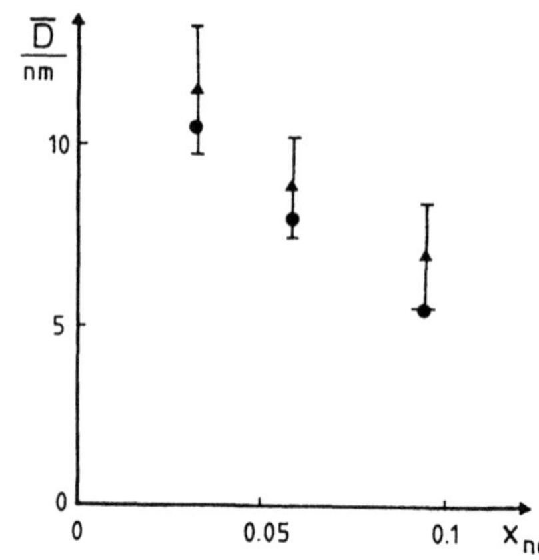

Fig. 22. Mean x-ray coherent thickness of the ISMP*s $\bar{D}$ in poly-oxi-methylene ($(CH_2)_4$: thermodynamics O; x-ray analysis: Δ

Final remarks

The defect-saturated inhomogeneous micro-phase is an appropriate equilibrium model for describing crystallization in eutectoid oligomers and copolymers. IMP*s or ISMP*s are inhomogeneous cooperative units. With lattice-incompatible units, like chain ends or *nc*-units squeezed into defect boundaries, the density drops to very low values – in case of oligomers down to about 45 % of the crystal density; in copolymers down to a density as in the melt. The existence of these "molecular diffuse" defect layers allows segregation of *nc*- units or chain-ends on a very local scale.

A thermodynamic treatment is made possible by defining micro-phases so that chains or *c*-sequences are uniquely assigned to IMP*s (or ISMP*) or to the melt. A fringed micelle structure results implicating that longitudinal "phase boundaries" of IMP*s and ISMP*s are in principle not sharply defined. Moreover, components belonging to neighboring micro-phases may interpenetrate, an effect which is pronounced in eutectoid copolymers. The boundaries of IMP*s and ISMP* are "diffuse interlayers".

For eutectoid copolymers we must consider *c*-sequences of different lengths as thermodynamic components in networks identical with the chains between crosslinks. This is a definition which breaks away from the classical conception of autonomous particles. Yet, in view of the results, especially those for PE-networks, *c*-sequences can locally always be exchanged between micro-phases so as to guarantee formation of ISMP*s as demanded by thermodynamics.

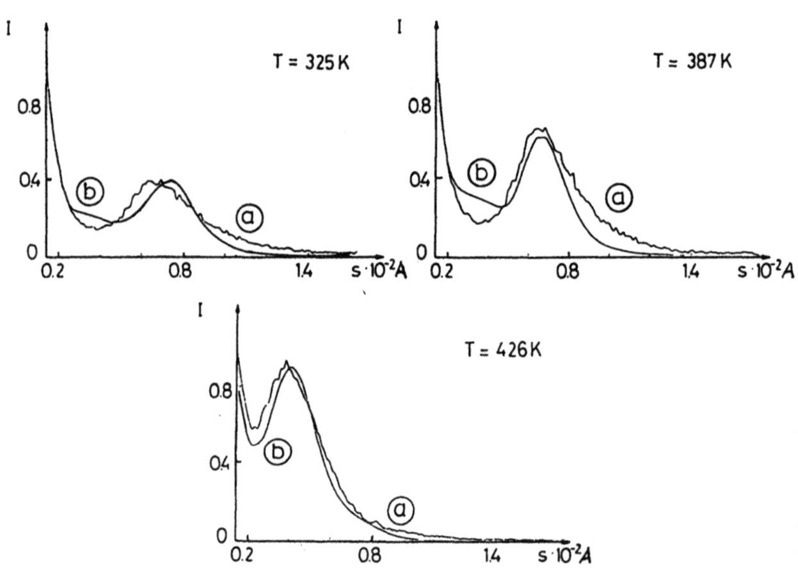

Fig. 23. (A) Synchrotron SAXS-pattern of the poly- oxi-methylene copolymer with $(CH_2)_4$–O–*nc*- units ($x_{nc} = 0.031$) at the temperature as indicated; (B) theoretical curves computed with the para-crystalline cluster model of Hosemann [37, 39, 40]

Thermodynamic equivalence of identical microphases is a fundamental symmetry which identifies eutectoid oligomers and copolymers as colloid-systems. Therewith it comes about that macroscopic and microscopic phenomena are "indistinguishable" provided that composition in molten regions does not fluctuate so much.

The typical colloid-structure of eutectoid oligomers and copolymers is characterized by the existence of larger thermodynamically equivalent subsystems. The intrinsic structure of the subsystems is, on the average, typified by clusters with a statistical structure related to the assembly of IMP*s or ISMP*s. These "primary structure units" are uniquely determined by the chains-structure of the molecules according to the chain-superstructure equivalence principle.

The IQS-subsystems must be so large as to guarantee that the same processes run off within each of them. The size of the subsystems depends therefore on the width chain- or c-sequence-length distribution. The minimum diameter was estimated to lie in the range of 10–100 nm. Diffusional motions which are necessary for adjusting composition and amount of the coexisting multi-component micro-phases, are thus confined to distances smaller than the radius of these IQS-subsystems. Because it is very likely that most processes occur by exchanging components between next neighboring microphases ("membrane effect"), microphase equilibrium can quickly be approximated within eutectoid oligomers and copolymers. In eutectoid low molecular weight multi-component systems multi-phase coexistence is far from being so perfect [47].

There are some fundamental problems. For example, variables must be well defined within microphases. The phase transition is bound to a large degree of cooperativity. Due to local fluctuations these conditions might become violated if the size of the nonhomogeneous layers is essentially low when there is no longer a well-ordered crystal core. To study such micro-phases more distinctly could give insights into the organization of defect structures in polymers with large concentrations of lattice-incompatible units.

There is no doubt that local crystallization segregation of ic- or nc-units leads to diffusive interlayers. Every understanding has therefore to deal with the delicate problem of characterizing these "defect-layers". Difficulties with the simple autonomy of ISMP*s as defined here should arise for strain-induced crystallization in networks: disordered chain segments in the outer regions of the diffusive interlayers are in fact coupled to the intrinsic deformation imposed onto the amorphous regions.

Appendix

The liquidus temperature $T_m(y)$ of eutectoid copolymers is given by [31, 32, 34, 40]:

$$T_m(y) = T_m(1 - 2\sigma_e/\Delta h(y, T))/N \qquad \text{(A1)}$$

where

$$N = 1 + (RT_m/\Delta h(y, T)) \ln[2y/(\Delta y + 2)] - \ln(x_y^{(m)}/x_c^{(c)}) + (y - 1)\, X^{(m)}/2 \,. \qquad \text{(A2)}$$

T_m is the melting temperature of the extended-c-sequence crystal of "infinite thickness". If the longitudinal interfacial free-enthalpy σ_e shows the same temperature dependence as the molar melting enthalpy, the ratio

$$\sigma_e/\Delta h(y, T) \qquad \text{(A3)}$$

is independent of temperature, while $\Delta h(T)$ is given by

$$\Delta h(T) = \Delta h(T_m) - \Delta C(T_m - T)\,. \qquad \text{(A4)}$$

ΔC is the molar thermal coefficient of the melting enthalpy. $X^{(m)}$ is the interaction parameter per unit in the multicomponent melt. $\Delta h(T, y)$ is defined as

$$\Delta h(y, T) = (1 - A/3)\, \Delta h(T)\, (y - y_k)\,. \qquad \text{(A5)}$$

The molar fraction of the c-sequences of the length y is given by

$$x_c^m = 0.5\, \Phi(y) \Big/ \sum_{\nu=1}^{y-1} \Phi(\nu) \qquad \text{(A6)}$$

in the ISMC of the thickness y

$$x_c^c = \Phi(y) \Big/ \sum_{\nu=y-\Delta y/2}^{y+\Delta y/2} \Phi(\nu) \qquad \text{(A7)}$$

in the melt.

Φ is the c-sequence-length distribution function, in the case of random copolymers defined by

$$\Phi(y) = x_{nc}^2\, x_c^{y-1}\,. \qquad \text{(A8)}$$

References

1. Wunderlich B (1973) Macromolecular Physics. Vol 1, Academic Press, New York London
2. Mandelkern L (1964) Crystallization of Polymers. McGraw-Hill Book Comp, New York San Francisco Toronto London
3. Flory PJ (1955) Trans Farad Soc 51:848
4. Wunderlich B (1980) Macromolecular Physics. Vol 3, Academic Press, New York London
5. Sanchez IC, Eby RK (1973) J Res Nat Bur Std 77A:353
6. Sanchez IC, Eby RK (1975) Macromolecules 8:638
7. Helfand E, Lauritzen JI Jr (1973) Macromolecules 6:631
8. Arakawa T, Wunderlich B (1966) J Polym Sci Phys Ed 4:53
9. Bodily D, Wunderlich B (1966) J Polym Sci Phys Ed 4:25
10. Patel GN, Keller A (1975) J Polym Sci Phys Ed 13:1181
11. Baltà Calleja FJ, Hidalgo A (1969) Koll Z Z Polym 229:21
12. Martinez de Salazar J, Baltà Calleja FJ (1980) J Cryst Growth 48:283
13. Kortleve G, Tuijnman CAF, Vonk CG (1965) J Polym Sci Phys Ed 10:123
14. Roe RJ, Geniewsky C (1980) J Cryst Growth 48:295
15. Glenz W, Renvanz B, Wilke W (1972) Koll Z Z Polymere 250:453
16. Alfonso GC, Fiorina L, Martuscelli E, Pedemonte E, Russo S (1973) Polymer 14:37
17. Willbourn AH (1959) J Polym Sci 34:569
18. Buckley GD, Ray NH (1952) J Chem Soc 3701
19. Wunderlich B, Poland D (1963) J Polym Sci Part A 1:357
20. Tanaka K (1960) Bull Chem Soc Jpn 33:1060
21. Baltà Calleja FJ, Rueda DR (1974) Polym J 3:216
22. Kalepky U, Ficher EW, Herchenröder P (1979) J Polym Sci 17:2117
23. Griskey RG, Foster GN (1970) J Polymer Sci Polym Chem Ed 8:1623
24. Ke B (1962) J Polym Sci 61:47
25. Flory PJ, Frij A (1963) Am Chem Soc 85:3548
26. Unseld K, Asbach GI, Kilian H-G (in preparation)
27. Dulfer, Asbach GI, Kilian H-G (in preparation)
28. Asbach GI, Geiger K (1979) Colloid Polym Sci 257:1094
29. Stracke F (1975) Thesis University of Ulm
30. Neppert BT (1985) Thesis Universtiy of Ulm
31. Kilian H-G (1968) Makromol Chem 116:219
32. Kilian H-G (1986) Progr Colloid Polym Sci 72:60
33. Hosemann R, Loboda-Cackovic J, Sassoi M, Weick D (1979) Progr Colloid Polym Sci 64:303
34. Glenz W, Kilian HG, Klattenhoff D, Stracke F (1972) Polymer 18:685
35. Vonk CG (1973) J Appl Cryst 6:819
36. Striebeck N, Ruland W (1978) J Appl Cryst 11:535
37. Heise B, Kilian H-G, Schmidt H (1981) Colloid Polym Sci 259:611
38. Kilian H-G, Unseld K, Jaeger E, Müller J, Jungnickel B (1985) Colloid Polymer Sci 263:607
39. Hosemann R, Bagchi SN (1962) Direct Analysis of Diffraction by Matter. North-Holland Publ Comp, Amsterdam
40. Rosenberger B, Rodriguez et al., to be published
41. Kilian H-G (1965) Koll Z Z Polym 202:97
42. Kilian H-G (1967) Koll Z Z Polym 215:131
43. Kilian H-G (1969) Koll Z Z Polym 231:534
44. Holl B (1987) Thesis, University of Ulm
45. to be published
46. Hoffmann M, Kröner H, Kuhn R (1977) Polymeranalytik I. Georg Thieme Verlag, Stuttgart
47. Chalmers (1964) Principles of Solidification. Wiley & Sons, New York London Sydney
48. Anderson FR (1965) J Polym Sci, Part C 8:275

Received June 25, 1988;
accepted July 19, 1988

Author's address:

Prof. Dr. H.-G. Kilian
Universität Ulm
Abt. Experimentelle Physik
Oberer Eselsberg
D-7900 Ulm, F.R.G.

Progress in Colloid & Polymer Science Progr Colloid Polym Sci 78:177–182 (1988)

Morphology and properties of blends of ethylene-propylene block copolymer with ethylene-vinylacetate copolymer

I. Hudec, J. Beniska, V. Khunová, and Ľ. Martinec[1])

Department of Plastics and Fibres, Faculty of Chemistry, Slovak Technical University, Bratislava, Czechoslovakia
[1]) Department of Materials Science, Faculty of Mechanical Technology of Slovak Technical University, Trnava, Czechoslovakia

Abstract: The statistical evaluation of the influence of ethylene-propylene block copolymer (EPB) and ethylene-vinylacetate copolymer (EVA) contents in blends, as well as time-temperature conditions of blending were studied in relation to phase morphology and mechanical properties of these blends.

The morphology and mechanical properties of binary EPB/EVA blends are markedly influenced by polymer composition. Desirable properties have been obtained by changing the blending conditions, for the same EVA content.

Key words: Polyolefins, morphology and properties of ethylene-propylene, morphology and properties of ethylene-vinylacetate, blending conditions.

Introduction

The properties of polymer blends are determined by the individual characteristics of the basic components, the manner of preparation, the phase morphology and the nature of the interphase [1].

Recently, ethylene-propylene block copolymers have often been used as construction materials, usually filled with mineral fillers [2, 3] or modified by elastomers [4–6]; they are also often used to increase compatibility of polyethylene polypropylene blends [7–9].

In our previous investigation [10] we studied the morphology of blends containing ethylene-propylene block copolymer (EPB) and EVA copolymer with low vinylacetate concentration. Our present work deals with the phase morphology, as well as the mechanical properties of blends of EPB and EVA copolymer with higher vinylacetate concentration (43,3 wt.%); their dependence on EVA copolymer concentration and conditions of blending were also studied.

Experimental

Materials

Ethylene-propylene block copolymer KD-703 used in this work was a commercial product (Slovnaft Bratislava), ethylene content 14,8 wt%, melt flow index 2,7 g · 10 min^{-1} measured by ČSN 640861 and density 0,903 g/cm^3. The EVA copolymer was used Levapren 450 N (Bayer, Leverkusen, FRG), vinylacetate content 43,3 wt.% and density 0,978 g/cm^3.

Blending technique

The blends were prepared on a laboratory twin-roller mill at different temperatures and mixing time, according to the planned experimental conditions. Temperature was varied from 170 °C to 220 °C and mixing time from 2 minutes to 10 minutes.

Methods and specimen preparation

For characterisation of the morphology of EPB and its blends, scanning electron microscopy [SEM] and optical microscopy techniques were used.

Mechanical properties, including tensile yield stress, yield strain and Charpy impact strength of notched specimens, were studied. Compression-moulded samples for morphological study and mechanical testing were prepared in a press at 200 °C.

Specimens used for SEM observations were fractured at liquid nitrogen temperature; the fractured surfaces of the specimens were etched with toluene for 10 minutes. Several samples were etched with 1 % solution of potassium permanganate in concentrated sulphuric acid for 2 hours at 25 °C. Gold was used for coating the etched surfaces. Micrographs were taken at several magnification ratios ranging from 500 : 1 to 5000 : 1 with a TESLA BS 300 scanning electron microscope.

Table 1. Mechanical properties bicomponent EPB/EVA blends

EVA content wt. %	Blending conditions Temperature °C	Time s	Tensile yield stress σ_y MPa	yield strain ε_y %	Charpy impact strength kJ · m^{-2} −20 °C	−40 °C
20	180	240	14.53	36.8	2.67	1.63
40	180	240	8.73	39.2	2.70	1.95
20	210	240	14.52	48.0	2.29	1.52
40	210	240	9.05	50.6	2.63	1.90
20	180	480	15.27	86.4	2.72	2.05
40	180	480	10.37	69.6	2.87	2.06
20	210	480	13.51	84.0	2.10	2.04
40	210	480	8.37	32.0	2.96	1.73
13.18	195	360	16.61	91.0	2.85	2.20
46.82	195	360	7.71	58.6	3.10	2.00
30	169.8	360	8.72	12.0	4.35	2.56
30	220.2	360	10.77	37.0	2.78	2.15
30	195	158	12.34	119.2	2.41	2.17
30	195	562	11.87	91.0	2.15	1.50
30	195	360	12.14	113.6	2.05	1.68
30	195	360	11.86	95.0	2.24	1.67
30	195	360	12.49	111.0	2.21	1.75
30	195	360	13.04	125.0	2.32	1.74
30	195	360	11.88	115.0	2.25	1.64
30	195	360	12.21	107.0	2.48	1.64
Standard error S_E			± 0,44	± 9,9	± 0,2	± 0,1

Results and discussion

The results obtained in the investigation of the influence of time-temperature conditions of blending and their correlation with the polymer concentration, phase morphology and mechanical properties are illustrated in the Figs. 1–3 and summarized in the Table 1.

The basic method of experimentation was a statistical three factor experiment as shown in Table 2.

From the regression of the experimental results obtained from 20 samples, second order polynomial was obtained (1 and 2).

$$\sigma_y = 12{,}252 - 2{,}666x_1 + 0{,}009x_2 + 0{,}002x_3 - 0{,}005x_1x_2 + 0{,}137x_1x_3 - 0{,}492x_2x_3 + 0{,}074x_1^2 - 0{,}779x_2^2 + 0{,}055x_3^2 \quad (1)$$

$$\varepsilon_y = 111{,}39 - 8{,}662x_1 + 1{,}805x_2 + 3{,}659x_3 - 4{,}375x_1x_2 - 9{,}225x_1x_3 - 7{,}825x_2x_3 - 14{,}874x_1^2 - 32{,}662x_2^2 - 4{,}158x_3^2 . \quad (2)$$

Table 2. Factor levels for the rotatable design

x_i	Factor	Coded factor levels − 1,682	−1,0	0	+1,0	+1,682	I_i
x_1	Concentration EVA, wt. %	− 13,18	20	30	40	46,82	10
x_2	Temperature, °C	169,8	180	195	210	220,2	15
x_3	Mixing time, s	158	240	360	480	562	120

Adequacy of these equations for tensile yield stress $-\sigma_y$ and yield strain $-\varepsilon_y$, respectively, has been verified by a statistical analysis. Standard error (S_E) of the regressed polynomials are given in Table 1.

Results of the statistical analysis of the mechanical properties revealed that the concentration of EVA influences most of the properties studied. In contrast the influence of temperature is less pronounced and the effect of mixing time is most obvious for the yield strain.

Comparison of the measured values of notched Charpy impact strength at −20 °C and −40 °C for the individual concentration levels shows that the impact strength increases with EVA concentration. Due to very small variations in the measured increase with values of impact strengths it was not possible to evaluate statistically a correlation between this property and the blending conditions.

It was evident from the evaluation of the properties studied, that yield strain shows most sensitivity to the blending conditions (Fig. 1a-1c). At low EVA content (13,18 wt. % – Fig. 1 a) the prolongation of the mixing time always leads an increase in the yield strain (ε_y), due to the improved degree of dispersion. The effect of temperature is more pronounced at higher mixing times.

As can be seen from Fig. 1 b at 30 wt. % EVA in the blend, the mixing temperature is the determining factor for ε_y. It is possible to reach the optimum condi-

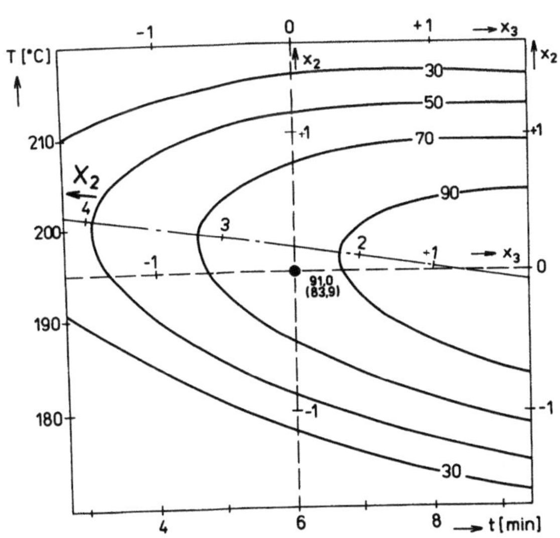

Fig. 1. Contour-line diagrams reflecting the dependence of yield strain (ε_y) on the time-temperature conditions at blending. Fig. 1 a.: The constant parameter: $x_1 = -1{,}682$; $c_{EVA} = 13{,}18\,\%$. Experimental point marked by dark points

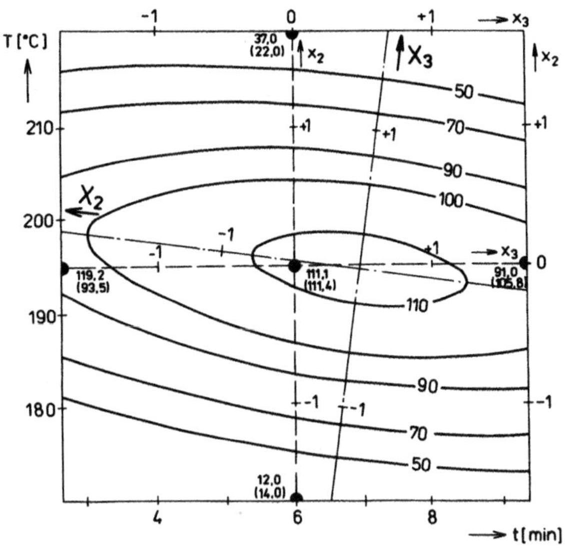

Fig. 1 b. The constant parameter: $x_1 = 0$; $c_{EVA} = 30\,\%$. Experimental points marked by dark points

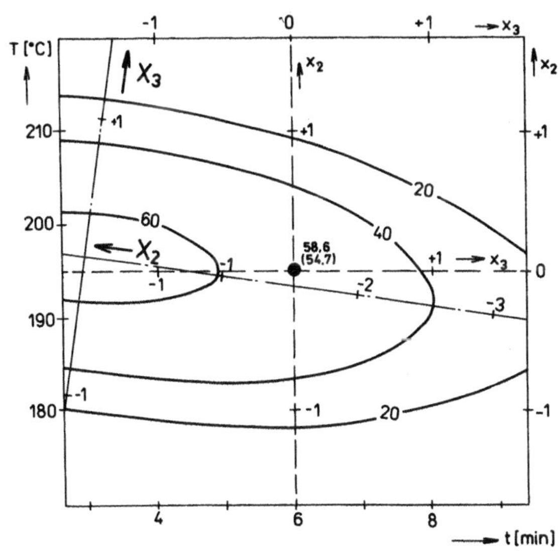

Fig. 1 c. The constant parameter: $x_1 = +1{,}682$; $c_{EVA} = 46.82\,\%$. Experimental point marked by dark points

tions for blend preparation – approximately 7 minutes mixing time at 195 °C. From the morphological study it can be confirmed that the optimum blending conditions exhibit maximum dispersion, as shown in Fig. 2.

It is interesting to note that the maximum EVA content (46,82 wt. %) enables the preparation of blends with the lowest yield strain (Fig. 1 c).

The observed profound change in the mechanical properties is probably connected with the marked changes in the phase morphology. The study of phase

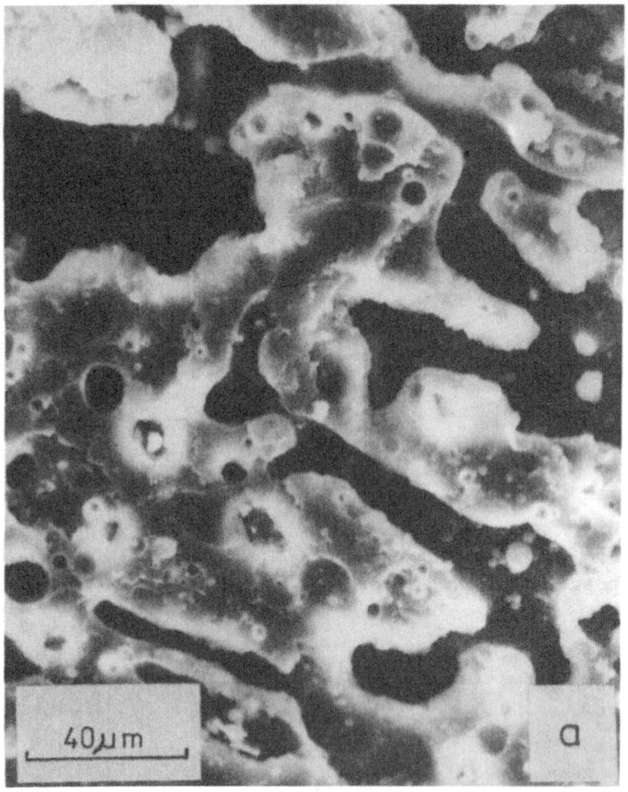

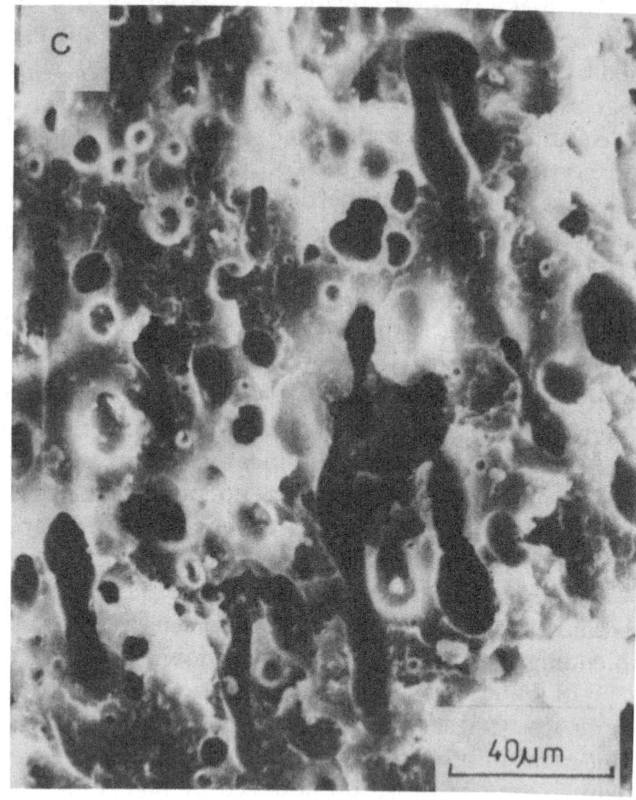

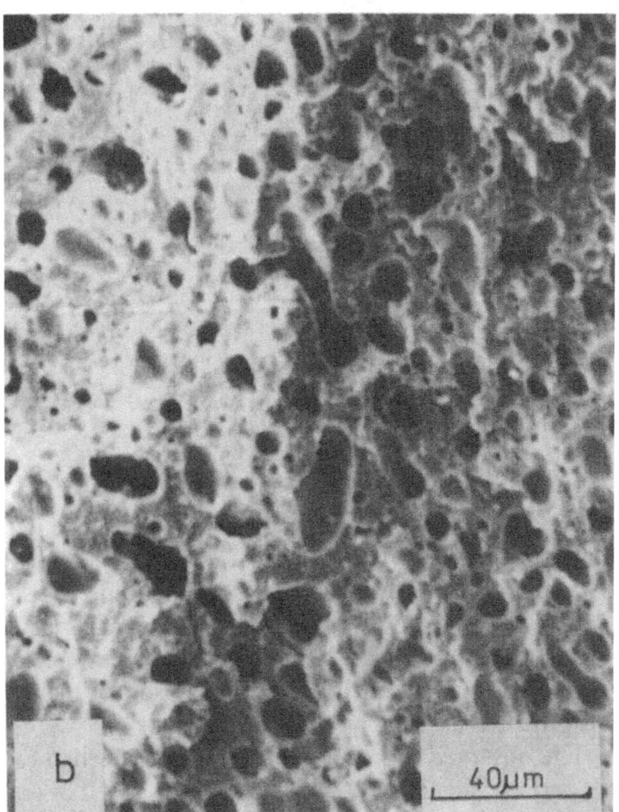

Fig. 2. SEM micrographs of EPB/EVA blends (70%/30%) prepared at different temperatures: a – 170 °C, b – 195 °C, c – 220 °C

morphology of EPB/EVA blends by means of SEM and optical microscopy proved, that increasing the EVA content in the blend not only enlarged the dispersed particles, but also changed the character of phase morphology (Fig. 3 a–3 d).

Whereas up to the 30 wt.% concentration a dispersion of EVA in EPB matrix was evident, increasing the EVA content up to 47 wt.% leads to a structure having two interpenetrating continuous phases (Fig. 2 d). The matrix is always formed by the ethylene-propylene copolymer, which influences the properties of the whole blend in a decisive manner.

Conclusions

The statistical evaluation, of the influence of EPB and EVA contents in blends as well as time-temperature conditions of blending, has shown that it is possible to markedly change the phase morphology and subsequently the mechanical properties of binary

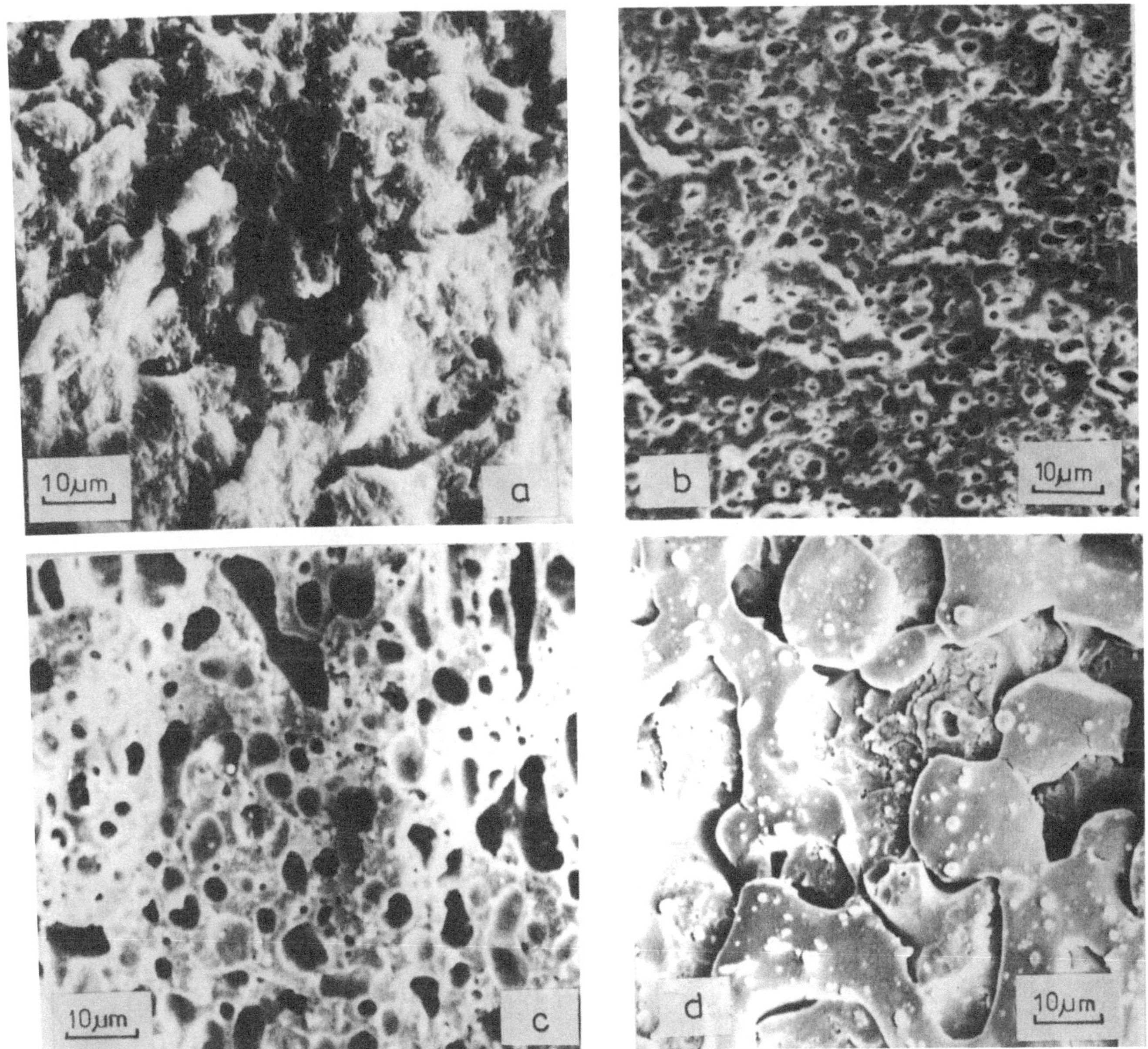

Fig. 3. SEM micrographs of EPB/EVA blends with different compositions: a) 100/0, b) 86,82/13,18, c) 70/30, d) 53,18/46,82. Blending conditions: 195 °C, 6 minutes

EPB/EVA blends, by varying the concentration of individual polymer components.

It was found that by changing the temperature and mixing time it is possible to influence the phase morphology and therefore the resulting mechanical properties in a desirable way, without changing the EVA concentration in the blend.

References

1. Chang DH (1981) Multiphase Flow in Polymer Processing. Acad Press, New York
2. (1984) Japan Patent 108 050
3. (1985) Eur Pat Appl EP 132 968
4. (1983) Japan Patent 195 758
5. Popov VP, Zareckaya EM, Belozerov VV (1984) Plast Massy (6):61

6. Glagoleva YuA, Vinogradova GA, Beloglazova GA, Šrabman ND, Buchgalter VI (1986) Plast Massy (3):25
7. Barlett DW, Barlow JW, Paul DR (1982) J Appl Polym Sci 27:2351
8. Ramsteiner F, Kanig G, Heckmann N, Gruber W (1983) Polymer 24:365
9. D'Orazio L, Greco R, Martuscelli E, Ragosta G (1984) In: Kryszewski M, Galeski A, Martuscelli E (eds) Polymer Blends: Processing, Morphology and Properties. Vol 2, Plenum Press, New York and London pp 111, 127
10. Hudec I, Beniska J (1985) In: Morphology of Polymers. 17th Europhysics Conference on Macromolecular Physics, Prague p 35–1

Received June 25, 1987;
accepted April 21, 1988

Authors' address:

Dr. I. Hudec
Department of Plastics and Fibres
Faculty of Chemistry, Slovak Technical University
Radlinského 9
Bratislava 81237, Czechoslovakia

Progress in Colloid & Polymer Science Progr Colloid Polym Sci 78:183–187 (1988)

On the morphology of high-modulus and high-strength polyethylene filaments

A. Schaper, E. Walenta, and E. Schulz

'Erich Correns' Institute of Polymer Chemistry, Academy of Sciences, Teltow-Seehof, G.D.R.

Abstract: Microstructural investigations of surface-grown and surface-grown/zone-drawn ultra-high molecular weight polyethylene (UHMWPE) fibers were carried out by scanning- and transmission-electron microscopy (SEM, TEM), by optical diffraction (OD) of the TEM images, and by small angle X-ray scattering (SAXS). The as-spun filaments show a "shish kebab" morphology characterized by a periodic arrangement of the overgrowths. During zone-drawing the shish kebabs are transformed into well-aligned needle crystals. At medium-draw ratios this is accompanied by considerable improvement of the mechanical properties, whereas draw-ratios exceeding approximately 2.5 lead to a reduction in strength and modulus due to "overdrawing" effects.

Key words: Surface-grown polyethylene, zone-drawing, fibrillar morphology, mechanical properties.

Introduction

In recent years several continuous processes of preparing high-strength and high-modulus PE fibers and films have been reported, such as solution and gel spinning and drawing, and the solid phase processing routes of tensile drawing, hydrostatic and ram extrusion and die drawing [1, 2]. Stretching of a molecular network of physical entanglements may be considered the basic principle of all the above processes [1, 3]. Except for pressure-crystallized melt extrudates [4], extension and alignment of the macromolecules have proved to be prerequisites for high strength and stiffness values. Numerous structural studies [4–18] suggest that, depending on the conditions of production, different morphologies occur ranging from interlocking shish kebabs [4] to extremely thin needle-shaped crystals [18]. The role topological defects play in determining the mechanical properties of the highly oriented materials has been stated [19].

In previous papers by Zenke et al. [20–22] the application of the surface-growth method to medium and high molecular weight PE was reported in conjunction with calorimetric measurements on the melting behavior of the as-spun and drawn fibers and with mechanical measurements, as well as some preliminary morphological observations. The main object of this and the accompanying paper [23] is to provide direct electron microscopic evidence of the real fibrous nature of surface-grown ultra-high molecular weight polyethylene (UHMWPE) before and after zone-drawing, and to characterize its structure more quantitatively. The influence of morphological defects on the mechanical properties will be demonstrateed as well.

Experimental

Filaments were surface-grown in a Couette type apparatus from a dilute solution (0.5 wt%) of Hostalen GUR (Hoechst; $\bar{M}_w = 1.5 \times 10^6$ g/mol, $\bar{M}_n = 10^5$ g/mol) in p-xylene at temperatures between 104 and 115 °C. A rotor speed of 100 min^{-1} and growth rates < 2m/min were employed. Zone-drawing by degrees was carried out by moving a narrow band heater along the loaded fiber sample at temperatures elevated from 130 to 145 °C. Moving speed was ca. 20 mm/min, the draw ratios varied between 1.2 and 2.7. For TEM studies only samples of $\lambda = 2.7$ have so far been used. Further investigations are in progress.

For investigating fiber surfaces we used scanning electron microscopy and Pt/C-replica transmission electron microscopy. Gold decoration [24] of thin, peeled-off sample portions, and staining [25] combined with ultramicrotomy were used to reveal the innermost parts of the filaments at high resolutions. Specimens

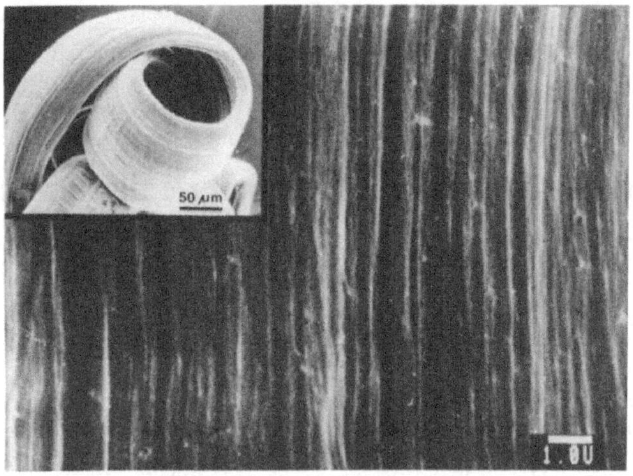

Fig. 1. The outward appearance of surface-grown UHMWPE filaments as revealed by SEM

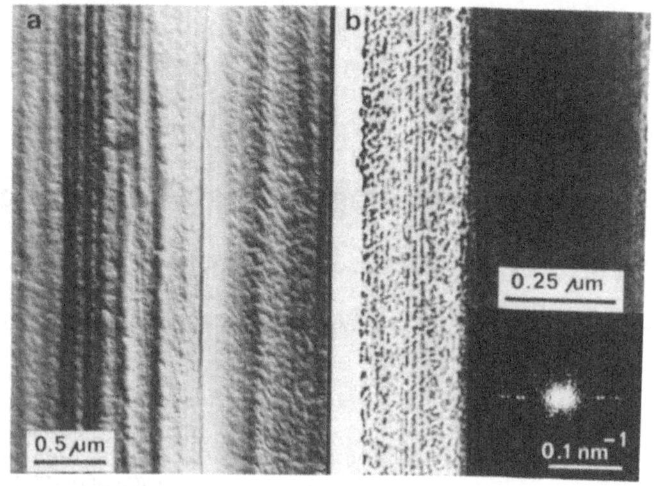

Fig. 2. Transmissioin electron micrographs of the fibrillar structure of as-spun filaments: a) replica, b) gold-decorated fiber fragment (optical transform inserted)

were chlorosulfonated at 65 °C for 11/2 to 3 h and treated with uranyl acetate, and besides sometimes with osmium tetroxide. With the fibers subsequently embedded in epoxy resin, longitudinal ultra-thin sections were prepared using a TESLA BS 490 A microtome and a glass knife. TEM studies were performed in a TESLA BS 500, and partly in a JEOL JEM 200 CX microscope. For evaluating the micrographs we used a 40 mW He-Ne laser optical bench [26] to record the optical transforms of the negatives.

SAXS was carried out using a Kiessig camera with Ni-filtered CuK_α-radiation to determine the meridional long period. Equatorial scattering was measured in a Kratky camera with a proportional counter attached. The scattered intensity was corrected for parasitic scattering and short-range electron density fluctuations. From the scattered intensity the fibril thickness was estimated by applying the Gaussian approach for the scattering from circular cylinders of radius R:

$I(s) \sim \exp(-\pi^2 R^2 s^2)$; $s = 2 \sin\theta \cdot \lambda^{-1}$; 2θ = scattering angle. The slope of the straight line in the Guinier plot yields the minimum fibril diameter ($\bar{D}_1$). Repeatedly applying the Guinier plot to the difference curves of the total intensity substracted by the respective linear part of the scattering (Gaussian analysis) yielded a measure of fibrillar clustering ($\bar{D}_2$, $\bar{D}_3$).

Results and discussion

Figure 1 shows scanning electron micrographs giving an impression of the outward fibrillar appearance of as-spun filaments. A closer inspection of the structure is possible by TEM via surface replicas (Fig. 2a) or by the aid of the gold decoration technique (Fig. 2b). Here, short interruptions along the contours of the fibrils occur which are indicative of overgrowth phenomena. Fibril transverse distances less than 20 nm are derived from Fig. 2b.

More reliable information is gained by direct TEM observations of stained sections. It should be noted that, due to the weak lateral interfibrillar cohesion, sectioning often appears similar to a longitudinal splitting rather than to a real cutting process. This is demonstrated by the micrographs of fiber fragments split off dur-

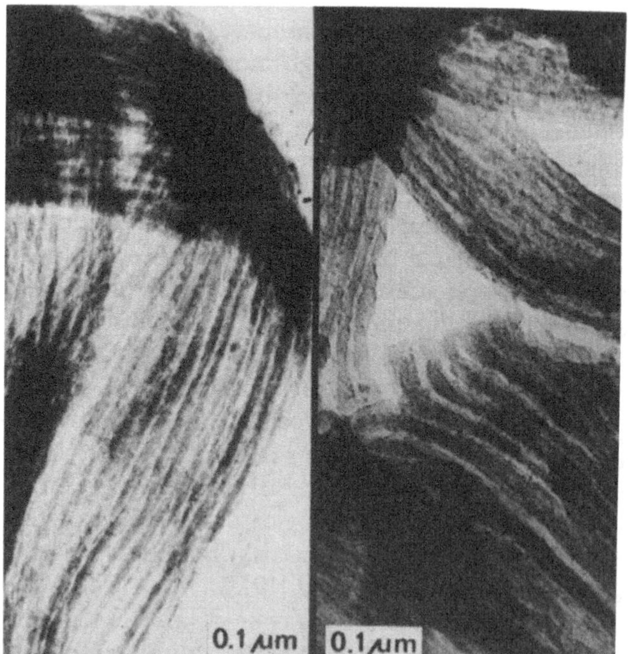

Fig. 3. Transmission electron micrographs of fibrillar fragments split off from stained fibers during sectioning

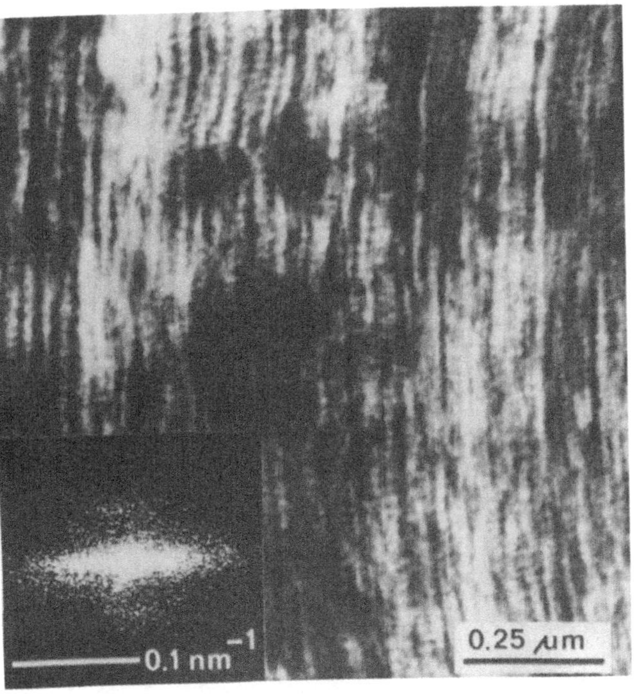

Fig. 4. "Shish kebab" morphology of a fiber surface-grown at 114.5 °C; stained ultra-thin section. Inset: optical transform (fiber axis vertical)

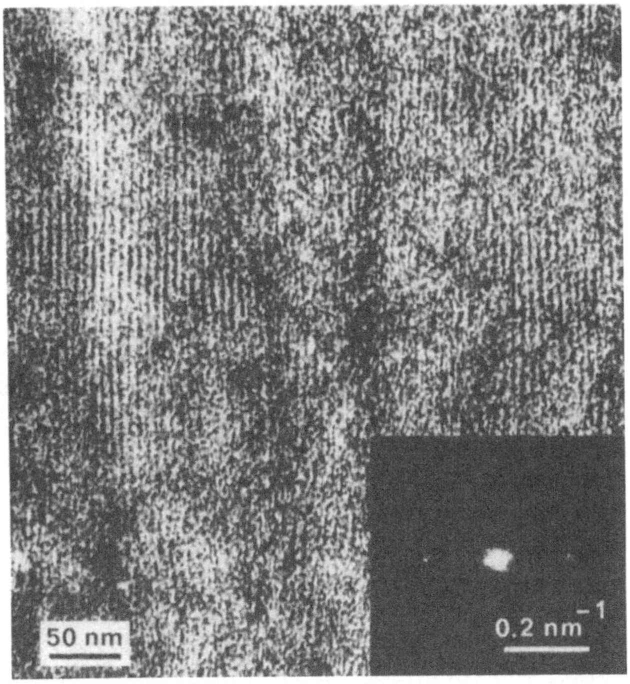

Fig. 5. Fibrillar morphology of a sample surface-grown at 104 °C and zone-drawn between 130 and 145 °C to $\lambda = 2.7$; stained ultra-thin section. Inset: optical transform (fiber axis vertical)

ing sectioning (Fig. 3). The microfibrillar structure of the ribbon-like fragments is clearly revealed. Figure 4 shows another view of the same 114.5 °C surface-grown filament. The micrograph evidences a shish kebab structure consisting of extended backbone fibrils periodically overgrown by small lamellar entities. This supermolecular order is reflected by the optical transform which displays faint, but unequivocal, meridional reflections in addition to the strong equatorial streak. As shown in [23] the equatorial scattering may as well approach definite maxima, disclosing transverse spacings of 15–35 nm. The results of EM and SAXS analyses are summarized in Table 1. The data agree well; only SAXS analyses reveals a larger tendency of the fibrils to coalesce laterally than does EM.

Table 1. Structural EM and SAXS characteristics of PE fibers surface-grown at different temperatures (a_s = lateral shish size, $\bar{D}$ = mean fibril/cluster diameter, $L_\parallel$ = long spacing, $L_\perp$ = transverse spacing)

Quantity	Method	104 °C	110 °C	114.5 °C
a_s [nm]	TEM$^{a)}$	< 12	–	< 12
$\bar{D}_1$ [nm]	SAXS	12.4	9.5	–
$\bar{D}_2$		22.4	20.3	–
$\bar{D}_3$		33.0	32.8	–
$L_\parallel$ [nm]	TEM$^{a)}$ + OD	12–24	–	12–24
	SAXS	17.6	–	–
$L_\perp$ [nm]	TEM$^{a)}$ + OD	15–35	–	15–35
	TEM$^{b)}$	–	11.5–20.4	–

$^{a)}$ data of stained sections; $^{b)}$ data of gold-decorated fiber fragments

The shish kebabs experience remarkable changes upon zone-drawing. The overgrowth disappears so that a very regular structure of smooth needle-like fibrils forms. Figure 5 shows a fiber zone-drawn up to $\lambda = 2.7$ exhibiting a needle diameter of about 4 ($\pm$ 0.5) nm. This and other micrographs of the same specimen and corresponding OD patterns displayed transverse spacings in the range 5.6–7.0 nm [23]. In the morphological transformation several molecular processes are involved [27, 28]: (i) melting of the folded-chain overgrowth, (ii) and its conversion into extended-chain material, (iii) backbone elongation, (iv) healing up of a fraction of crystal defects, and (v) distribution along the fibril axes of remaining defects, e.g., trapped entanglements, chain ends, and twist disclinations [29].

Table 2. Mechanical properties of surface-grown and surface-grown/zone-drawn PE filaments

Quantity	104 °C	110 °C	114.5 °C	zone-drawn
strength [GPa]	1.7	1.8	2.2	1.2– 6.0
modulus [GPa]	47.5	66.3	74.7	38.0–135.0
strain at break [%]	9.5	4.3	5.2	8.0– 3.0

A mechanical characterization of our samples is given in Table 2. As to the large variety of results achieved by zone-drawing, the inferior mechanical values correspond to the highest draw ratios exceeding 2.5; they are caused by gross topological defects, such as those shown in Figs. 6 and 7. Here (Figs. 6 and 7), boundary layers and longitudinal voids are revealed that separate fibrillar aggregations laterally (Fig. 6), as well as distortions and terminations of whole fibril bundles (Fig. 7). Some of these defects arise from the surface growth process, whereas others are the result of what can be termed overdrawing (compare Ref. [28]), during which new defects may develop by chain scission and accretion of small discontinuities.

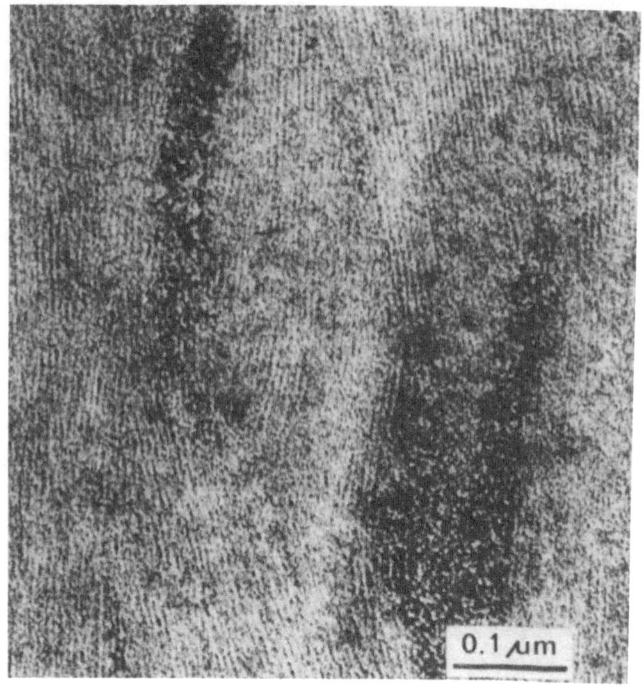

Fig. 7. Another view of the same fiber as in Figs. 5 and 6 showing distorted and terminated bundles of microfibrils

Conclusions

TEM observations of stained ultra-thin sections parallel to the fiber axis enable us to precisely visualize the elementary fibrillar structure of surface-grown and zone-drawn polyethylene filaments. TEM combined with optical diffraction yields structural parameters which are comparable to small angle X-ray scattering results. In zone-drawing the increased perfection of the fibrillar structure under non-optimum drawing conditions is opposed by the occurrence of gross morphological defects which deteriorate the fiber properties.

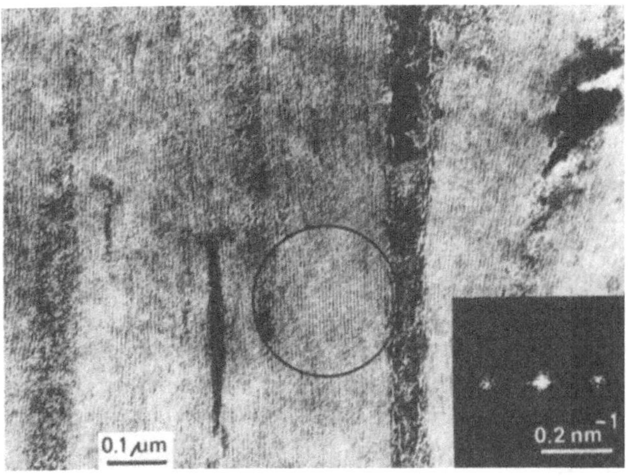

Fig. 6. Boundary layers and voids within the same fiber as used for Fig. 5. Inset: optical transform of the marked area

Acknowledgements

The authors thank Drs. D. Zenke and R. Hirte for providing the as-spun filaments and for the helpful disscussions, and Dr. Z. Pelzbauer from the ICM Prague for his collaboration in electron microscopic work. The provision of research facilities in the Institute of Solid State Physics and Electron Microscopy in Halle (Saale) is appreciated.

References

1. Ward IM (1985) Adv Polym Sci 70:1
2. Lemstra PJ, Bastiaansen CWM, Meijer HEH (1986) Angew Makromol Chem 145/146:343
3. Grubb DT (1983) J Polym Sci Polym Phys Ed 21:165
4. Odell JA, Grubb DT, Keller A (1978) Polymer 19:617
5. Bashir Z, Hill MJ, Keller A (1986) J Mater Sci Lett 5:876
6. Zwijnenburg A, van Hutten PF, Pennings AJ, Chanzy HD (1978) Colloid Polym Sci 256:729

7. Smook J, Torfs JCM, van Hutten PF, Pennings AJ (1980) Polym Bull 2:293
8. van Hutten PF, Pennings AJ (1980) J Polym Sci Polym Phys Ed 18:927
9. van Hutten PF, Koning CE, Smook J, Pennings AJ (1983) Polymer Commun 24:237
10. Grubb DT (1977) In: Misell DL (ed) The Institute of Physics. Proc EMAG, 12–14 Sept, Glasgow, Bristol London, Conf Ser 36:399
11. Hill MJ, Barham PJ, Keller A (1980) Colloid Polym Sci 258:1023
12. Grubb DT, Hill MJ (1980) J Cryst Growth 48:321
13. Frye CJ, Ward IM, Dobb MG, Johnson DJ (1979) Polymer 20:1310
14. Frye CJ, Ward IM, Dobb MG, Johnson DJ (1982) J Polym Sci Polym Phys Ed 20:1677
15. Sherman ES, Porter RS, Thomas EL (1982) Polymer 23:1069
16. Adams WW, Briber RM, Sherman ES, Porter RS, Thomas EL (1985) Polymer 26:17
17. Smith P, Boudet A, Chanzy H (1985) J Mater Sci Lett 4:13
18. Petermann J, Gohil RM (1979) J Mater Sci 14:2260
19. Smook J, Hamersma W, Pennings AJ (1984) J Mater Sci 19:1359
20. Zenke D, Schulz E, Hirte R, Weigel P (1985) Acta Polym 36:110
21. Zenke D, Schaper A, Schulz E, Nitzsche K, Hirte R (1984) Intern Conf on Man-Made Fibres, Siofok, May 6–9, Hungary
22. Marichin VA, Mjasnikova LP, Zenke D, Hirte R, Weigel P (1984) Polym Bull 12:287
23. Schaper A, Hirte R, Zenke D, Hillebrand R, J Mater Sci Lett, to be published
24. Bassett GA, Blundell DJ, Keller A (1976) J Macromol Sci B1:161
25. Kanig G (1974) Kunststoffe 64:470
26. Hillebrand R, Scheerschmidt K, Neumann W, Werner P, Pippel A (1984) Beitr Forschungstechn 11:87
27. van Hutten PF, Pennings AJ (1980) Makromol Chem, Rapid Commun 1:477
28. Smook J, Torfs JCM, Pennings AJ (1981) Makromol Chem 182:3351
29. Pennings A (1979) Makromol Chem Suppl 2:99

Received June 25, 1987;
accepted April 21, 1988

Authors' address:

Dr. A. Schaper
Institut für Polymerenchemie "Erich Correns"
Akademie der Wissenschaften der DDR
Kantstraße 55
DDR-1530 Teltow-Seehof, G.D.R.

Influence of filler on the polypropylene structure

V. Khunová, V. Smatko, I. Hudec, and J. Beniska

Department of Plastics and Fibres, The Chemical Faculty of Slovak Technical University, Bratislava, Czechoslovakia

Abstract: The nucleating effect of different kinds of fillers in polypropylene composite materials were studied using a dilatometric method and the composites were examined by scanning electron microscopy to relate the composite structure to physical properties. Filler concentration was varied for each composite.

A positive influence due to the nucleating effect, of the incorporated fillers on the polypropylene structure, was achieved only with low bentonite concentrations. For all types of filler employed, an increase in the filler content raises the rate of crystallization, reduces the crystalline phase content, and diminishes the size of supermolecules formed. Particle size and concentration of filler influence the physical properties of polypropylene. In highly filled composites agglomerates are formed and disintegration of filler particles occurs as a result of intensive mechanical strain.

Key words: Polyolefins, nucleating agent, crystallinity, filler.

Introduction

The influence of fillers on polymer structure, on kinetics of crystallisation, and on the ultimate properties of composite polymer materials, is indisputable. The complexity of this problem is given by a number of factors concerning the filler as well as polymer [1].

In the case when a filler acts as a heterogeneous nucleating agent, its nucleating effect is controlled by the filler concentration, by the size and shape of the filler particles, the chemical composition of the filler and also by the distribution of the filler particles in the polymer matrix. It is also affected by the process of addition of filler into the polymer, modification of the filler surface and the mutual interaction of the polymer and the filler [2].

One of the reasons which stimulates our interest is the Czechoslovak origin of the raw material used in this work. The investigation deals with the phase structure of the polypropylene composite materials filled with bentonite, zeolite, $CaCO_3$, and china clay.

Experimental

Materials

Powdered polypropylene (a commercial product of Slovnaft, n.p. Bratislava) and calcined fillers (bentonite, zeolite, $CaCo_3$, china-clay) with particle size $d_{95} = 5\mu m$ (commercial products from Hnusta and Rohoznik of Czechoslovakia).

Procedure

Composite materials were prepared by mixing in a laboratory two-roll mill for 6 min at 190 °C.

The kinetics of crystallisation were studied by dilatometry at 137 °C. From the results obtained the degree of crystallinity was calculated and a half time for crystallinity ($\tau_{1/2}$) was determined graphically from crystallization isotherms. We also obtained the Avrami exponent (n) and the crystallisation rate constant (K) using the Avrami equation:

$$\log\log\frac{h_o - h_\infty}{h_t - h_\infty} = \log\frac{K}{1-\lambda_\infty}\log e + n\log t \quad (1)$$

where h_o, h_t and h_∞ are the mercury column heights at time $t = 0$, $t = t$ and $t = \infty$, respectively; $1 - \lambda$ is the weight fraction of the

Table 1. Crystallisation kinetics of china clay filled polypropylene composites

% wt filler	n	$K \cdot 10^7$	% cryst	$\tau_{1/2}$ min
0	2.76	1.6	54.3	63.8
0.5	3.38	3.7	53.8	62.1
1.0	3.52	25.0	53.1	43.8
3.0	3.64	218.2	50.6	42.2
5.0	3.04	653.1	48.8	31.9
15.0	2.96	3117.9	42.9	16.3

Table 3. Crystallisation kinetics of zeolite filled polypropylene composites

% wt filler	n	$K \cdot 10^7$	% cryst	$\tau_{1/2}$ min
0.5	2.69	4.4	52.4	44.7
1.0	3.21	24.9	49.4	30.1
3.0	3.11	453.7	50.3	32.5
5.0	2.43	804.9	52.2	35.9
15.0	2.51	3934.4	40.1	17.2

crystalline phase at the end of the primary crystallisation process.

The structures of the composite materials studied were examined using a scanning electron microscope (SEM). Specimens used for SEM observations were fractured at −20 °C. Gold was used for coating the fractured surfaces. Micrographs were taken at several magnifications ranning from 1:1000 to 1:10000 with a TESLA BS 350 scanning electron microscope.

Compression-molded samples for morphological study and mechanical testing were prepared in a press at 200 °C.

Results and discussion

The experimental results from the study of the influences of china clay, $CaCO_3$, and zeolite on the crystallisation kinetics of polypropylene are demonstrated in Tables 1, 2, and 3. These tables show the degree of crystallinity (% cryst.) and half time of crystallisation ($\tau_{1/2}$) as well as values of n and K in Eq. (1).

By increasing the filler content (using china clay, $CaCO_3$, and zeolite), the extent of crystallinity was decreased. This polymer amorphosity is related to the reduction of the mobility and change in the density of the macromolecules at the polymer filler interface. These macromolecules due to their conformational restrictions, do not take part in the process of crystallisation. Increasing filler content, therefore, favors kinetic and not thermodynamic factors.

From the values of the Avrami exponent n calculated, it is evident that the presence of filler has no influence on the mechanism of supernuclear structure of the polymer. In the studied cases (china clay, $CaCO_3$ and zeolite) both two and three dimensional growths of crystals take place simultaneously [3].

The half time for crystallisation decreases with increasing filler content, suggesting an increase in the rate of crystallisation. A similar conclusion was also drawn from the values determined for rate the constant K. Filler does not influence the mechanism of growth, but it does influence the rate of nucleation. An increase in the number of nucleation centers leads to reduced dimensions of the crystalline supermolecular structure.

It is evident from experimental results that incorporation of very low concentrations of bentonite filler increases the degree of crystallinity in the polypropylene matrix. The maximum increase in crystallinity was recorded at 1% wt. bentonite content, with a simultaneous maximum in the tensile strength of polypropylene (Table 4). Higher concentrations of bentonite decrease the crystallinity of the polypropylene phase, with a corresponding reduction in tensile strength. This trend was similar to that observed for other fillers. Values of the Avrami exponent were

Table 2. Crystallisation kinetics of $CaCO_3$ filled polypropylene composites

% wt filler	n	$K \cdot 10^7$	% cryst	$\tau_{1/2}$ min
0.5	2.89	2.6	50.0	61.8
1.0	3.75	269.5	48.2	46.9
3.0	3.73	350.1	49.3	31.6
5.0	2.63	959.4	44.3	26.6
15.0	2.86	3404.7	40.4	17.3

Table 4. Degree of crystallinity and tensile strength of bentonite filled polypropylene composites

% wt. filler	0	0.6	0.8	1.0	5.0	10.0	20.0
% cryst.	71.2	73.6	74.2	74.8	66.8	65.4	60.2
tensile strength MPa	38.8	41.6	41.2	42.8	40.3	38.7	35.6

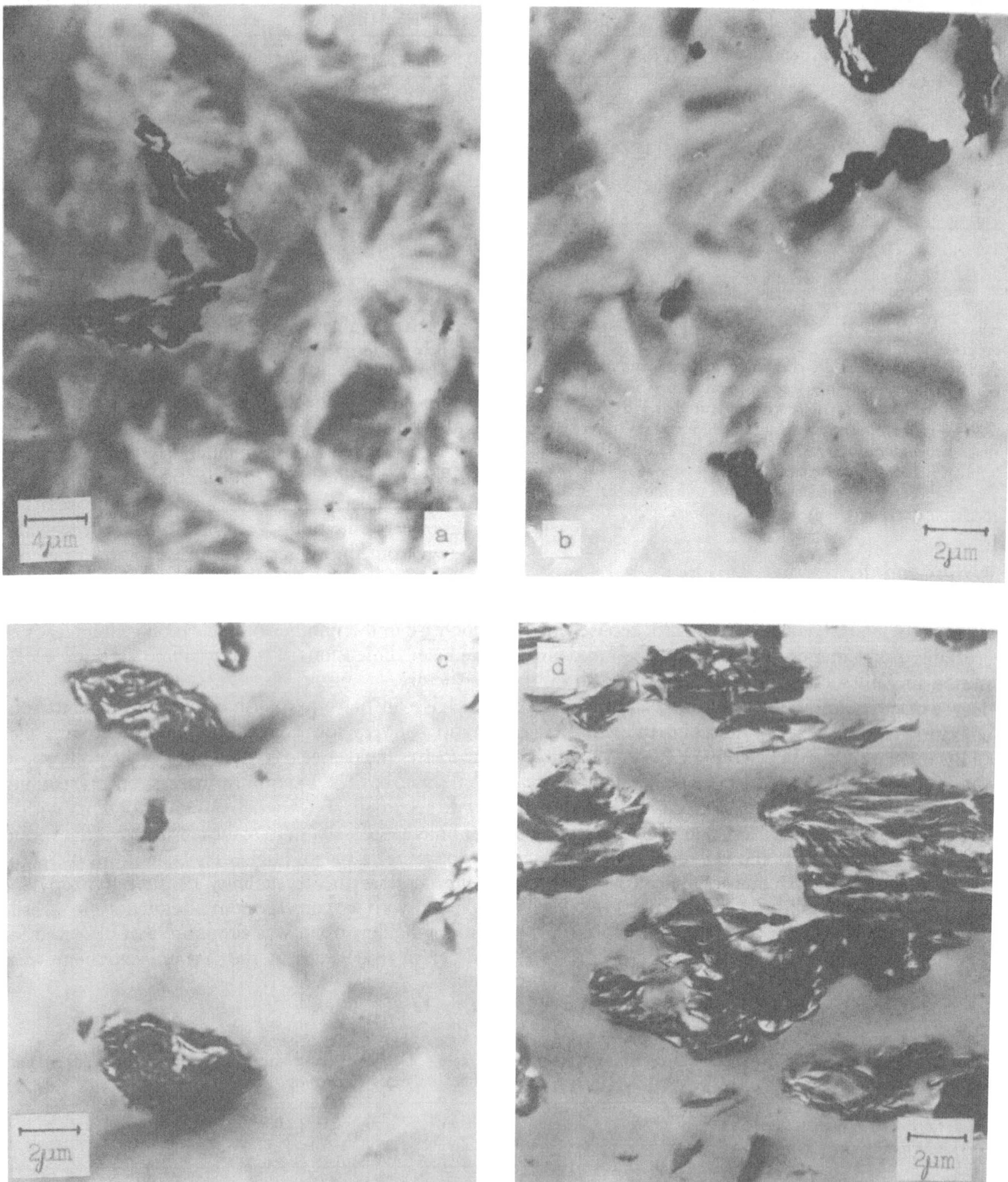

Fig. 1. SEM micrographs of polypropylene-bentonite composite. Bentonite content: (a) = 4 %wt; (b) = 10 %wt; (c) = 20 %wt; (d) = 60 %wt

Fig. 2. SEM of polypropylene composite containing 50 % wt zeolite shows phase structure with: (a) = agglomerated particles; (b) = broken particles

greater than 3 in all cases [4]. This leads to the presumption of three dimensional growth for the supermolecular structure of polypropylene.

Micrographs from the SEM study confirmed our presumption of three dimensional growth in the crystalline phase. Figure 1a shows the microstructure of polypropylene containing 4 % wt. bentonite. It is evident that the distribution of filler in the spherulite structure is random. Furthermore, the dilatometric study reveals that the increase in the filler content reduces the spherulite size. An excellent spherulite structure for polypropylene-bentonite composite was found, using a higher magnification, at 10 % wt. bentonite content (Fig. 1b). In composites containing more than 20 % wt. filler, the identification of individual spherulites was difficult at any magnification (Fig. 1c). However, a reduction in the spherulite dimension was evident in the latter case. On the other hand, at 60 % wt. bentonite content the identification of a structure was not possible (Fig. 1d).

We also studied the structure of the composite materials to elucidate the reason for deterioration of tensile strength at higher filler concentrations (50 % wt.). We found that in this concentration range the heterogeneity of the composites increases significantly, as a result of creation of agglomerates (Fig. 2a), as well as disintegration (Fig. 2b) of the filler particles. In both cases the filler acts as a stress concentrator. This effect is influenced by the particle size of the filler used.

Conclusion

A positive influence due to the nucleating effect of incorporated fillers on the polypropylene structure was only achieved at low bentonite concentrations. For all types of filler employed, increases in the filler content raises the rate of crystallisation, reduces the crystalline phase content and diminishes the size of supermolecules formed.

Particle size and concentration of filler influence the physical properties of polypropylene. In highly filled composites agglomerates are formed and disintegration of filler particles occours as a result of intensive mechanical strain.

References

1. Lipatov YS (1977) Physical Chemistry of Filled Polymers. Chimija, Moscow, p 64
2. Solomko VP (1980) Filled Crystalline Polymers. Naukova dumka, Kyev, p 212
3. Krevelen DWV (1972) Properties of Polymers Correlations with Chemical Structure. Elsevier, Amsterdam London New York, p 386
4. Khunova V (1984) J Electrical Insulation and Cable Technology 36(3):141

Received August 25, 1987; accepted April 21, 1988

Authors' address:

Dr. Viera Khunová
Department of Plastics and Fibres
The Chemical Faculty of Slovak Technical University
Bratislava 81237, Czechoslovakia

Progress in Colloid & Polymer Science Progr Colloid Polym Sci 78:192–195 (1988)

Water in some DNA model systems

M. Neubert and J. Nedbal

Charles University, Department of Polymer Physics, Prague, Czechoslovakia

Abstract: In order to facilitate the interpretation of the DNA dielectric spectrum, dielectric properties of poly ~ (*N*-methyl-capryl-amide) PMCA and modified poly-(caprolactam) PKMG were measured in a range of $-50° - 90$ °C, and 10^{-1} Hz $- 10^4$ Hz. All measurements were performed under defined humidity conditions. Several absorption regions were observed. Only the low temperature region γ at -140 °C was found to be humidity independent. The non-Arrhenius region α is connected with the glass transition of the amorphous phase. We assume that the hydrogen bonds linking the polyamide groups act as temporary physical crosslinks. The sorption of water probably diminishes the effective network density and thus the glass transition temperature. The polyamide groups pairing is seriously destroyed at 5 wt% of water. If the hydration goes on, the water molecules act merely as a plasticizing agent via lowering the internal viscosity. The relaxation region β is interpreted as local motion of water-polyamide groups complex and the region β' (PKMG) as the motion of side groups with water molecules attached. Both types of kinetic units move in surroundings of similar internal viscosity.

Key words: Dielectric relaxation, polyamides, DNA, humidity, plasticizing effects.

1. Introduction

Deoxyribonucleic acid (DNA) has drawn the attention of many physicists because of its important role in the processes of proteosynthesis and genetic information transfer. The dielectric properties of DNA were measured first in solutions and then occasionally in solid state [1, 2, 3]. The influence of water content on the complex permitivity showed to be very outstanding and intricate.

In order to facilitate the interpretation of the DNA dielectric spectrum, suitable model polymer systems were sought to be studied under identical conditions. These polymers had to be polar enough to sorb sufficient quantities of water.

Two particular polyamide systems were selected because their dielectric properties had been studied extensively and their properties have been thoroughly reviewed, e.g., [4, 5].

2. Experimental

The influence of water content on dielectric properties of polar polymers was studied for poly (N-methyl-capryl-amide) PMCA

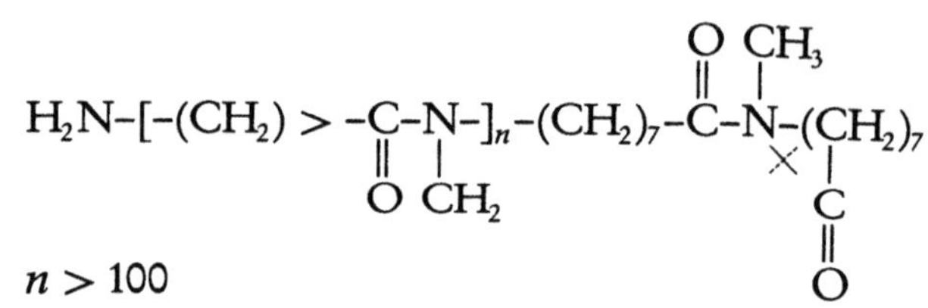

and modified poly (caprolactam) PKMG with four types of side groups randomly distributed on 40% of nitrogen atoms:

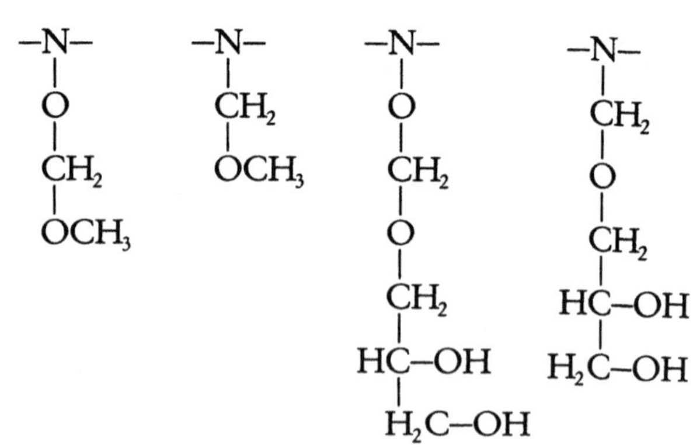

Thick self-supporting layers of these polymers (0.1 mm) were casted from ethanol solution. The samples were inserted into a

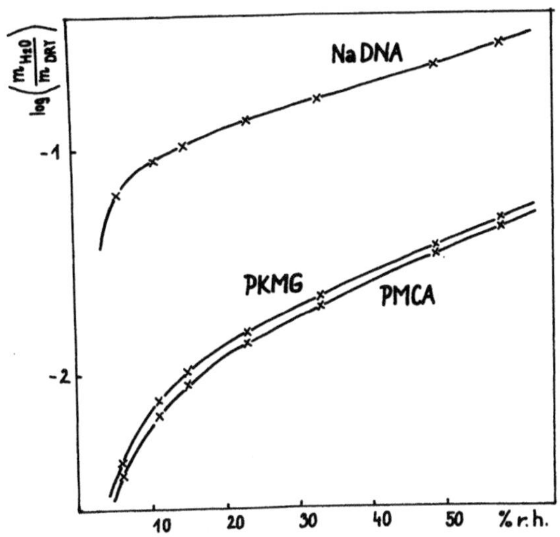

Fig. 1. Sorption characteristics for PMCA, PKMG and NaDNA

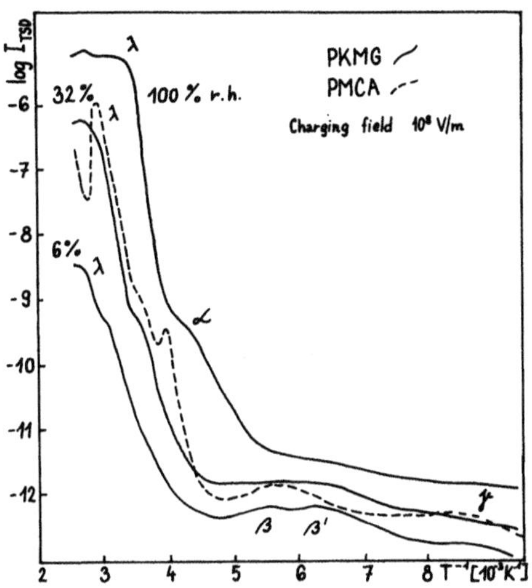

Fig. 2. TSD curves for PKMG (6%, 32%, 100% r.h.) and PMCA (32% r.h.)

three-electrode condenser with a guard ring, all electrodes were made from stainless steel and separated electrically from each other by teflon. The measuring electrode diameter was 10 mm. A liquid ultrathermostat with temperature stability 0.1 °C enabled us to measure the dielectric properties in the temperature interval −50 °–90 °C. Construction of the sample holder made it possible to cool the sample with liquid nitrogen and to raise the temperature at a rate of 4 K/min. All measurements were performed under defined humidity conditions; relative humidity of the ambient was controlled by inserting Petri discs with several saturated salt solutions into the double vacuum cover of the condenser.

In the range 10^{-1} Hz – 10^4 Hz a homemade bridge with an automatic symmetrization of the applied voltage was used. The temperature dependence of the complex permitivity at $\omega = 10^4 s^{-1}$ and thermostimulated depolarization currents (TSD) were measured using Tesla BM 484 bridge and electrometer Keithley 610 C.

3. Results and conclusions

Sorption characteristics obtained for PMCA, PKMG, and natrium salt of DNA are presented in Fig. 1. It can be seen that the amount of water sorbed into the PMCA and PKMG is an order lower than in the case of NaDNA (2 wt% at 16% r.h.). This result is consistent with the conception of specific sorption points which may be occupied by water molecules. It can be concluded from the small difference between PMCA and PKMG that there are more sorption points in PKMG, probably near the oxygen atoms of side groups, as compared to PMCA. The sorption points in PMCA are assumed to be merely near the polyamide groups of the main chain which are interconnected by hydrogen bonds, leading to a high degree of crystallinity in polyamides without side groups.

TSD results for PKMG with three different water contents are shown in Fig. 2. Similar to other polymers [4, 5], several absorption regions were observed. Only the low temperature region γ at −140 °C was found to be humidity independent. Other absorption regions were shifted to lower temperatures after the water content was raised. While this effect was only moderate in the case of region β, the temperature shifts of regions λ and α were very pronounced. On the other hand, the magnitude of region α was humidity independent unlike that of region β. The latter consisted of two close peaks, denoted β and β', disappeared in the atmosphere of phosphorus dioxide.

The dependence of TSD peak position on water content can be seen in Fig. 3. Two features of region α behavior should be emphasized. The dry value 71 °C can be well compared with the glass transition temperature of other polyamides (65 °C for nylon 6). The initial large drop of this temperature caused by adding water to the sample (30 K per first 1 wt%) becomes small when the water content exceeds 5 wt%. But the regions' λ, β, and β' temperature shifts at low water content were small. The TSD peak λ temperature began to fall at 5 wt% of water.

Figure 4. offers the temperature dependence of the a. c. conductivity for PKMG with different water content at 50 Hz and 1492 Hz. The temperature dependence of all regions exhibits the same behavior as in

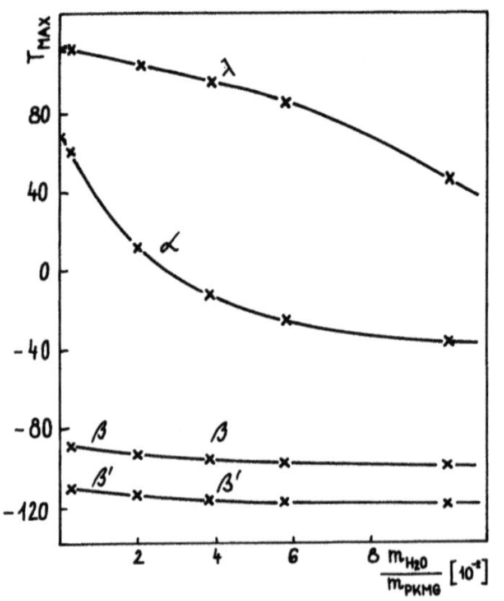

Fig. 3. Humidity dependence of the TSD peaks position in PKMG

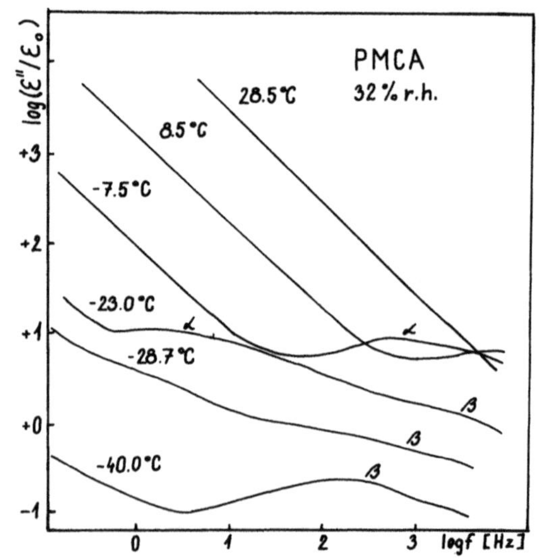

Fig. 5. Frequency dependence of the imaginary part of the complex permitivity for PMCA (32 % r.h.) at different temperatures

TSD. In addition, a strong increase in a.c. conductivity can be detected at high temperatures when the water content reaches 5 wt%.

As to PMCA, only small differences were found in comparison to PKMG. The TSD curve for PMCA at 32 % r.h. is presented in Fig 2. (dashed line). The absorption region consists of only one distinct peak that can be found at −95 °C. Its position and magnitude are water dependent. Extrapolating the position to zero water content, the value −84 °C was obtained.

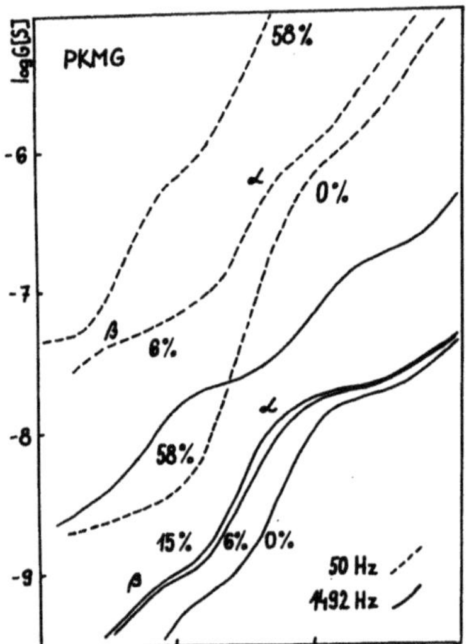

Fig. 4. Temperature dependence of a. c. conductivity for PKMG

This absorption region corresponds to the high temperature part of region β in PKMG. The second difference is connected with better separation of regions α and λ.

The frequency dependence of the complex permitivity imaginary part measured at several temperatures for PMCA with water sorbed at 20 °C and 32 % r.h. is shown in Fig. 5. The absorption regions α and β are quite recognizable. For these regions in both studied polymers PMCA and PKMG Arrhenius plots were drawn and activation energies were determined for regions β and β' (14.5 and 13.5 kcal/mol). The influence of water content on activation energies was only moderate. The values in samples equilibrated at 100 % r.h. were 0.5 kcal/mol lower. The frequency shifts of the peak α fulfilled the WLF equation.

There is another strong absorption region at low frequencies with similar temperature behavior as α region. Both the real and imaginary parts of the complex permitivity rise sharply on lowering the frequency with a slope −.95 in a log–log plot. We connect this strong low frequency polarization with λ peak in TSD. It can be observed even in dry samples. Similar low frequency dispersion has been often observed by many researchers in polymer systems containing ions, hydrogen bonds, and water. Two explanations are usually offered. The first [6] assumes that the abovementioned behavior is a bulk property of the material and the concept of an anomalous dispersion is used to describe a nonconductive long-range charge transport. The second explanation stresses the role of the sample-electrode interface and the theory of space-charge polarization [7] is taken into account. This problem will be an object of our next study.

The non-Arrhenius region α is obviously connected with the glass transition of the amorphous phase in polyamides. It is well known [5] that the hydrogen bonds linking the polyamide groups are nearly all realized at room temperature. Nevertheless, their creation and break are very quick and the motions of large parts of the main chains are possible. We can assume that the hydrogen bonds act as temporary physical crosslinks. The sorption of water probably diminishes the hydrogen bond time of life, the effective network density and, thus, the glass transition temperature. Approximately 9% and 43% of polyamide grous in the amorphous phase are hydrated in samples of PMCA with 1 wt% and 5 wt% of water content, respectively. It can be concluded that the polyamide groups pairing is seriously destroyed at 5 wt% of water. If the hydration goes on (Fig. 3.), the water molecules act merely as a plasticizing agent and they lower the glass transition temperature T_g slightly via lowering the internal viscosity. A similar humidity dependence of the glass transition temperature can be found in polyelectrolytes where the ions act as physical crosslinks. Plasticization of ethyl acrylate ionomers [8] with water depresses the T_g. It varies linearly with water content, but with different slopes above and below two water molecules per ion pair. The ion clusters are assumed to change their nature or even to fall apart. On the other hand, the same effect cannot be expected in NaDNA where water molecules help to maintain the regular quasicrystalline structure. Water desorption causes profound disturbance in DNA base pairing and the hydrogen bonds break.

The relaxation region β is commonly interpreted [9] as the movement of the polymer-water complex. We suppose that these local motions are connected with polyamide groups. This idea explains the increase in the magnitude of absorption and the plasticizing effect of water. The influence of water molecules on the internal viscosity in polyamides can be seen in the region β' behavior, too. We propose that the region β' observed in PKMG is due to the motion of the side groups with water molecules attached. These kinetic units move in similar surroundings as those responsible for region β.

The problem of interactions in polymer systems with small molecules remains open and furter investigation must follow.

References

1. Ando V, Fukada E (1976) J Polym Sci Polym Phys Ed 14:63
2. Mascarenhas S et al. (1978) In: Fukada E (ed) Riken Symposium on Electrical Properties of Polymers. Saitama, Japan
3. Neubert M, Bakule R, Nedbal J (1985) In: Proc 5th Intern Symp Electrets, Heidelberg, IEEE, p 825
4. Hedvig P (1977) Dielectric Spectroscopy of Polymers. Akademiai Kiado, Budapest
5. Seanor DA, Link GL, Parker TG (1972) In: Jenkins AD (ed) Polymer Science. North-Holland Publishing Company, Amsterdam
6. Shablakh M, Dissado LA, Hill RM (1984) J Biol Phys 12:63
7. MacDonald JR (1973) J Chem Phys 58:4982
8. Matsuura H, Eisenberg A (1976) J Polym Sci Polym Phys 14:773
9. Curtis J (1961) J Res Nat Bur Std 65A:185

Received June 23, 1987;
accepted April 26, 1988

Authors' address:

Dr. M. Neubert
Department of Polymer Physics
Faculty of Mathematics and Physics
Charles University
V Holešovičkách 2
180 00 Prague 8, Czechoslovakia

Note on the solution of the soliton problem in trans-polyacetylene

L. Valenta[1]), L. Wojtczak[2], and B. Rolov[3])

[1]) Faculty of Mathematics and Physics, Charles University, Prague, Czechoslovakia
[2]) Department of Solid State Physics, University of Łódź, Łódź, Poland
[3]) Department of Theoretical Physics, Latvian State University, Riga, USSR

Abstract: A soliton solution in the continuum approximation is suggested for the Su-Schrieffer-Heeger model [1] of the trans-polyacetylene. Contrary to [1], the solitons described in this paper are nontopological.

Key words: Polymers, solitons.

Introduction

Su et al. [1] started an avalanche of many papers dealing with the soliton problem in polyacetylene polymeric chains. They discussed the idea of a topological soliton-domain wall and its consequences. They used a plausible trial function to describe it since they did not have a solution corresponding to their Hamiltonian. We later showed, and reported on it on several occasions, that there may be a nontopological soliton solution. Since the problem remains we think it may be of interest, and we present our results.

Model Hamiltonian

In [1] it was assumed that the alternating single and double bonds in the polymeric CH_x chain are as shown in Fig. 1. The C and H atoms are expected to create a planar configuration. The choice of the model Hamiltonian is not quite obvious. Su et al. used

$$\hat{H} = -\sum_{ns} t_{n+1,n} \left(C^{+}_{n+1,s} C_{ns} + C^{+}_{ns} C_{n+1,s}\right) - \frac{1}{2}\sum_{n} K \left(u_{n+1} - u_n\right)^2 + \frac{1}{2}\sum_{n} M \dot{u}_n^2 \quad (1)$$

c^{+}_{ns}, c_{ns}, denoting the creation and annihilation operators of the π electrons of spin $s(\pm 1/2)$ on the n-th CH group; M is the total mass of the CH group, K is the effective σ spring constant and

$$t_{n+1,n} = t_0 - \alpha(u_{n+1} - u_n) \quad (2)$$

is the hopping integral, where t_0 is the hopping integral for the undimerized chain and α is the electron-lattice displacement coupling constant. The configuration coordinates u_n describe the translation of the n-th CH group as a whole along the helical symmetry axis of the chain. [1] contains the kinetic energy of the CH groups, their elastic energy in harmonic approximation, and the hopping term where the hopping integral is linearly dependent on the distance of the nearest neighbor CH groups. The explicit Coulomb interaction between the π electrons is missing from $\hat{H}$, being only partially included in the screened values of t_0 and α.

It is, of course, not evident that such a Hamiltonian is sufficient and its usefulness must be tested. A helpful critical contribution to the problem may be found in [2], where the author, Springborg, states, "we still

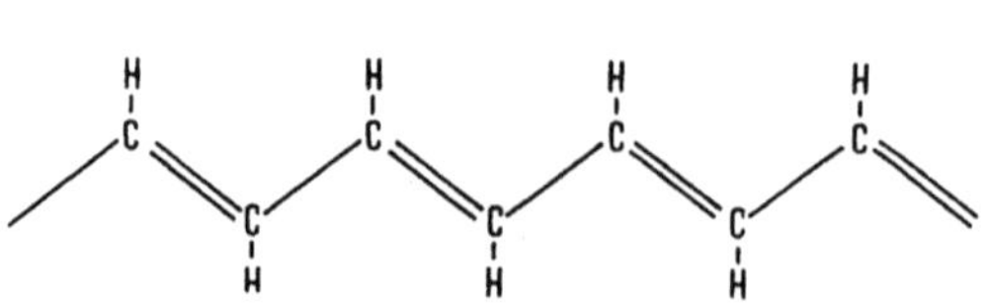

Fig. 1. Model of the trans-polyacetylene assumed in [1]

believe that the Su et al. Hamiltonian can be used to describe trans-polyacetylene, but it has to be interpreted differently". The statement "differently" means that the operators $c^+_{n,s}$, $c_{n,s}$ should not be restricted to the π electrons only, but they also should take into account the σ electrons. We shall, therefore, also accept the model with alternating single and double bonds, but we shall start with a more general Hamiltonian that we believe is more realistic.

In infinite chains there are two equivalent ground states with single and double bonds mutually exchanged and a combination of two semi-infinite chains with violation of the order of regularly alternating single and double bonds, even leads to the introduction of the transition region, i.e., domain wall or topological soliton. In our research we found evidence to suggest that a chain in a real system may be in only one of the possible phases, and that in both cases (namely, that discussed in [1] and that in this paper), coexist in real systems which, of course, wave a finite length.

As shown in Fig. 2, we subdivide the CH groups into two "sublattices": the "odd" and the "even". Denoting the position of a CH group by R_{2n} ("even") or R_{2n+1} ("odd"), $n = 0, \pm 1, \pm 2 \ldots$, where $R_m = R_{m_0} + u_m = ma + u_m$, $m = 0, \pm 1, \ldots$ and a being the equilibrium lattice constant, we assume the following Hamiltonian

$$\begin{aligned}\hat{H} = & -\sum_{ns} [t_0 - \bar{\bar{\alpha}}(u_{2n+1} - u_{2n})] (C^+_{2n+1,s} C_{2n,s} \\ & + C^+_{2n,s} C_{2n+1,s}) + \frac{1}{2} U \sum_{ns} \hat{n}_{2n,s} \hat{n}_{2n,-s} \\ & - \sum_{ns} [t_0 - \bar{\alpha}(u_{2n} - u_{2n-1})] (C^+_{2n,s} C_{2n-1,s} \\ & + C^+_{2n-1,s} C_{2n,s}) + \frac{1}{2} U \sum_{ns} \hat{n}_{2n+1,s} \hat{n}_{2n+1,-s} \\ & + \frac{1}{2M} \sum_{n} [p^2_{2n-1} + p^2_{2n}] \\ & + \frac{1}{2} K \sum_{n} (u_{2n+1} - u_{2n})^2 \\ & + \frac{1}{2} K \sum_{n} (u_{2n} - u_{2n-1})^2 . \end{aligned} \tag{3}$$

As before, the c^+_{ns}, c_{ns} create and destroy the electrons of spin $s\left(= \pm \frac{1}{2}\right)$ on the *n-th* CH group, M is the total

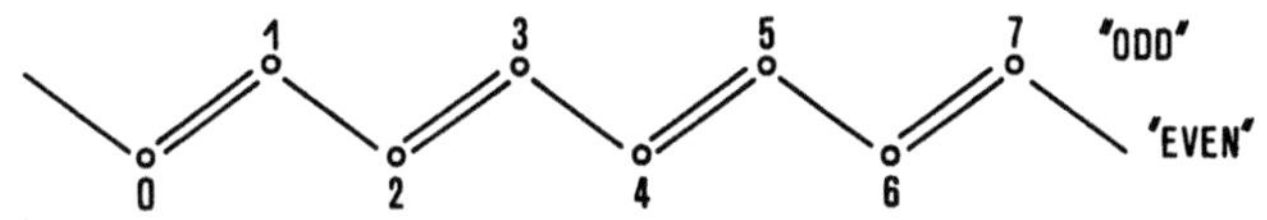

Fig. 2. Subdivision of the CH groups in two sublattices

mass of the CH group, $\hat{n}_{ns} = c^+_{ns} c_{ns}$ are the number of particles operators and u_n, p_n denote the coordinates and their conjugated momenta. We take them as classical numbers (c-numbers). The first and third terms in Eq. (3) are again the hopping terms, where we assume two different electron-phonon coupling constants $\bar{\alpha}$ and $\bar{\bar{\alpha}}$ according to the single or double bond, i.e.,

$$t = \begin{cases} t_0 - \bar{\alpha}(u_{2n} - u_{2n-1}) & \text{(single)} \\ t_0 - \bar{\bar{\alpha}}(u_{2n+1} - u_{2n}) & \text{(double)}. \end{cases} \tag{4}$$

The second and the fourth terms in (3) correspond to the Coulomb energy of the pairs of electrons with opposite spins. One could think of different $\bar{K}$ and $\bar{\bar{K}}$ as being similar to $\bar{\alpha}$ and $\bar{\bar{\alpha}}$. However, that is not of special interest for infinite chains investigated in this paper. And a remark more on the use of u_n and p_n. One could treat them as operators (q-numbers), but there is no essential difference and we shall therefore use the Hamilton equations of classical mechanics in what follows.

Equations of motion

We assume that the wave function to (3) may be approximately represented by

$$|\psi(t)\rangle = \frac{1}{N} [1 + \sum_{ns} A_{ns}(t)\, C^+_{ns}]\, |0\rangle \tag{5}$$

$$NN^* = 1 + \sum_{ns} A^*_{ns}(t)\, A_{ns}(t) = |N(t)|^2 \tag{6}$$

A test function similar to (5) may be found in papers dealing with similar problems, e.g., [3–5]. It is evidently not a general ansatz since the states of the type $c^+c^+|0\rangle$, $c^+c^+c^+|0\rangle$, etc., are not included. However, it turned out to be useful.

We need to find the time dependence of $A_{ns}(t)$. It may easily be verified using the anticommutation rules for the c.., c^+.., that

$$\frac{A_{ns}(t)}{|N(t)|^2} = \langle \psi(t) | C_{ns} | \psi(t) \rangle = \phi_{ns}(t) \tag{7a}$$

and

$$\frac{A^*_{ns}(t)}{|N(t)|^2} = \langle \psi(t) | C^+_{ns} | \psi(t) \rangle = \phi^*_{n,s}(t). \quad (7b)$$

Similar calculations show that

$$\langle \psi(t) | C^+_{is} C_{ks'} | \psi(t) \rangle = \phi^*_{is} \phi_{ks'} |N|^2 \quad (8)$$

and

$$\langle \psi(t) | \hat{n}^1_{2n,s} \hat{n}^1_{2n,-s} | \psi(t) \rangle = 0. \quad (9)$$

Hence, if we write the equations of motion for the $c_{..}$ and $c^+_{..}$, it means the equations for the time derivatives of $c_{..}$, $c^+_{..}$, we obtain a system of difference-differential equations for the $\phi_{ns}(t)$. These equations lead after the averaging procedure using the Eqs. (7–9) to

$$i\hbar \frac{d\phi_{2n,s}}{dt} = -[t_0 - \bar{\bar{\alpha}}(u_{2n+1} - u_{2n})]\, \phi_{2n+1,s} - [t_0 - \bar{\alpha}(u_{2n} - u_{2n-1})]\, \phi_{2n-1,s} \quad (10a)$$

and

$$i\hbar \frac{d\phi_{2n-1,s}}{dt} = -[t_0 - \bar{\alpha}(u_{2n} - u_{2n-1})]\, \phi_{2n,s} - [t_0 - \bar{\bar{\alpha}}(u_{2n-1} - u_{2n-2})]\, \phi_{2n-2,s}. \quad (10b)$$

Equations (10a, 10b) have the same structure with only the interchanged role of $\bar{\alpha}$ and $\bar{\bar{\alpha}}$.

Now, we calculate the average value of the Hamiltonian

$$\langle \hat{H}(t) \rangle = \langle \psi(t) | \hat{H} | \psi(t) \rangle$$

with respect to Eqs. (6–9). We obtain

$$\langle \hat{H}(t) \rangle = -\sum_{ns} [t_0 - \bar{\bar{\alpha}}(u_{2n+1} - u_{2n})](\phi^*_{2n+1,s}\phi_{2n,s} + \phi^*_{2n,s}\phi_{2n+1,s})|N|^2 - \sum_{ns} [t_0 - \bar{\alpha}(u_{2n} - u_{2n-1})](\phi^*_{2n,s}\phi_{2n-1,s} + \phi^*_{2n-1,s}\phi_{2ns})|N|^2 + \frac{1}{2M}\sum_n [p^2_{2n-1} + p^2_{2n}] + \frac{1}{2}K\sum_n [(u_{2n+1} - u_{2n})^2 + (u_{2n} - u_{2n-1})^2]. \quad (11)$$

$H(t) = H$ may serve as a Hamilton function in the sense of classical mechanics (see, e. g. [3])

$$\langle \hat{H}(t) \rangle = \varkappa(u_m, p_m)$$

and we have then

$$\dot{p}_m = -\frac{\partial H}{\partial u_m}, \quad \dot{u}_m = \frac{\partial H}{\partial p_m}. \quad (12)$$

It follows

$$\dot{u}_m = \frac{p_m}{M}$$

and

$$M\ddot{u}_m = -\frac{\partial H}{\partial u_m}. \quad (13)$$

Eq. (13) leads to

$$M\ddot{u}_{2n} = \sum_s [\bar{\bar{\alpha}}(\phi^*_{2n+1,s}\phi_{2n,s} + \phi^*_{2n,s}\phi_{2n+1,s}) - \bar{\alpha}(\phi^*_{2n,s}\phi_{2n-1,s} + \phi^*_{2n-1,s}\phi_{2n,s})]|N|^2 + K[u_{2n+1} + u_{2n-1} - 2u_n] \quad (14a)$$

and

$$M\ddot{u}_{2n-1} = \sum_s [-\bar{\bar{\alpha}}(\phi^*_{2n-1,s}\phi_{2n-2,s} + \phi^*_{2n-2,s}\phi_{2n-1,s}) + \bar{\alpha}(\phi^*_{2n,s}\phi_{2n-1,s} + \phi^*_{2n-1,s}\phi_{2n,s})]|N|^2 + K[u_{2n} + u_{2n-2} - 2u_{2n-1}]. \quad (14b)$$

Equations (10a,b) and (14a,b) represent a system of coupled equations to be solved simultaneously. And again, Eqs. (14a,b) for even and odd CH groups are similar in their structure, but not identical. The expressions containing $\phi^+_{..}$ and $\phi_{..}$ are always sums of two complex conjugated products, i.e., they are real numbers, which together with real $\bar{\alpha}$, $\bar{\bar{\alpha}}$ leads to real $\ddot{u}_{..}$ as necessary.

Experience with such systems of equations shows that there is some hope in solving them according to the continuum approximation. We therefore change Eqs. (14 a, b) to

$$M\ddot{u}_{2n} = \sum_s [(\bar{\bar{\alpha}}\, \phi^*_{2n+1,s} - \bar{\alpha}\, \phi^*_{2n-1,s})\, \phi_{2n,s} + \phi^*_{2n,s}\, (\bar{\bar{\alpha}}\, \phi_{2n+1,s} - \bar{\alpha}\, \phi_{2n-1,s})]\,|N|^2 + K[u_{2n+1} + u_{2n-1} - 2u_n] \qquad (15\,\mathrm{a})$$

and

$$M\ddot{u}_{2n-1} = \sum_s [(\bar{\alpha}\, \phi^*_{2n,s} - \bar{\bar{\alpha}}\, \phi^*_{2n-2,s})\, \phi_{2n-1,s} + \phi^*_{2n-1,s}\, (\bar{\alpha}\, \phi_{2n,s} - \bar{\bar{\alpha}}\, \phi_{2n-2,s})]\,|N|^2 + K[u_{2n} + u_{2n-2} - 2u_{n-1}]\,. \qquad (15\,\mathrm{b})$$

The continuum approximation means that we use, e.g.,

$$\phi_{2n\pm1,s} \cong \phi_{2n,s} \pm a \frac{\partial \phi_{2n,s}}{\partial x} + \frac{1}{2} a^2 \frac{\partial^2 \phi_{2n,s}}{\partial x^2}\,,$$

$$u_{2n\pm1} \cong u_{2n} \pm a \frac{\partial u_{2n}}{\partial x} + \frac{1}{2} a^2 \frac{\partial^2 u_{2n}}{\partial x^2}$$

and analogously for the other groups of indices, a is one half of the lattice constant. As a consequence, e.g., Eq. (15 b) changes to

$$M\ddot{u}_{2n-1} = \sum_s \Big\{(\bar{\alpha} - \bar{\bar{\alpha}})\, \phi^*_{2n-1,s}\, \phi_{2n-1,s} + a(\bar{\alpha} + \bar{\bar{\alpha}}) \frac{\partial^2 \phi^*_{2n-1,s}}{\partial x}\, \phi_{2n-1,s} + \frac{1}{2} a^2 (\bar{\alpha} - \bar{\bar{\alpha}}) \frac{\partial^2 \phi_{2n-1,s}}{\partial x^2}\, \phi_{2n-1,s}\Big\}\, |N|^2 + C.C. + K\, a^2 \frac{\partial^2 u_{2n-1}}{\partial x^2}\,.$$

This equation sufficiently illustrates the situation if we do the same with the remaining Eqs. (10 a, b) and (15 a). There is an evident substantial simplification of the mathematical difficulties, namely, if we come back to the SSH model putting $\bar{\alpha} = \bar{\bar{\alpha}} = \alpha$. We accept it in the following considerations, despite the fact that some special features are lost. We expect, however, to get some insight into the problem and, moreover, we expect that despite this simplification, the electron-phonon interaction will still be able to impress its influence on the creation of the soliton state.

Using the continuum approximation we find that the equations for the odd and even CH groups are now the same and independent of the spin index s.' This sounds quite reasonable, since there is no longer any difference in the physical conditions for the even and odd groups and our Hamiltonian does not contain any terms which would distinguish between spins up or down.

Equations (10 a, b) are then reduced to

$$i\hbar \frac{\partial \phi_A}{\partial t} = -2t_0\, \phi_s + 2\alpha\, a \frac{\partial u}{\partial x}\, \phi_s - t_0\, a^2 \frac{\partial^2 \phi_s}{\partial x^2} + \alpha\, a^3 \frac{\partial}{\partial x} \left(\frac{\partial u}{\partial x} \frac{\sum o_s}{\partial x}\right) \qquad (16)$$

and Eqs. (15 a, b) to

$$M \frac{\partial^2 u}{\partial t^2} = 2a\, \tilde{\alpha} \sum_s \frac{\partial}{\partial x} (\phi^*_s\, \phi_s) + K a^2 \frac{\partial^2 u}{\partial x^2} \qquad (17)$$

where $u = u(x, t)$ and $\phi = \phi(x, t)$. $\tilde{\alpha} = \alpha \cdot |N|^2$, and, moreover, we assume that N, given by Eq. (6), corresponds to a permanent geometrical shape moving with constant velocity and we may treat it as a constant. Looking for solutions describing some excitations moving with constant velocity, it is useful to assume

$$u(x, t) = u(x - vt) = u(y)$$
$$\phi(x, t) = \phi(x - vt) = \phi(y)\,. \qquad (18)$$

Denoting

$$\frac{du}{dy} = u'\,, \quad \frac{d^2u}{dy^2} = u''$$

and

$$\frac{d\phi_s}{dy} = \phi'_s\,, \quad \frac{d^2\phi_s}{dy^2} = \phi''_s\,,$$

Equation (17) becomes

$$u'' = \frac{2\alpha a}{Mv^2 - Ka^2} \sum_s \frac{d}{dy} (\phi^*_s\, \phi_s) \qquad (19)$$

which may immediately be integrated to

$$u' = \frac{2\alpha a}{Mv^2 - Ka^2} \sum_s \phi_s^* \phi_s + \text{konst } (= C). \quad (20)$$

Now we can u' and u'' from Eqs. (19) and (20) and insert it in Eq. (16). It may be seen, however, that the terms on the right side of Eq. (16) are multiplied successively by a^0, a^1, a^2, a^3, where a, now the lattice constant of the Su et al. model, is a small quantity of the order 10^{-8} cm. It seems plausible, if we neglect the last term with a^3, assuming a reasonable behavior of ϕ_s and its derivatives.

By inserting for u from Eq. (20), the index s may be omitted

$$i\hbar\,\dot{\phi} = 2(-t_0 + \alpha\, a\, C)\,\phi + \frac{(2\,\alpha\, a)^2}{Mv^2 - K\,a^2}\, 2|\phi|^2\,\phi - t_0\, a^2\, \phi''. \quad (21)$$

We used $\sum_s \phi_s^*\,\phi_s = 2\phi_s^*\,\phi_s$ due to the expected symmetry $\phi_\uparrow = \phi_\downarrow = \phi$. Equation (21) belongs to the family of the so-called nonlinear Schrödinger equation and its solutions and their properties are known (see e.g. [6, 7]). Let us write Eq. (21) as

$$i\dot{\phi} + t_0\, a^2\, \ddot{\phi} - 2(t_0 + \alpha\, a\, C)\,\phi + \frac{8\,\alpha^2\, a^2}{Ka^2 - Mv^2}\, |\phi|^2\,\phi = 0. \quad (22)$$

It is known that Eq. (22) has a soliton solution if the coefficient

$$G = \frac{8\,\alpha^2\, a^2}{Ka^2 - Mv^2} > 0,$$

i.e., if

$$K\,a^2 > Mv^2. \quad (23)$$

If we take $K = 21$ eV/Å like in [8] and $a = 1{,}22$ Å [1] we see that v must obey the condition

$$v < 10^6 \text{ cm/sec}.$$

The limit 10^6 cm/sec is comparable with the electron velocity in the Bohr hydrogen atom orbit 1s. Denoting in Eq. (22)

$$2t_0 + 2\alpha\, a\, \mathscr{C} = \lambda, \quad t_0\, a^2 = J \quad (24)$$

the normalized soliton solution may be written as

$$\phi(x,t) = \frac{\sqrt{\dfrac{G}{8J}}\, \exp\left\{i\left[\dfrac{\hbar v}{2J}(x - x_0) - wt\right]\right\}}{\cosh\left[\dfrac{G}{4J}(x - x_0 - vt)\right]} \quad (25)$$

where

$$\hbar w = \lambda + \left(\hbar^2\, v^2 - \frac{1}{4}\, G^2\right)\Big/ 4J.$$

There are also other solutions of Eq. (22) for $K\,a^2 < Mv^2$ and also some periodic solutions which may be easily derived from Eq. (22). We do not intend, however, to discuss them in this paper (see e.g. [3], [7] form more details).

$\phi(x,t)$ given by Eq. (25) represents no domain wall (topological soliton) like that suggested in [1]. In our case, as shown e.g. in [3], it is something more like a tangle of atoms with a changed density of CH groups per unit of length of the chain. The shape of the probability distribution of the electrons (see Eqs. (7a, b)) is determined by the cosh( ). It moves along the chain with the velocity v.

Conclusions

It was our aim to show that in a system like trans-polyacetylene, solitons of a nontopological nature (Davydov-like solitons) may exist. It seems that the answer is probably positive. Some uncertainty, however, still persists, partly due to the model itself, and partly due to the approximations accepted during our calculations. We saw what should be done respecting the single and double bonds. It should, perhaps, be one of the next steps towards a more satisfactory theory. It remains to draw physical consequences from the expected existence of the nontopological soliton, compare it with the Su et al. topological-domain wall-soliton, etc. This should be a subject of future investigations.

Acknowledgements

We would like to express our thanks to Dr. Vojtěch Kapsa for his helpful assistance in checking our calculations.

References

1. Su WP, Schrieffer JR, Heeger AJ (1980) Phys Rev B 22:2099
2. Springborg M (1986) Phys Rev B 33:8475

3. Davydov AS (1985) Solitons in Molecular Systems. Reidel, Dordrecht
4. Davydov AS (1986) Ann der Physik 43:93
5. Lindner U, Fedyanin VK (1978) phys stat sol (b) 89:123
6. Davydov AS, Kislukha NI (1976) phys stat sol (b) 75:735
7. Lindner U, Fedyanin VK (1979) phys stat sol (b) 95 K 83
8. Stafström S, Chao KA (1984) Phys Rev B 29:2266

Received February 1, 1988;
accepted April 21, 1988

Authors' address:

Prof. Dr. L. Valenta
Faculty of Mathematics and Physics
Charles University
Ke Karlovu 3
121 16 Prague 2, Czechoslovakia

Author Index

Subject Index

GPSR Compliance

The European Union's (EU) General Product Safety Regulation (GPSR) is a set of rules that requires consumer products to be safe and our obligations to ensure this.

If you have any concerns about our products, you can contact us on ProductSafety@springernature.com

In case Publisher is established outside the EU, the EU authorized representative is:

Springer Nature Customer Service Center GmbH
Europaplatz 3
69115 Heidelberg, Germany

Batch number: 10578403

Printed by Printforce, the Netherlands